Management of Marine Protected Areas

Management of Marine Protected Areas

A Network Perspective

Edited by Paul D. Goriup

WILEY Blackwell

Registered Office
John Wiley & Sons Ltd, The Atrium, Southern Gate, Chichester, West Sussex, PO19 8SQ, UK

Editorial Offices
111 River Street, Hoboken, NJ 07030, USA
9600 Garsington Road, Oxford, OX4 2DQ, UK
The Atrium, Southern Gate, Chichester, West Sussex, PO19 8SQ, UK
Boschstr. 12, 69469 Weinheim, Germany

For details of our global editorial offices, customer services, and more information about Wiley products visit us at www.wiley.com.

Wiley also publishes its books in a variety of electronic formats and by print-on-demand. Some content that appears in standard print versions of this book may not be available in other formats.

Library of Congress Cataloging-in-Publication data has been applied for

ISBN: 9781119075776

Cover image: © Ultramarinfoto/gettyimages
Cover design: Wiley

Set in 10/12pt Warnock by SPi Global, Pondicherry, India

Printed in Singapore by C.O.S. Printers Pte Ltd

10 9 8 7 6 5 4 3 2 1

Contents

List of Contributors

Alexandrov, Boris
Institute of Marine Biology
National Academy of Sciences of Ukraine
Odessa
Ukraine
Email: borys.aleksandrov@gmail.com

Arkin, Sinan
Institute of Marine Science
Middle East Technical University
Erdemli
Mersin
Turkey

Beal, Stephen
NatureBureau
Newbury
UK
Email: stephen_beal@hotmail.co.uk

Bitetto, Isabella
COISPA Tecnologia & Ricerca
Bari
Italy

Boero, Ferdinando
Università del Salento – CNR-ISMAR
Italy
Email: boero@unisalento.it

Braun, Daniel
Research Group of Prof. Dr. Detlef
Czybulka
Faculty of Law
University of Rostock
Germany
Email: daniel.braun@o2online.de

Breil, Margaretha
Fondazione Eni Enrico Mattei (FEEM) and
    Euro-Mediterranean Center for Climate
    Change
Venezia
Italy

Cebrian, Emma
Centre d'Estudis Avançats de Blanes
(CEAB-CSIC)
Girona
Spain;
Department of Environmental Sciences
University of Montilivi
Girona
Spain

Claudet, Joachim
National Center for Scientific Research
CRIOBE
Perpignan
France

Dominguez-Carrió, Carlos
Institut de Ciències del Mar (ICM-CSIC)
Barcelona
Spain

Dudley, Nigel
Equilibrium Research
Bristol
UK;
School of Geography
Planning and Environmental Management
    at the University of Queensland
Brisbane
Australia
Email: nigel@equilibriumresearch.com

Evans, Julian
Department of Biology
Faculty of Science
University of Malta
Malta

Fach, Bettina A.
Institute of Marine Science
Middle East Technical University
Erdemli
Mersin
Turkey

Galil, Bella
The Steinhardt Museum of
   Natural History
Israel National Center for Biodiversity
   Studies
Tel Aviv University
Tel Aviv
Israel
Email: bella@ocean.org.il

Garcia-Rubies, Antoni
Centre d'Estudis Avançats de Blanes
   (CEAB-CSIC)
Girona
Spain
Email: tonigr@ceab.csic.es

Gili, Josep Maria
Institut de Ciències del Mar (ICM-CSIC)
Barcelona
Spain

Goriup, Paul D.
NatureBureau
Newbury
UK
Email: paul.goriup@fieldfare.biz

Haynes, Thomas
NatureBureau
Newbury
UK

Hockings, Marc
School of Geography
Planning and Environmental Management
   at the University of Queensland
Brisbane
Australia

Keskin, Çetin
Faculty of Fisheries
Istanbul University
Beyazıt
Istanbul
Turkey

Macpherson, Enrique
Centre d'Estudis Avançats de Blanes
   (CEAB-CSIC)
Girona
Spain

March, David
SOCIB – Balearic Islands Coastal
   Observing and Forecasting System
Palma
Spain

Markandya, Anil
Basque Centre for Climate Change (BC3)
Bilbao
Spain;
Ikerbasque Foundation
Bilbao
Spain

Melià, Paco
Dipartimento di Elettronica
Informazione e Bioingegneria
Politecnico di Milano
Milano
Italy;
Consorzio Nazionale Interuniversitario per
   le Scienze del Mare
Roma
Italy
Email: paco.melia@polimi.it

Minicheva, Galina
Institute of Marine Biology
National Academy of Sciences of Ukraine
Odessa
Ukraine

Ojea, Elena
Basque Centre for Climate Change (BC3)
Bilbao
Spain;
University of Vigo
Spain
Email: elenaojea@uvigo.es

Öztürk, Ayaka Amaha
Faculty of Fisheries
Istanbul University
Beyazıt
Istanbul
Turkey

Öztürk, Bayram
Faculty of Fisheries
Istanbul University
Beyazıt
Istanbul
Turkey
Email: ozturkb@istanbul.edu.tr

Pascual, Marta
Basque Centre for Climate Change (BC3)
Bilbao
Spain;
Ikerbasque Foundation
Bilbao
Spain

Requena, Susana
Institut de Ciències del Mar (ICM-CSIC)
Barcelona
Spain

Sanders, Natalie
NatureBureau
Newbury
UK
Email: Natalie@NatureBureau.co.uk

Sardá, Rafael
Centre d'Estudis Avançats de Blanes
    (CEAB-CSIC)
Girona
Spain
Email: sarda@ceab.csic.es

Schachtner, Eva
Leibniz Institute of Ecological Urban and
    Regional Development
Dresden
Germany
Email: eva.coconet@gmx.de

Schembri, Patrick J.
Department of Biology
Faculty of Science
University of Malta
Malta
Email: patrick.j.schembri@um.edu.mt

Topaloğlu, Bülent
Faculty of Fisheries
Istanbul University
Beyazıt
Istanbul
Turkey

Webster, Chloë
Mediterranean Protected Areas Network
    (MedPAN)
Marseilles
France
Email: chloe.webster@medpan.org

Zaitsev, Yuvenaliy
Institute of Marine Biology
National Academy of Sciences of Ukraine
Odessa
Ukraine

Foreword

So much of our national and international effort to protect nature has been concentrated on terrestrial protected areas. Even the international agreements under the Convention on Biological Diversity (CBD) have accorded less prominence to protecting coastal and marine areas. Indeed, it was not until the agreement of the CBD's Aichi targets for 2020 in 2012 that marine protection began to gather real momentum with a target of 10% coverage compared with 17% for the terrestrial environment. This refocusing is a welcome recognition of the importance of looking after coastal and marine ecosystems in the longer term, especially in the light of the progressive degradation as a result of human activities at sea and on land and the relatively uncertain effects of global climate change. Now, there is a need to concentrate greater effort on strategic planning in the Mediterranean and Black Seas, including the establishment, protection and enforcement of Marine Protected Areas (MPAs) in these two naturally, culturally, economically and socially important seas. So this book is a timely reminder of what we know, what problems need to be addressed, what progress has been made, what can be learnt from other parts of the world, what actions are being taken and what more needs to be done using MPA mechanisms and processes to sustain life in and around the seas in the longer term.

It is obvious that the coastal and marine environment cannot be considered in isolation to what happens on land, especially the effects of infrastructure development on many parts of the coastline of the two seas to exploit the favourable weather conditions and shoreline situations and the delivery of water, nutrients and pollutants (and consequential eutrophication) into the seas from the surrounding rivers.

It is also obvious that looking after the marine environment of the two seas cannot just focus on nature and be a top-down process focusing on the protection of species and habitats. Both seas have a long history of human occupation and human passage in all directions and there are many internationally important cultural artefacts reflecting this long history. And there are many communities still dependent on the seas for the provision of natural resources for human survival, especially fish. The question of which comes first – nature or people – is an often-posed one in this book. The answer is both as they are really indivisible – hence the development of new approaches to looking after nature, including MPAs, which stress the importance of societal engagement throughout rather than the more traditional western approach of leaving it to the experts in nature. This does not mean that understanding and maintaining and, where necessary, restoring natural processes is not important: it is vital for the future of nature itself and for the survival of human societies.

The development of protection of the coastal and marine environment has to be seen from the perspective of nation states which have a stake. In all, 21 nations have coasts on the Mediterranean Sea, and whilst there are six on the Black Sea coast, another 10 nations make inputs through the rivers flowing into the sea.

With these points in mind, why do the Mediterranean and Black Seas require MPAs? The simple answer is that there are international and regional agreements requiring signatory states to protect the marine environment. More fundamentally, there remain many conflicts, for example, between fishermen and conservation to ensure that fish stocks are in a healthy biological state for the future, between tourism development and coastal pollution, between waste disposal through the river systems and the cleanliness of the marine environment, and between over-exploitation of key species and water pollution and their gradual loss and in some cases extinction. And, there is the potential inequality between those nations which exploit more resources and those which have a lesser environmental footprint. It is for these reasons that formal conventions have been long established for each of the seas: the Barcelona Convention for the Mediterranean and the Bucharest Convention for the Black Sea. Within these multilateral structures, many protocols for the protection of the seas have been developed, including systems of protected areas. Of particular note are the Specially Protected Areas of Mediterranean Importance and the Special Areas of Protection in the EU Member States. But protection is not just about designation of sites and areas, as there are too many so-called parks which exist only on paper. It is more fundamentally about the perpetual protection of nature and natural processes within the context of changing societal values, availability of new scientific information, and implementation of effective processes of engagement for all stakeholders.

Only through these approaches can the effectiveness of protection be secured and be assured for the future.

Much good progress has been made, as the chapters in this book illustrate. Of particular note are the sanctuaries and no-take zones to allow fish stocks to recover from over-exploitation and for the spawning biomass to increase to a state of biological sustainability and therefore allow fishing to recommence. There are important 'spillovers' of young fish from these protected areas into the wider seas which indicate that fish stocks are recovering. Also of note are the interactive processes established between the nation states and also, for example, between the MPA managers under the MedPAN initiative. A great deal is known about how the seas operate naturally – the water flows and the current patterns at all levels in the water column – and therefore where there are more likely to be pollution sinks and lack of water interchange which create negative conditions for marine life. Within the territorial seas of the EU Member States the Marine Strategy Framework Directive, with its target of achieving Good Environmental Status in all EU Waters by 2020, is a testing and very welcome target to stimulate action.

But more needs to be done bearing in mind that only 0.012% of the Mediterranean Sea is fully protected with effective MPAs and only about 1.7% of the Black Sea has protected area status.

In the former, greater action over the whole sea and coastal area, rather than just within the EU Member States' jurisdiction, is needed; but this has to recognize the relatively weaker economies, especially in North Africa and the Middle East, and therefore the limited resources available to address these issues. The learning of lessons from the various EU initiatives, and the EU states continuing to help the non-EU states to do more through technical aid and financial support, would be a very worthwhile effort. Also, means of cooperation through informal networks, such as the IUCN Centre for Mediterranean Cooperation, are important for sharing knowledge and experience of what works and what is less successful.

The key issue, seen from an external perspective, is to ensure that all of the nation states around the two seas are fully committed to working together and within their own territorial seas to achieve protection and restoration of the natural environment for the benefit of their own citizens now and in the future. This requires political will which is not always forthcoming and is often placed well behind other pressing priorities. Maybe arguing for acting in the nation's own interests and at the same time acting in the interests of 'the commons of the seas' might have some effect. Certainly new laws and protocols take a long time to get agreement and implement, so softer approaches are worthwhile in the shorter term.

Taking the long view is key if the measures implemented are to be effective in safeguarding nature and natural processes and providing benefit to human communities. Inevitably, this may mean reductions in income in the short term, for example for fishermen while stocks are allowed to recover, or increases in the costs to developers to reduce environmental side effects. That surely is a price worth paying for the longer term interests of nature and society jointly.

The diversity of the seas, the challenges due to the varying depths and nature of the water columns, and the variation in the human impacts all suggest to me the need for tailor-made measures for protection within the general approaches laid down in the two conventions and in the EU Natura programme for EU Member States. There is no need, however, to reinvent the wheel as there is plenty of international experience,

some of which is cited in the book, on which to base improvements in the protected mechanisms used. The work under the IUCN Marine Protected Areas Programme is a classic source for ideas and approaches and what works in different situations which would merit greater attention and use by practitioners in both seas. Adoption of protected area practices from the terrestrial sphere is, however, very unlikely to be helpful as they are less dynamic, and are rarely three dimensional – with the exception of the learning from best practice examples globally of connecting individual protected areas into networks especially in recognition that nature does not recognize site boundaries imposed upon it for administrative convenience. Clear management objectives and means of measuring effectiveness of implementation and feeding back into reviews of management are critically important; the IUCN Management Effectiveness Evaluation approach is well tried and tested around the world for this purpose.

Engagement of key stakeholders throughout the process of development, implementation and review of effectiveness of MPAs is absolutely necessary – we know from experience around the world that imposed top-down solutions do not work. Given the diversity of cultural histories and modern culture around these seas, recognition has to be given to ensuring that representatives of these aspects are factored into the process of design and management of MPAs. Hence, the IUCN work on governance types and mechanisms is a very helpful toolkit as are the methods of ecosystem-based management described in the book. It also means ensuring that expertise on negotiation and conflict resolution are part of the armoury of those involved in seeking agreement on strategies and action plans, otherwise disputes will continue and there will be no meeting of minds on what really needs to be done.

I hope that all of those who read the chapters in this book will be encouraged by what has been achieved through implementing MPAs in the Mediterranean and Black Seas. More importantly, I hope that readers will be stimulated to engage in further improving the quality of the coastal and marine areas for the benefit of present and future generations. Remember this means making sure that nature is allowed to function effectively, otherwise human society in the future will not benefit.

Professor Roger Crofts CBE, FRSE,
FCIEEM, FRSGS, FRGS, FRSA
WCPA Regional Vice-Chair Europe
2001–08, WCPA Emeritus

Editor's Preface

The genesis of this book lies in a large-scale collaborative research project, 'Towards coast to coast networks of marine protected areas (from the shore to the high and deep sea), coupled with sea-based wind energy potential' (CoCoNet), funded by the European Community's 7th Framework Programme from 2012 to 2016 (Grant Agreement No. 287844). Led by the Italian National Inter-University Consortium for Marine Science, under the management of Professor Ferdinando Boero, CoCoNet was one of the largest multi-disciplinary environmental projects ever to cover both the Mediterranean and Black Seas simultaneously. With 39 partners from 22 countries, CoCoNet strove to apply a holistic ecosystem approach to the management of the two basins, with a particular focus on developing a scientifically rigorous system for establishing ecologically coherent networks of Marine Protected Areas (MPAs). In addition to the science, CoCoNet also investigated associated socio-economic issues (led by NatureBureau, UK), and the technical potential for integrating offshore wind farms with the marine environment (led by the Hellenic Centre for Marine Research, Greece).

As the project reached maturity and the results were both being shared among the research teams and being widely published in journals (in more than 140 papers), it became apparent that some of the key cross-cutting ideas risked being fragmented or diluted unless they could be drawn together in a single volume. Through the good offices of Bob Carling, NatureBureau approached Wiley with a proposal for a book, mainly drawing on the results of CoCoNet, but also containing contributions from other world authorities in MPAs to more fully elaborate the concepts presented. Wiley accepted the proposal and this book came to fruition.

Not only does this book share a lot of new knowledge about the state of the marine environment in the Mediterranean and Black Seas, it does so around the central theme of networks of MPAs. Of course, the notion that individual MPAs would be ecologically more useful and effective if they are linked in networks (or systems) has been around for a long time. Indeed, there are a plethora of manuals and guidelines explaining how this should be done. However, there is surprisingly little evidence that networks deliver significantly more than the sum of their parts. To a large degree, this is due to an inevitable lack of comparative controls against which to assess how a particular network is faring. But other important reasons include the absence of a strong theoretical basis for their design, and imprecision about what an 'ecologically coherent' MPA network actually constitutes, as well as how such networks can be built, managed and monitored as discrete entities.

Such problems are explored in this book and some potential advances proposed: how to locate ecologically coherent MPA networks within 'cells of ecosystem functioning' (CEFs), that can be defined as 'a marine volume with coherent oceanographic, biological and ecological features, leading to higher degrees of connectivity than with nearby CEFs'; that networks can come in a variety of mutually interacting types that, once made explicit, can be used to design effective 'network-aware' management strategies; and that the adoption of the Marine Strategy Framework Directive (2008/56/EC) and Maritime Spatial Planning Directive (2014/89/EU) by EU Member States has put in place a progressive legislative system for achieving Good Environmental Status of EU marine waters, a system in which MPAs, as individual sites and as networks of them, will have a strong role to play in the two seas.

The network perspective of the book, based on the typology described in Chapter 3, can be used to analyse its 14 chapters as shown in the matrix. Each chapter addresses two or more of the network types identified in that chapter (namely Conservation, Connectivity, Socio-economic, Geographic, Collaborative, Cultural and Transnational) in order to give full coverage of the wide variety of forms and functions of MPA networks.

Paul D. Goriup
Chairman of the Board
NatureBureau, Newbury, UK

Matrix to show the relationships between book chapters and MPA network types.

Chapter number	Chapter title	Author(s)	Conservation	Connectivity	Socio-economic	Geographic	Collaborative	Cultural	Transnational
1	From Marine Protected Areas to MPA Networks	Ferdinando Boero	■						
2	Ecological Effects and Benefits of Mediterranean Marine Protected Areas: Management Implications	Antoni Garcia-Rubies, Emma Cebrian, Patrick J. Schembri, Julian Evans and Enrique Macpherson	■			■	■		
3	Typology, Management and Monitoring of Marine Protected Area Networks	Stephen Beal, Paul D. Goriup and Thomas Haynes	■			■	■		
4	Marine Protected Area Governance and Effectiveness Across Networks	Nigel Dudley and Marc Hockings	■				■		
5	Marine Protected Areas as Spatial Protection Measures under the Marine Strategy Framework Directive	Daniel Braun	■						■
6	Socioeconomic Impacts of Networks of Marine Protected Areas	Elena Ojea, Marta Pascual, David March, Isabella Bitetto, Paco Melià, Margaretha Breil, Joachim Claudet and Anil Markandya	■		■				
7	Multi-criteria Decision-Making for Marine Protected Area Design and Management	Paco Melià	■			■	■		
8	Ecosystem-Based Management for Marine Protected Areas: A Systematic Approach	Rafael Sardá, Susana Requena, Carlos Dominguez-Carrió and Josep Maria Gili	■			■	■		
9	Developing Collaboration among Marine Protected Area Managers to Strengthen Network Management	Chloë Webster	■				■		
10	Eyes Wide Shut: Managing Bio-Invasions in Mediterranean Marine Protected Areas	Bella Galil	■						
11	Marine Protected Areas and Marine Spatial Planning, with Special Reference to the Black Sea	Eva Schachtner				■			■
12	Black Sea Network of Marine Protected Areas: European Approaches and Adaptation to Expansion and Monitoring in Ukraine	Boris Alexandrov, Galina Minicheva and Yuvenaliy Zaitsev				■			■
13	Prospects for Marine Protected Areas in the Turkish Black Sea	Bayram Öztürk, Bettina A. Fach, Çetin Keskin, Sinan Arkin, Bülent Topaloğlu and Ayaka Amaha Öztürk				■			■
14	Marine Protected Areas and Offshore Wind Farms	NatureBureau, Newbury, UK				■	■		■

1

From Marine Protected Areas to MPA Networks

Ferdinando Boero

Università del Salento – CNR-ISMAR, Italy

The Ecology of Beauty

Just like terrestrial National Parks, Marine Protected Areas (MPAs) were first established at places where biodiversity had some prominent features. In the Mediterranean Sea, for instance, the first MPAs were established at places that were perceived as 'beautiful' by scuba divers who started to explore marine landscapes and singled out the most scenic ones (see Abdulla *et al.*, 2008 for a review on Mediterranean MPAs). The European Landscape Convention (ELC) (Council of Europe, 2000) is in line with this approach to site selection. The ELC, in fact, states that 'The sensory (visual, auditory, olfactory, tactile, taste) and emotional perception which a population has of its environment and recognition of the latter's diversity and special historical and cultural features are essential for the respect and safeguarding of the identity of the population itself and for individual enrichment and that of society as a whole'.

What is perceived as valuable in a given environment, then, is part of the heritage of the resident population and contributes to its culture. The positive impressions described in the ELC simply identify beauty, defined as follows in a popular dictionary: 'a combination of qualities, such as shape, colour, or form, that pleases the aesthetic senses, especially the sight'.

The perception of beauty, however, is directly linked to cultural paradigms and can change with them. Cetaceans, for instance, were once perceived as evil 'monsters' that brave sailors had to exterminate, as Melville's story of Moby Dick tells us. Nowadays, they are worshipped as gods. Even white sharks (*Carcharodon carcharias*), again depicted as terrifying beasts in movies like Spielberg's *Jaws*, are now considered as highly valuable, deserving strict protection.

Following this aesthetic approach, large vertebrates or, in alternative, beautiful and scenic habitats (i.e. the charismatic expressions of nature) are usually identified as deserving protection, whereas important ecological actors are simply ignored. Everybody wants to save the whales, but nobody wants to save the bacteria, even if bacteria are indispensable for ecosystem functioning (and also for our own body functions), whereas whales are not. On the one hand, our impact on bacteria is not so huge: they become rapidly resistant to antibiotics and are not affected much by our influence, being able to evolve rapidly so as to cope with environmental changes. On the other hand we could easily exterminate cetaceans, if only we intended to do it.

The preservation of beautiful portions of the environment, and of the fauna and flora inhabiting them, has been instrumental in the understanding of the value of nature. This approach to the defence of nature is shared by almost all environmentalist movements who evoke charismatic portions of nature in their logos, full of dolphins and panda bears. The growth of human population, with the adoption of economic paradigms aimed at the continuous growth of the economic capital, as if resources were infinite, has led to an alarming erosion of the planet's natural capital. Habitat destruction, both on land and in the seas, and climate change show that we need more than beauty to preserve nature. Protected areas, in this framework, have been some sort of surrogate that justified the destruction of nature where protection was not directly enforced. Focusing on the unique and beautiful facets of nature, often perceived as the sole expression of 'biodiversity', led to protection of natural structures, while disregarding natural functions that are not restricted to charismatic species and habitats.

Beauty is important, but the conservation of nature requires more than aesthetics.

From Landscapes to Habitats

The European Landscape Convention is centred on the way the culture of a population perceives and modifies nature, somehow 'improving' it with wise management. This is particularly evident in countries like Italy, where millennia of agriculture and architecture have led to unique landscapes that are considered of paramount importance in Article 9 of the Italian Constitution. In this sense, the landscape is the result of human interventions that led to changing a 'wild'

expression of nature into a 'gentler' one. Usually the products of these interventions are aesthetically valid, and the result is beauty. However, a beautiful landscape might be limited in the expression of biodiversity (especially if agriculture is involved), calling for the need of preserving nature *per se*, and not its modifications, whatever their aesthetic value. It can happen, furthermore, that a local 'culture' adopts some behaviours that are against the integrity of nature, as happened in Region Apulia with date mussel (*Lithophaga lithophaga*) consumption. The harvesting of date mussels from rocks caused extensive denudation of Apulian rocky bottoms (Fanelli *et al.*, 1994). The destruction of hard bottom habitats came to an end only after a long process of generating public awareness, together with the enforcement of new laws.

To cope with an overly anthropocentric approach to our interactions with the environment, the EU Habitats Directive (92/43/EEC) embraced a completely different perspective: habitats of community importance must be protected, even if this goes against the aspirations of the resident populations!

Sites protected under the Habitats Directive do not necessarily comprise beautiful landscapes, and the low level of ocean literacy in almost every country is often a source of conflict between the expectations of lay people and the preservation of natural capital. The resident communities are puzzled when they are prevented from building a new harbour just because there is a seagrass meadow on the bottom. Local populations often label as 'algae' the phanerogam *Posidonia oceanica*, whose presence can lead to the establishment of a protected site, and consider it as a nuisance. The decomposing leaves that accumulate on the beach repel tourists, who complain about their appearance and smell. The recognition of the ecosystem service of these accumulations of leaves is not part of local cultures, who do not realize that stranded leaves protect the beach from erosion. The stranded leaves are removed, sometimes with bulldozers, and huge quantities of sand are removed with them. Lacking a buffer of amassed leaves, wave action starts to erode the beach. Beaches are a source of income, and the wider they are, the higher the income, since more tourists can be crammed onto them. Beach erosion reduces incomes, and this is redressed by beach replenishment. Without the protection of *Posidonia* leaves, however, the newly placed sand is also rapidly eroded and often accumulates on the seagrass meadow, smothering it. *Posidonia* meadows are bioconstructions, since the new rhizomes grow over the old ones, raising the bottom of the sea and making it more stable. The death of the meadow is a catastrophe for the coast, since its role of erosion buffer ceases to protect the shore. Once the protection from erosion is completely gone, due to unwise management of coastal systems, physical defences are built in order to protect the beach, with a radical change of the whole landscape.

It is undeniable that some 'cultures' have a vague understanding of the functioning of nature, and the Habitats Directive is an attempt to bring a more objective approach to our relationship with natural systems.

Our land-based culture, however, still biases the Habitats Directive because although it considers marine habitats that are not necessarily 'beautiful', they are invariably benthic. For the Habitats Directive, the marine space is bi-dimensional, just as the terrestrial one. The third dimension, on land, is occupied just by the size of bodies, and by the temporary presence of flying organisms in the air, so it is right to speak about 'areas'. In marine systems, however, the water column is a three-dimensional habitat for a host of organisms that have almost no interactions with the sea bottom. Since oceans cover over 70% of the Earth, the water column is the most widespread habitat of the planet, and it is a volume.

Many marine organisms live their whole life suspended in the water, and even benthic ones derive their food from currents, not to mention the spread of propagules. A Habitats Directive which includes the marine biome but does not consider the third dimension of the water column is fundamentally flawed.

Protecting beautiful places, and managing the habitats of European Community importance, is a first step towards recognizing the significance of the marine environment, inviting science to design an approach to its management and protection that goes beyond the biases of the current 'culture'. Indeed, it calls for actions aimed at developing the 'ocean literacy' to alter our scant perception of the values of the oceans that is linked to our terrestrial history.

From Hunting and Gathering to Farming

If we were just like all the other species on the planet, when our populations increase to above the carrying capacity (i.e. the maximum number of individuals of a species an ecosystem can bear), overly eroding the natural capital that sustains us, our numbers should decrease due to a shortage of resources. This would lead to the re-constitution of the natural capital, according to the popular prey–predator model developed by Lotka and Volterra (Gatto, 2009), in which we are the predators and the rest of nature is the prey. But we are not like the other species. When confronted with a shortage of natural resources, we abandoned hunting and gathering and invented agriculture (Diamond, 2002). We domesticated a restricted set of animal and plant species, and started to culture them so as to satisfy our needs. Agriculture leads to the eradication of all competing species from a piece of land so as to rear just the domesticated one. The terrestrial animals we rear as food are almost invariably herbivores or, in some cases, omnivores, and we cultivate the plants we feed them with. This leads to habitat modification, and what the ELC considers as precious is often just the eradication of natural diversity and its substitution with agricultural systems.

In terrestrial systems there are no natural populations of both animals and plants that can provide massive amounts of resources. In the seas, by contrast, we can still extract resources from natural populations, and fishing is just a form of hunting. In recent decades, however, we have been rapidly passing from harvesting fish, crustaceans, molluscs and so on to aquaculture. What happened on land is now happening in the seas: wild populations cannot feed us all, and our pressure on them is leading several species towards commercial extinction, meaning the benefits from fishing are less than the costs incurred. Increasing the efficiency of fisheries, furthermore, is giving little hope of saving the remaining fish. The transition from fisheries to aquaculture is the final stage in the shift from hunting and gathering to farming. In the sea, contrary to what we do in terrestrial systems, we tend to rear carnivores rather than herbivores.

The Western world, in fact, is fed with farmed carnivorous species, such as sea bream (*Dicentrarchus labrax*) and salmon (*Salmo salar* and *Oncorhynchus* spp.), fed with smaller fish caught from surviving natural populations. This is clearly an unsustainable operation, since it exacerbates the overexploitation of natural populations: after having destroyed the populations of the larger fish, we culture them and we feed them with smaller fish caught from natural populations. Emerging countries cannot afford such costly forms of aquaculture and eat lower quality, but also less impacting, farmed herbivorous species such as tilapia (*Tilapia* spp.) and pangasius (*Pangasianodon hypophthalmus*).

The awareness of the impact of industrial fishing did induce some management of natural populations resulting in the protection of target species from overexploitation (Pikitch *et al.*, 2004). This has been done by restricting fishing activities at important places and during important periods. The relevance of these spaces and times depends on the biology of the species under management. Spawning grounds, nursery areas, and feeding grounds are identified species by species, and fisheries are restricted in order to allow for successful recruitment of the managed species. The ban of industrial fishing, *per se*, is a measure of protection and its positive impact, albeit temporal, is another form of marine conservation even though the aim is just to relieve fish from our excessive pressure, so as to continue to exploit their populations.

The reproductive rates of many fish species are so high that populations can be restored in reasonable time, as the abundance of fish in well-managed MPAs demonstrates (Guidetti *et al.*, 2008). Since the environmental impact of farming carnivorous species is higher than that of simply fishing, the survival of sustainable natural fisheries is a measure of the health of marine systems, and fisheries science must lead to better results, in conjunction with conservation science.

Landscapes, Habitats and Fish are Not Enough

The introduction of concepts such as 'ecosystem-based management', 'ecosystem approach' and 'integrated coastal zone management' is the clear expression of a broader view in the way we interact with the rest of nature (Pikitch *et al.*, 2004; Heip *et al.*, 2009). Ecosystems are not just structures, they also function through myriad processes, as their name implies. Knowledge of the connections among the different structures is crucial for managing what we intend to exploit, and to conserve what we want to protect. The link between biodiversity (structure) and ecosystem functioning (function) is the conceptual tool that guides a proper understanding of how the natural world works (Heip *et al.*, 2009). In a strategic document, the European Marine Board identified the adoption of holistic understanding as the greatest challenge for marine scientists worldwide (Arnaud *et al.*, 2013). It is obvious, for instance, that fish do not proliferate as isolated entities from the rest of the environment: they need to be considered as part of ecosystems throughout their life cycle, from the fertilized egg to the adult. This, for instance, should oblige fisheries scientists to consider the impact of predators of fish eggs and larvae, such as gelatinous plankton, in their models of fish population dynamics (Boero, 2013). The match (or mismatch) of a bloom of the by-the-wind sailor (the hydrozoan *Velella velella*) with the spawning of fish species that deliver floating eggs, for instance, can have (or not have) devastating effects on the fisheries yields of the subsequent months (Purcell *et al.*, 2015). However, the cause–effect relationship is usually not perceived since the impact (fewer fish) becomes apparent only when the cause (increased *Velella* predation and/or competition) is over, the lapse of time depending on the growth rate of the fish species concerned. If larval mortality is treated as a constant in fisheries models, fisheries management cannot be effective. The causes of potential failures in fish recruitment (resulting from depressed larval development) must be ascertained and fisheries science must overcome the almost complete separation from gelatinous plankton science (Boero *et al.*, 2008).

Similarly, the quality of the various habitats that fish frequent during their whole lifespan can have a crucial impact on fisheries yields, determining more or less successful recruitment. Yet, the scientists who study fish populations in MPAs are usually not directly involved in traditional fisheries science, even if their research tends to show that MPAs often improve fish yields due to spillover effects (Planes *et al.*, 2000). Fisheries scientists, though, usually disregard the role of MPAs and propose other management measures to promote sustainable exploitation of fish populations. Fisheries scientists are probably right, since the total surface of MPAs is scant, if compared with the vastness of the oceans, and the protected environments are almost invariably coastal and restricted to the sea bottom. While the current extent of protected marine space can improve local conditions, it is nowhere near sufficient to manage the entirety of fish populations. Furthermore, fisheries are just one of the manifold threats to the marine environment, and a more integrative approach to conservation is badly needed.

Good Environmental Status

Of course, a solution might be to increase the size and the density of MPAs, encompassing the SLOSS debate (Single Large Or Several Small) (Olsen *et al.*, 2013) with the Several Large approach. The increase in both the number and the size of MPAs, however, would cause conflicts between national and local authorities and the resident communities that, usually, are resistant to any limitation of their 'freedom' of (ab) using the environment.

Networks of MPAs seem the best solution for this conundrum (Olsen *et al.*, 2013). The Marine Strategy Framework Directive (MSFD, 2008/56/EC) sets the target of reaching Good Environmental Status (GES) in all EU waters by 2020. The situation of the European Seas will improve significantly if this strategic goal can be achieved, or at least if the trend towards its achievement triggers effective conservation measures.

The MSFD includes 11 descriptors of GES, which in their synthetic formulation are:

- Descriptor 1: Biodiversity is maintained
- Descriptor 2: Non-indigenous species do not adversely alter the ecosystem
- Descriptor 3: The population of commercial fish species is healthy
- Descriptor 4: Elements of food webs ensure long-term abundance and reproduction
- Descriptor 5: Eutrophication is minimised
- Descriptor 6: The sea floor integrity ensures functioning of the ecosystem
- Descriptor 7: Permanent alteration of hydrographical conditions does not adversely affect the ecosystem
- Descriptor 8: Concentrations of contaminants give no effects
- Descriptor 9: Contaminants in seafood are below safe levels
- Descriptor 10: Marine litter does not cause harm
- Descriptor 11: Introduction of energy (including underwater noise) does not adversely affect the ecosystem.

As Boero *et al.* (2015) remarked, pursuing GES based on these measures represents a real revolution in the management of marine ecosystems. In the past, the precise measurement of key environmental variables (temperature, salinity, nutrients, pollutants of any kind) was considered to be sufficient to evaluate the state of the environment. This led to the establishment of sophisticated observation systems that check these variables through the use of satellites, buoys, gliders, and a vast array of sensors. The collected data are then stored in huge databases that contain the 'history' of environmental systems. The factors that should inform us about the quality of the environment, however, do not represent the real state of any habitat. From the perspective of GES, these variables acquire a meaning only when they affect the living

component: if some of these variables change but this does not lead to any change in the biological component of ecosystems, then the change is irrelevant. The individual stressors, furthermore, do not act in isolation from each other. Instead, they interact with each other, with cumulative effects that might lead to misinterpretations of the quality of the environment. If considered in isolation from each other, these variables can have values that are below the threshold that is known to affect the living component of the environment. These effects are often assessed by laboratory experiments, under controlled conditions, in which only one variable is altered, whereas the others remain constant. The ensuing tolerance curves assess the impact of each stressor on selected species. However, even if the values of each stressor are below the thresholds, it can happen that biodiversity loses vigour, and many key species show signs of distress due to cumulative impacts (Claudet and Fraschetti, 2010).

To cope with this shortcoming, the MSFD defines GES while considering the status of both biodiversity and ecosystem functioning. The first descriptor of GES is just the status of biodiversity, whereas all the other descriptors regard the impact of specific stressors on biodiversity, ecosystem functioning and, in the case of Descriptor 9, human health.

Once a stress is identified, in terms of biodiversity and/or ecosystem function perturbation, then it can be addressed so as to mitigate its impact.

The logic of this approach is impeccable, but its application is far from straightforward. It is very simple to produce sensors that measure physical and chemical variables; even biogeochemistry can be assessed with automated instruments. Moreover, the geological features of the sea bottom can be mapped and assessed with very powerful tools. The descriptors of GES, however, consider biodiversity and ecosystem functioning, and the currently available instruments do not measure these features: they mostly consider abiotic features or measure some simple biotic variable, such as chlorophyll concentrations.

A new way of looking at the quality of the environment is then required, and the study of MPAs is somehow 'pre-adapted' to tackle this problem. Marine Protected Areas have been instituted to protect biodiversity and to enhance ecosystem functioning, and so adhere, at least in theory, to all the specifications of GES. The assessment of the efficacy of MPA management should consider the attainment of GES. If the requirements prescribed by some descriptors are not met, management should be changed in order to remove impediments to the attainment of GES.

Connectivity

The Marine Strategy Framework Directive of the European Union does not require the attainment of GES in MPAs only: GES is to be reached in all EU waters by 2020. This expectation is very ambitious, since GES is not reached even in the best-managed MPAs, but its logic is flawless. It is futile to hope for GES at any one place, if the surrounding environment is not in good condition as well. Marine Protected Areas are not like islands, separated from each other by the sea: the sea connects them.

Every individual living at a specific location produces propagules (the life cycle stages that propagate the species, whether as eggs, larvae, fragments, adults, etc.) that are taken away by the currents, to colonize other sites. Each site is a source of propagules for downstream sites that are reached by the current passing in its vicinity, and is a sink of propagules coming from the organisms living at upstream sites. Connectivity, then, is the degree of connection across sites within a given area. The very concept of connectivity teaches us that it is pointless to

manage specific sites (e.g. MPAs) without managing the systems in which they are nested in terms of connectivity. This insight is leading to a paradigm shift in conservation biology: from MPAs to networks of MPAs (Olsen *et al.*, 2013).

Connectivity is a very general concept: the connections among various parts of a given water body cannot be measured in a way that represents all living beings. Some species have a higher vagility (i.e. propensity to move from one place to another) than others and the differences greatly affect connectivity at a micro level. Grantham *et al.* (2003) tackled the problem of dispersal distances in a suite of habitats, considering just marine invertebrates, and reached the conclusion that the ensuing connections are very varied and that MPAs must therefore be designed based on the specific habitats that are going to be protected. Accordingly, networks of MPAs should encompass this problem, providing protection over large scales. However, it is also important to design MPA networks so as to respect complex connectivity patterns, in order to achieve a compromise that covers the different scales of vagility of the species assemblages that are going to be protected and/or managed. Knowing the basic biology of species, however, is not enough: ecological constraints and habitat availability can restrict the colonization of localities that can be reached by a given species but that are not suitable for its existence. For example, Johannesson (1988) considered two species of the mollusc genus *Littorina* with opposite dispersal strategies (planktonic versus brooding). The species with planktonic larvae should be a better colonizer than the brooding one. However, the brooder species had a higher propensity than the one with planktonic larvae to persist at a sink habitat widely separated from source areas. The 'paradox of Rockall' (Johannesson, 1988) shows that larval dispersal is not the sole factor responsible for connectivity. Sink areas that are distant from propagule sources tend to be colonized by low dispersal species that can reach them by rafting and that re-colonize the area without dispersing their propagules. In this regard, Boero and Bouillon (1993), analysing the distribution of more than 300 hydrozoan species of the Mediterranean Sea, showed that species with a long-lived medusa stage do not have a wider distribution than that of brooding species, brooders often being more widespread than highly vagile species (Shanks *et al.*, 2003).

As a result of such studies, it is clear that the levels of connectivity across an area are better studied by at least four methods:

1) The reconstruction of the oceanographic framework that potentially connects the various sites
2) The search for propagules (including asexual ones, and rafters) in the plankton collected in the connecting currents
3) The similarity of species assemblages across the considered area (so-called beta diversity)
4) The similarity in the genetic composition of a suite of species that represent a vast array of taxa.

The integration of the results of these different analyses leads to a more reliable representation of the degree of actual connectivity, helping to design more ecologically coherent networks of MPAs.

Networking According to Nature or to Bureaucracy?

The application of coherent policies of management and conservation of MPAs is particularly well developed in the Mediterranean area. The management entities of many Mediterranean MPAs are part of MedPAN (Webster, this volume) and, through it, the best practices evolved by the directors of each MPA are disseminated and improved, so as to find increasingly better ways of protecting nature. It is undeniable that issues regarding nature conservation have to be addressed over vast scales, and that the comparison of the efficacy of measures at different places is conducive to increasingly better ways of protecting the environment. It is also true, however, that there is not a one-size-fits-all way of solving the problems stemming from our relationship with the rest of nature. Special measures are necessary to protect remarkable properties of the marine environment, such as the presence of unique expressions of biodiversity in terms of either species (e.g. monk seals *Monachus monachus*, or cetaceans) or habitats (e.g. bioconstructions of any kind). Defending unique structures, however, is not enough: connectivity calls for a more integrated approach than just a structural one. Structures must be coupled with the ecosystem functions that allow for, if not underpin, their existence, and this approach calls for the expansion of management far beyond the boundaries of MPAs.

It is crucial, in this framework, to identify the units of conservation, namely the portions of marine space that are highly connected with each other and whose features are more dependent on each other than on those of sites that belong to other units. The identification of these units leads to the construction of networks of MPAs that are based not only on the enforcement of protection measures through bureaucratic imperatives, but also on the recognition of ecological principles that rule the functioning of the managed environments, just as the definition of GES prescribes.

These units might be based on climatic and biogeographic features, comprising areas where species compositions are similar due to shared climatic conditions; or on oceanographic features, where current patterns determine propagule transport; or on geological features of the sea bottom; or, indeed, on geo-political features that might be conducive to common management by various states. In such politically fragmented seas as the Mediterranean and the Black Seas, this approach requires development of and adherence to international agreements since it is highly unlikely that a single state will cover the whole extension of ecologically coherent conservation units.

It is evident, however, that the identification of these units of conservation must be holistic, covering most of the features that the single disciplines making up the complex of marine sciences now study in isolation. To satisfy this need, Boero (2015a) proposed to treat the marine environment as a living super-organism made of cells: the 'cells of ecosystem functioning' (CEFs). The exercise of dividing the marine space into larger conservation units than MPAs is not novel (see Olsen *et al.*, 2013 for a review), and its necessity is shared throughout the scientific community and among decision-makers.

Towards a Holistic View of Marine Systems

The previously mentioned quest for integrated, ecosystem-based, and holistic approaches to marine conservation requires a complex representation of marine spaces based on the assemblage of the available knowledge in an ecologically coherent fashion.

The physical background is the backbone of ecosystem description. The discovery of the oceanic conveyor belt (Broecker, 1991), with the recognition of the crucial role of polar regions as surface sites of deep water formation, marked a revolution in physical oceanography that parallels the discovery of continental drift to explain the current disposition of continental masses. The oceans are in fact one, the global ocean, and all are connected by horizontal and vertical currents. The cold and dense surface waters of the poles tend to sink and to become the deep waters of non-polar portions of the

ocean system, pushing up the spent waters of the deep. Everything is connected, in the oceans, and life is running on an apparently perpetual conveyor belt that distributes nutrients and propagules throughout the world. The single, interconnected oceanic system, however, can be divided into coherent portions, defined by the disposition of continental masses.

The Mediterranean Sea, in particular, due to its geological, oceanographic and bio-ecological features, is a miniaturized replica of the world ocean and, due to its smaller size, responds more quickly to the drivers of change that affect the whole planet (Lejeusne *et al.*, 2010). It is convenient, thus, as a first approach to the identification of coherent conservation units, to focus on the Mediterranean Sea so as to set up a feasible rationale that could possibly apply to whole oceanic systems.

From a physical oceanography point of view, the Mediterranean conveyor belts (Pinardi *et al.*, 2004) can be considered as analogous to the large oceanic conveyor belt (see Figure 1.1).

The Mediterranean Sea has a higher salinity than the Atlantic Ocean since freshwater inputs are lower than evaporation rates. The superficial Gibraltar Current enters from the Gibraltar Strait and brings Atlantic waters into the Mediterranean Sea, compensating the water deficit due to excessive evaporation. The Gibraltar Current crosses the Sicily Channel and flows into the Eastern Mediterranean, to flow back at about 500 m depth as the Levantine Intermediate Current that returns to the Atlantic, through the deepest part of the Gibraltar Strait. Since the average depth of the Mediterranean Sea is 1500 m, and the deepest part of the basin, in the Ionian Sea, exceeds 5000 m, the water renewal of the upper 500 m is not enough to bring oxygen to the depths of the Mediterranean Sea, where plants and other primary producers do not have enough light to perform photosynthesis and produce

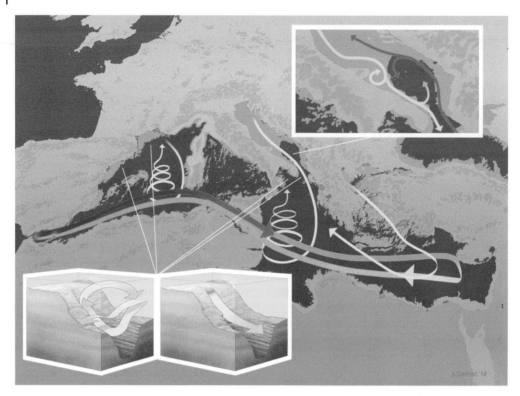

Figure 1.1 Circulation patterns in the Mediterranean Sea. A surface current enters the basin from the Gibraltar Strait, flows through the Sicily Channel and reaches the Levant Basin. The Gibraltar Current flows back at about 500 m depth as the Levantine Intermediate Current. Water renewal below 500 m occurs through the 'cold' engines in the Gulf of Lions for the Western Basin and in the Northern Adriatic and Northern Aegean Seas for the Eastern Mediterranean. In the cold engines, cold, oxygen-rich water flows through canyons (bottom left inset) with a 'cascading' process. The canyons outside cold engine areas can trigger upwelling events (bottom right inset). Other patterns of circulation regard the formation of gyres (top inset). Artwork: Alberto Gennari.

oxygen. Without photosynthesis, deep-sea animals would rapidly consume the oxygen dissolved in the water, leading to anoxic conditions that are not favourable to metazoan life. Without an oxygen supply from the surface, the Mediterranean deep-sea biodiversity would be much reduced and just a few simple life forms would survive, as happens in the Black Sea below 300 m depth.

The 'cold engines' of the Gulf of Lions, the Northern Adriatic and, from time to time, the North Aegean are crucial to the existence of deep-sea life in the Mediterranean Sea. At these sites, northern winds enhance evaporation and lower the temperature, causing a marked density increase in the well-oxygenated surface waters. The thermo-haline differences of the water masses of the cold engines in respect to the surrounding waters result in the so-called cascading of dense oxygenated waters that cross the continental shelf and, then, reach the deep sea through marine canyons. The cold engine of the Gulf of Lions renews the deep waters of the Western Mediterranean Basin, whereas the Northern Adriatic engine, sometimes replaced by the North Aegean one, refreshes the depths of the Eastern Mediterranean Basin.

The Gibraltar and the Levantine Intermediate currents join the various parts of the basin, defining the Mediterranean Sea as a single and very large unit. The cold engines produce vertical thermo-haline exchanges that define the Western and the Eastern Mediterranean as two large sub-units that, based on coastal morphology, are in their turn divided into the well-known 'seas' that make up the Mediterranean system.

Oceanographic conditions determine further sub-divisions of the seas that make up the Mediterranean. In the Adriatic Sea, for instance, the cold engine causes a thermo-haline current that flows southwards across the continental shelf and along the Italian coast to the Ionian Sea through the Bari Canyon. To balance this outflow, an incoming current enters the Adriatic Sea from the eastern coast of the basin, and reaches the Gulf of Trieste, where the circle is closed. The presence of headlands such as those at Istria, Conero and Gargano leads to the formation of a northern, a central and a southern gyre, with horizontal currents that connect the western and the eastern coasts of the basin, along which the currents flow in opposite directions. In this way, the Adriatic Sea could be divided into three coherent oceanographic cells, where ecosystems might function in distinct fashions, while being anyway connected by the northward current along the eastern Adriatic coast and the southward current along the western coast.

The Adriatic Sea is shallow and does not have canyons in its central and northern part, but canyons leading to the deep sea from the coast are a common feature of the rest of the Mediterranean shelf. Some are involved in the cascading phenomena generated by the cold engines but, in the majority of the canyons, the currents that flow parallel to the coast tend to sink offshore, bringing oxygen to the deep sea. These offshore downwellings push deep waters through the canyon, resulting in upwelling currents that connect the deep sea with the coastal areas (Hickey, 1995). There are about 500 Mediterranean canyons that, presumably, play the role of auxiliary engines to the three main cold engines, and underpin the survival of life in the deepest part of the Mediterranean Sea through vertical water exchanges. The upwellings, furthermore, bring nutrients towards the shore, enhancing primary production such as the spring phytoplankton bloom. Based on these oceanographic patterns and on the presence of a higher concentration of resting stages of both phyto- and zooplankton than outside the canyons, Della Tommasa *et al.* (2000) proposed that marine canyons are reservoirs of propagules (in this case resting stages of planktonic organisms) that are injected towards the coast together with the nutrients, so triggering the phytoplankton and zooplankton blooms that are at the base of the functioning of all oceanic systems.

The hydrodynamic patterns, generated by a combination of wind energy, changes in salinity and temperature, and interactions of currents with bottom and coastal morphology, define the physical framework that leads to the formation of masses of water that are more connected within their boundaries than they are with neighbouring masses, while remaining part of a coherent water body. The main sub-units can be further divided into smaller units according to the presence of fronts, gyres, eddies, upwellings and downwellings, defining what Boero (2015a) called the cells of ecosystem functioning, CEFs, mentioned earlier. With this metaphor, the Mediterranean Sea is a body (which is anyway dependent on other bodies, in this case the Atlantic Ocean) that can be divided into increasingly smaller functional parts, from wide ecological regions *sensu* Longhurst (2010) to CEFs as the smallest functional units.

The Cells of Ecosystem Functioning

Oceanographic conditions shape the associated ecological processes. The ensemble of areas where physical processes connect different portions of the environment might be considered a CEF. However, the long-term observation of oceanographic features shows high variability, including sudden and radical changes, as happened with the Eastern Mediterranean Transient (Pinardi *et al.*, 2004). Phenomena such as El Niño, the North Atlantic Oscillation, and, in recent decades, global warming, lead to a suite of multiple states that might not overlap in space. Eddies and gyres, furthermore, can have variable strengths, and even invert their rotation. Upwellings are stronger in some seasons and weaker in others. Extreme events such as the occurrence of very hot or very cold periods can have huge impacts on biological features, with effects that persist for a long time after the occurrence of the episodes. Rivetti *et al.* (2014), for instance, showed that the deepening of the summer thermal stratification caused large-scale mass mortalities of resident species of cold-water affinity. Temperature increases, furthermore, have favoured the massive expansion of non-indigenous species that continue to enter through the Suez Canal, establishing viable populations in the Mediterranean Sea.

The strong annual (seasonal) and inter-annual fluctuations and variations of the physical drivers determine the bio-ecological features that represent an integration of these fluctuations over the long term (Boero, 1994), with episodic events adding variability to this complex situation (Boero, 1996).

The interactions among species assemblages (the expression of biodiversity) and the physical variables lead to the formation of ecosystems and determine their functioning (Boero and Bonsdorff, 2007).

The inter-annual variability of planktonic communities is well known from long-term series (Boero *et al.*, 2014), whereas only recently has the long-term response of benthic communities to important physical changes, mainly due to global climate change, started to be quantified (Puce *et al.*, 2009). It is important, then, to establish not only the potential CEFs, in terms of physical features, but also the tangible CEFs in terms of biodiversity and ecosystem functioning: a CEF is defined by a higher level of internal connectivity compared with connections to nearby CEFs. It can happen, however, that cells that appear physically separated, at least temporarily, such as the central and southern Adriatic cells, defined by two adjacent gyres, might have such connected biological populations that a single, larger cell and, hence, a single large conservation unit, should be defined.

Obviously, these multiple physical states, leading to multiple ecosystem states, can be revealed only through continuous observation and cannot be predicted by current modelling techniques. No model, for instance, predicted the occurrence of the Eastern Mediterranean Transient.

Moreover, the approaches followed so far to assess the quality of the environment are more focused on structure than on function. The evaluation of ecosystem functioning in large marine ecosystems has been assessed only rarely (e.g. Godø *et al.*, 2012).

Mapping the Seas

Mapping benthic communities is relatively easy and, with state-of-the-art technologies, can be accomplished in reasonable time frames. Benthic communities can be subject to strong seasonal variation, especially in coastal areas, but their areas of occurrence are generally rather stable in space. Maps can be made from time to time and compared so as to ascertain changes in habitat distribution.

The Habitats Directive, with the associated Natura 2000 network, applies a terrestrial approach to the marine realm. The description of habitats, furthermore, is based on the features of vegetation and on the concept that the dynamics of communities leads to climax conditions after a series of deterministic seres. These concepts apply only partially to the marine domain. In marine systems the water column is the most crucial component, being the habitat of both plankton and nekton, whose temporal variability is very high if compared with that of the benthos. The connections between the sea bottom and the water column are so intimate that the functioning of their communities cannot be understood if they are considered as separate entities (Boero *et al.*, 1996).

Terrestrial habitat maps are bi-dimensional and consider the vegetation as a descriptor of diversity. Maps of marine habitats resulting from the application of the Habitats Directive are similar to terrestrial ones, since they consider just the benthic realm. However, marine habitat maps would be far more complex if the water column was taken into account. What is happening at the surface does not necessarily reflect the rest of the water column, and temporal patterns are very distinct, so the same physical space has different ecological features in different periods of the year, usually changing from year to year. The dimensions are four: the two of the surface area of the sea bottom, the third one of the volume of the water column (and its diversity through its entire depth), and finally the time dimension.

As a result, CEFs are fuzzy units that cannot always be sharply defined (due to their temporal instability) but nevertheless are more internally coherent than they are with neighbouring cells. Some cells may be relatively distinct, such as the northernmost part of the Adriatic Sea, whereas others can be alternately separated or joined, as occurs in the two gyres that characterize the central and the southern Adriatic Sea. According to the source and sink approach (Pulliam, 1988), some cells are a source for other cells that receive their products as sinks and, in their turn, can be sources or sinks for other cells, but the roles can be inverted according to different situations. The cold current generated in the Northern Adriatic, flowing southwards along the Italian coast, brings nutrients that support the white coral formations that thrive in the depths of the Southern Adriatic and Ionian Seas. The Northern Adriatic, in its turn, receives propagules from the current that enters the Adriatic from the Ionian Sea and that flows northwards along the coasts of Greece, Albania, Croatia and Montenegro, reaching Slovenia and then Italy.

It is clear, in this framework, that connectivity is not only a matter of propagules (of any kind) but also of food and nutrients, becoming almost a representation of ecosystem functioning, from the base of trophic networks (in terms of nutrients for phytoplankton, due to terrestrial runoffs and bacterial and fungal decomposition) to their very apex, namely the nekton.

The features of CEFs must be georeferenced, but the maps do not need to be overly accurate. These features of the environment, being very variable in time, cannot be found again with absolute precision, based on their representation on a map. An area where a gyre is enhancing primary production cannot be mapped with the same precision as an area covered by a seagrass *Posidonia oceanica* meadow. For sea grasses, the accuracy of the map can be tested by repeating the observation, and checking if the mapped feature is exactly in the place reported by the map. But this is not feasible for an area where fish forage, reproduce or spawn, or where phytoplankton and zooplankton bloom. It is however possible to identify some stable features that can be mapped with high accuracy. Canyons, as mentioned, can generate upwelling currents that bring nutrients from the deep sea to the coast. These upwellings provide a nutrient supply that favours primary production, and this ecosystem feature can be mapped. The 500 canyons that indent the continental shelf of the Mediterranean Sea (with the exclusion of those that are influenced by cascading phenomena generated by the cold engines) should be considered putative CEFs due to vertical currents. Their presence could lead to the testable hypothesis that the upwellings they generate foster ecosystem functioning in terms of phytoplankton production. Merging the representation of these vertical currents with the horizontal currents generated by both the winds and the configuration of the coast (i.e. gyres, eddies, fronts, etc.) should lead to maps that reflect the functioning of ecosystems in space. The multiple states of ecosystem features should be referred to these spaces, with maps that allow for temporal variability, supplemented with the distribution of habitats on the bottom, and of the behavioural patterns of important fish species (in terms of nursery, foraging and spawning areas).

Such maps are not available yet, and their realization is a compelling challenge, leading to the integrated, holistic and ecosystem-based approach that, in spite of being continuously invoked, has been rarely accomplished, so far.

Upgrading the Observation Systems and Managing the Networks

As mentioned earlier, the enforcement of the MSFD, so as to reach and maintain GES, calls for observation systems that assess the quality of the environment according to 11 descriptors of GES, which in turn are based on two main pillars: biodiversity and ecosystem functioning in all its facets. Current

observation systems must be upgraded, so as to cover all the relevant variables. Marine Protected Areas are the perfect places to perform continuous observation of the descriptors of GES, in terms of biodiversity (structure) and ecosystem functioning (function). The personnel of MPAs must be instructed on how to make these measurements, building on the experience of several marine stations that have been constantly monitoring the features of the water column for decades (Boero *et al.*, 2014).

Marine Protected Areas, however, are not enough and it is important to observe also control sites that are not under special protection regimes so as to be sure that GES is reached not only at already protected sites but throughout the sea. This calls for continuous evaluation of the features of biodiversity inhabiting both the sea bottom and the water column, with the establishment of long-term series of observations; this approach has tended to be unwisely dismissed due to the illusion of measuring the quality of the environment through the use of automatic devices. While current sensors can provide physical, chemical and biogeochemical information, they cannot measure either biodiversity or ecosystem functioning, and are therefore inadequate for the purposes of the MSFD (Boero *et al.*, 2015).

The continuous observation of ecosystem features should have two goals:

1) Assess the attainment of GES
2) Measure the efficiency of management.

Based on the definition of CEFs, and on the continuous check of their features through upgraded observation systems, the managers of MPAs must collaborate across networks of MPAs, leading to the definition of common policies within each network, based on the integrated study of the marine environment so as to perform efficient management and protection: the MedPAN structure, in the Mediterranean Sea, already represents a partnership of MPA managers.

Marine stations and other research institutions, furthermore, must be involved in a science-based management of the networks, leading to collaboration among states, in order to design regulations that will be tailored on the ecological conditions of the managed area, and not on the contingencies of political or bureaucratic situations.

Some goals of the MSFD might be difficult to reach through local management. Descriptor 2 of GES, for instance, requires that non-indigenous species do not affect the ecosystems in a negative way. It is undeniable that some aliens are real pests that impair the functioning of ecosystems. The case of the alien ctenophore *Mnemiopsis leidyi*, for instance, led to a disaster in Black Sea fisheries (Boero, 2013), although, in this case, the management of a hypothetical network of MPAs might have had little responsibility for an event that was mediated by species transport in ballast waters.

The early detection and risk assessment of non-indigenous species (NIS) is essential to determine appropriate action to prevent their spread, or to identify routes of arrival and to control them, whenever possible. Ship-driven introduction of alien species is particularly important (Boero, 2002) and is amenable to control measures. The recent doubling of the size of the Suez Canal, however, is likely to ease the arrival of more species of Lessepsian immigrants (Galil *et al.*, 2015; Galil, Chapter 10, this volume), and is much more difficult to control. This will probably aggravate the impact of non-indigenous species on the functioning of Mediterranean ecosystems, so worsening the situation required by Descriptor 2 of GES.

The observation systems, thus, will have to be set up also at the gateways to the Mediterranean Sea, with a particular focus on the Suez Canal, both in the Mediterranean and in the Red Sea.

Marine Protected Areas, *per se*, do not offer protection from NIS invasion, even though healthy ecosystems such as those ensuing from effective protection might be more resistant to invasions. In some cases the prohibition of human activities might enhance the chances of success for an invasive NIS. In such cases protection can be suspended and eradication measures might be taken, resulting from careful scientific assessments.

Human Capacity Building

The reliance on automated and physically oriented measurement of environmental quality has led to the perception that 'simple observation' as performed by the old naturalists is obsolete. In particular, the importance of describing species and understanding their roles, having taxonomic expertise at its base, has been disregarded and taxonomic expertise is vanishing across Europe and the Western world in general. In the era of biodiversity, the science of naming species (taxonomy) is in distress (Boero, 2010a), a rather paradoxical situation.

The definition of biodiversity (the first descriptor of GES) without taxonomy is simply flawed. The second descriptor, furthermore, covers the impact of alien species on ecosystem functioning. This requires knowing not only the resident species but also the species that might reach places where they have never been found. This means knowing all species, at a planetary level, since the introduction by shipping can bring species from any part of the world.

Furthermore, species identification is not enough. We must also assess the impact of alien species on the functioning of the ecosystems, and this means understanding their roles and their relationships with other species (Piraino *et al.*, 2002).

This level of knowledge requires a revival of traditional natural history (Boero, 2010b), while exploiting the most advanced techniques to tackle these very difficult problems. We need to create new expertise that is able to integrate the expert observation of nature with the so-called 'next generation' instruments (Boero and Bernardi, 2014; Boero *et al.*, 2015).

Marine Protected Areas, together with marine research stations, are the best places to build the new expertise required to manage the environment in a holistic way, as required by the MSFD. This will have to be accomplished by the collaborative effort of consortia of European universities and natural history museums.

Taxonomy, for instance, cannot just concern naming specimens by reference to already known species, or the description of new species based on some preserved specimens or on some genetic sequence. The knowledge of both phenotypes and genotypes is necessary but not sufficient. It is also necessary to elucidate the life cycles and life histories of species, and to define their ecological niche at least in terms of 'who eats whom', so as to ascertain the roles of species. Trait analysis, for instance, is often performed, ascribing the same traits to species that resemble each other, extending the knowledge acquired for one species to a whole group of species.

A new kind of biodiversity expertise is badly needed, if the requirements of GES are to be achieved. Moreover, it is also urgent to train 'integrative scientists' who are able to bridge the various disciplines, in order to reach the holistic approach so often invoked and yet so rarely achieved. Mathematical modelling leading to predictions of the kind 'if the situation is A at time 0, it will be B at time 1' is of course to be encouraged but the complexity of the highest levels of organization of nature does not produce the same results as those that have been reached at the lowest ones. The intertwining between biodiversity and ecosystem functioning cannot be treated as interactions of subatomic particles or black holes. When life enters the game, the number of variables becomes too high to handle with the tools of simpler disciplines. It is not by chance that the insights provided by the work of Charles Darwin cannot be translated into algorithms (Boero, 2015b) and the 'natural history' approach (upgraded with all the next generation technologies) is conducive to better insights about the functioning of complex natural systems (Ricklefs, 2012; Tewksbury *et al.*, 2014).

Extinction in the Mediterranean Sea

The re-building of taxonomic expertise, in the light of current concepts of ecology, biogeography and conservation, will probably show that current data on the distribution of species and habitats, as well as the models ensuing from them, should be treated with great caution. It is often the case that the distribution of biota is reconstructed by assembling data derived from different sampling methods and periods, lumping together very old records with recent ones. This leads to mistakes in evaluating the current state of biodiversity, since a species recorded from some place several decades ago might not still be present at the same place. Hence, a distribution map constructed by assembling new and old records does not account for the actual distribution of a given species (or habitat), and any conservation measure based on such data will prove ineffective.

This matter has become particularly salient in recent decades, since global change is rapidly modifying the physical features of the seas, especially as far as temperature is concerned. This is leading to radical modifications of biota, with increasingly widespread signs of stress for species that are adapted to temperate conditions and cannot withstand temperatures that reach values above their limits of tolerance (Rivetti *et al.*, 2014). It would be not surprising if some species have become extinct due to such changes in physical conditions, as well as to the arrival of more competitive aliens, pre-adapted to the new, warmer conditions.

The analysis of the distribution in time and space of a well-known group of Mediterranean invertebrates demonstrates the shortcomings of taking simplistic approaches to represent the distribution of biodiversity. Stemming from recent monographic work (Bouillon *et al.*, 2004), Gravili *et al.* (2015) divided the records of Mediterranean species of non-siphonophoran Hydrozoa into time intervals. Out of the 398 known species, only 162 (41%) have been reported in the last decade, while 53 (13%) were not recorded in the literature for at least 41 years. According to the Confidence of Extinction Index (Boero *et al.*, 2013), 60% of the 53 missing species are extinct, and 11% are probably extinct from the basin. From a biogeographical point of view, the missing species are 34% endemic, 19% boreal, 15% Mediterranean-Atlantic, 11% Indo-Pacific, 11% circumtropical, 4% cosmopolitan, 2% tropical-Atlantic, and 4% non-classifiable. Fluctuations in species composition in a certain area cause high variability in the expression of both structural and functional biodiversity. As a consequence, regional biodiversity should be analysed through its temporal evolution, to detect changes and their possible causes. This approach has profound implications for biodiversity assessments and also for the compilation of red lists of species that are in danger of extinction. Such analyses require a detailed knowledge of the literature covering a given taxon, so as to ascribe records to different periods. In spite of continuous claims of biodiversity crises, extinction has rarely been proven in the Mediterranean Sea, or indeed in any other oceanic system (Boero *et al.*, 2013). Nevertheless, the example of the Mediterranean non-siphonophoran Hydrozoa suggests that biodiversity is changing at a fast pace, and that current species lists are the result of adding new records (usually made up of non-indigenous species) to the old ones, leading to an apparent steady increase of the species pool of a basin. This artefact, furthermore, is biased by the distribution of sampling effort, the distribution of species often directly corresponding to the distribution of specialists and of their sampling effort, which is often concentrated around their institutional location.

All this calls for regular monitoring of species diversity at key locations, with all-species inventories, in order to produce solid estimates of the extant species pools and to observe their evolution in time. Puce *et al.* (2009), for instance, comparing recent and 25-year-old assessments of the phenology and the species pool of hydrozoans at a specific location, found substantial changes that suggest a great influence of global change on biodiversity expression. It is rather unfortunate, in this respect, that long-term series are not being maintained in most countries and that they run the risk of being dismissed even where they have been carried out over a long period (Boero *et al.*, 2014).

Conclusion and Recommendations

Marine systems are still generally in such conditions that, with fisheries, we can extract resources from natural populations, but this will not last for long if we do not enforce appropriate measures of both management and conservation of the natural capital. All governments and nations concur in recognizing the value of biodiversity, and the integrity of nature has been the object of a recent Encyclical by Pope Francis (Bergoglio, 2015), with full recognition of the central role of science in the preservation of nature, since it is impossible to protect something that is ignored. Increasing our understanding of complex natural objects such as the oceans, and the life therein, is still a 'great challenge' (Arnaud *et al.*, 2013) and there are no shortcuts that will improve our knowledge with little effort. Naming all species is probably a feasible accomplishment, with adequate investment (Costello *et al.*, 2013) and this should be the first step towards the inventory of biodiversity (the bulk of the natural capital). A second step is the understanding of the roles of species (Piraino *et al.*, 2012), and then the link between the diversity of species and the functioning of ecosystems (Boero and Bonsdorff, 2007). This will require an understanding of the geographic distribution of ecosystems in the marine space, and the concept of CEFs probably deserves further consideration; at present it represents only a scientific hypothesis and it needs to be tested at multiple places. The identification of management and protection units, however, is crucial to enforce efficient policies and this has not been accomplished yet.

The 11 descriptors of GES of the MSFD cover the most important features of the environment, but their principles need to be translated into action through the enforcement of policies, and these, to be effective, must be science-based.

The need for new observational approaches developed by new types of expertise is the logical outcome of a century of extreme reductionism and specialization. We have built a series of very solid bricks of knowledge. Now they have to be assembled so as to acquire a conceptual continuity. This challenge cannot be avoided.

Acknowledgements

This chapter stems from the author's recent work as coordinator of the European Community's 7th Framework Programme (FP7/2007–2013) project CoCoNet (Grant Agreement No. 287844, http://www.coconet-fp7.eu/) and thorough discussions throughout the project consortium. Financial support was also provided by the project PERSEUS (Grant Agreement No. 287600, http://www.perseus-net.eu/site/content.php) also of the 7th Framework Programme, and by the flagship Italian project RITMARE (http://www.ritmare.it/). Support by the European Union for VECTORS (http://www.marine-vectors.eu/) and from the Italian Ministry of Research (PRIN) is also acknowledged.

References

Abdulla, A., Gomei, M., Maison, E. and Piante, C. (2008) *Status of Marine Protected Areas in the Mediterranean Sea*. IUCN, Malaga and WWF, France. 152 pp.

Arnaud, S., Arvanitidis, C., Azollini, R. *et al.* (2013) *Navigating the Future IV*. Position Paper 20 of the European Marine Board, Ostend, Belgium. 203 pp. ISBN 9789082093100.

Bergoglio, J. [aka Pope Francis] (2015) *Lettera Enciclica Laudato si' del Santo Padre Francesco sulla cura della casa comune*. Vatican City. 191 pp.

Boero, F. (1994) Fluctuations and variations in coastal marine environments. *P.S.Z.N.I: Marine Ecology*, **15** (1), 3–25.

Boero, F. (1996) Episodic events: their relevance in ecology and evolution. *P.S.Z.N.I: Marine Ecology*, **17**, 237–250.

Boero, F. (2002) Ship-driven biological invasions in the Mediterranean Sea, in 'Alien marine organisms introduced by ships in the Mediterranean and Black Seas'. *CIESM Workshop Monographs*, **20**, 87–91.

Boero, F. (2010a) The study of species in the era of biodiversity: a tale of stupidity. *Diversity*, **2**, 115–126.

Boero, F. (2010b) Marine Sciences: from natural history to ecology and back, on Darwin's shoulders. *Advances in Oceanography and Limnology*, **1** (2), 219–233.

Boero, F. (2013) Review of jellyfish blooms in the Mediterranean and Black Sea. *GFCM Studies and Reviews*, **92**. 53 pp.

Boero, F. (2015a) The future of the Mediterranean Sea ecosystem: towards a different tomorrow. *Rendiconti Lincei*, **26**, 3–12.

Boero, F. (2015b) From Darwin's Origin of Species towards a theory of natural history. *F1000 Prime Reports*, **7**, 49. doi:10.12703/P7-49

Boero, F. and Bernardi, G. (2014) Phenotypic vs genotypic approaches to biodiversity, from conflict to alliance. *Marine Genomics*, **17**, 63–64.

Boero, F. and Bonsdorff, E. (2007) A conceptual framework for marine biodiversity and ecosystem functioning. *Marine Ecology – An evolutionary perspective*, **28** (Supplement 1), 134–145.

Boero, F. and Bouillon, J. (1993) Zoogeography and life cycle patterns of Mediterranean hydromedusae (Cnidaria). *Biological Journal of the Linnean Society*, **48** (3), 239–266.

Boero, F., Belmonte, G., Fanelli, G. *et al.* (1996) The continuity of living matter and the discontinuities of its constituents: do plankton and benthos really exist? *Trends in Ecology and Evolution*, **11** (4), 177–180.

Boero, F., Bouillon, J., Gravili, C. *et al.* (2008) Gelatinous plankton: irregularities rule the world (sometimes). *Marine Ecology Progress Series*, **356**, 299–310.

Boero, F., Carlton, J., Briand, F. *et al.* (2013) Marine extinctions: patterns and processes. *CIESM Workshop Monographs*, **45**, 5–19.

Boero, F., Kraberg, A.C., Krause, G. and Wiltshire K.H. (2014) Time is an affliction: why ecology cannot be as predictive as physics and why it needs time series. *Journal of Sea Research*. doi:10.1016/j.seares.2014.07.008

Boero, F., Dupont, S. and Thorndyke, M. (2015) Make new friends, but keep the old: towards a transdisciplinary and balanced strategy to evaluate Good Environmental Status. *Journal of the Marine Biological Association*, **95** (6), 1069–1070.

Bouillon, J., Medel, M.D., Pagès, F. *et al.* (2004) Fauna of the Mediterranean Hydrozoa. *Scientia Marina*, **68** (Supplement 2), 1–449.

Broecker, W.S. (1991) The great ocean conveyor. *Oceanography*, **4** (5), 79–89.

Claudet, J. and Fraschetti, S. (2010) Human impacts on marine habitats: a regional meta-analysis in the Mediterranean Sea. *Biological Conservation*, **143**, 2195–2206.

Costello, M.J., May, R.M. and Stork, N.E. (2013) Can we name Earth's species before they go extinct? *Science*, **339**, 413–416.

Council of Europe (2000) *European Landscape Convention*. http://www.coe.int/t/dg4/cultureheritage/heritage/Landscape/Publications/Convention-Txt-Ref_en.pdf

Della Tommasa, L., Belmonte, G., Palanques, A. *et al.* (2000) Resting stages in a submarine canyon: a component of shallow–deep-sea coupling? *Hydrobiologia*, **440**, 249–260.

Diamond, J. (2002) Evolution, consequences and future of plant and animal domestication. *Nature*, **418**, 700–707.

Fanelli, G., Piraino, S., Belmonte, G. *et al.* (1994) Human predation along Apulian rocky coasts (SE Italy): desertification caused by *Lithophaga lithophaga* (Mollusca) fisheries. *Marine Ecology Progress Series*, **110**, 1–8.

Galil, B., Boero, F., Fraschetti, S. *et al.* (2015). The enlargement of the Suez Canal and introduction of non-indigenous species to the Mediterranean. *Limnology and Oceanography Bulletin*, **24** (2), 41–43.

Gatto, M. (2009) On Volterra and D'Ancona's footsteps: the temporal and spatial complexity of ecological interactions and networks. *Italian Journal of Zoology*, **76**, 3–15.

Godø, O.R., Samuelsen, A., Macaulay, G.J. *et al.* (2012) Mesoscale eddies are oases for higher trophic marine life. *PLoS ONE*, **7** (1), e30161. doi:10.1371/journal.pone.0030161

Grantham, B.A., Eckert, G.L. and Shanks, A.L. (2003) Dispersal potential of marine invertebrates in diverse habitats. *Ecological Applications*, **13** (1) Supplement, S108–S116.

Gravili, C., Bevilacqua, S., Terlizzi, A. and Boero, F. (2015) Missing species among Mediterranean non-Siphonophoran Hydrozoa. *Biodiversity and Conservation*, **24**, 1329–1357.

Guidetti, P., Milazzo, M., Bussotti, S. *et al.* (2008) Italian Marine Reserve effectiveness: does enforcement matter? *Biological Conservation*, **141**, 699–709.

Heip, C., Hummel, H., van Avesaath, P. *et al.* (2009) *Marine Biodiversity and Ecosystem Functioning.* Printbase, Dublin, Ireland. 91 pp.

Hickey, B.M. (1995) Coastal submarine canyons, in *Topographic Effects in the Ocean* (eds P. Müller and D. Henderson). SOEST Publication, University of Hawaii Manoa. pp. 95–110.

Johannesson, K. (1988) The paradox of Rockall: why is a brooding gastropod (*Littorina saxatilis*) more widespread than one having a planktonic larval dispersal stage (*L. littorea*)? *Marine Biology*, **99**, 507–513.

Lejeusne, C., Chevaldonné, P., Pergent-Martini, C. *et al.* (2010) Climate change effects on a miniature ocean: the highly diverse, highly impacted Mediterranean Sea. *Trends in Ecology and Evolution*, **25**, 250–260.

Longhurst, A.R. (2010) *Ecological Geography of the Seas* (2nd edition). Elsevier Academic Press, Burlington, USA.

Olsen, E.M., Johnson, D., Weaver, P. *et al.* (2013) *Achieving Ecologically Coherent MPA Networks in Europe: Science Needs and Priorities* (eds K.E. Larkin and N. McDonough). Marine Board Position Paper 18. European Marine Board, Ostend, Belgium. 88 pp.

Pikitch, E.K., Santora, C., Babcock, E.A. *et al.* (2004) Ecosystem-based fishery management. *Science*, **305**, 346–347.

Pinardi, N., Zavatarelli, M., Arneri, E. *et al.* (2004) The physical, sedimentary and ecological structure and variability of shelf areas in the Mediterranean Sea, in *The Seas* (eds A.R. Robinson and K.H. Brink). Harvard University Press, Harvard. pp. 1245–1331.

Piraino, S., Fanelli, G. and Boero, F. (2002) Variability of species' roles in marine communities: change of paradigms for conservation priorities. *Marine Biology*, **140**, 1067–1074.

Planes, S., Galzin, R., Garcia Rubies, A. *et al.* (2000) Effects of marine protected areas on recruitment processes with special reference to Mediterranean littoral ecosystems. *Environmental Conservation*, **27** (2), 126–143.

Puce, S., Bavestrello, G., Di Camillo, C.G. and Boero, F. (2009) Long-term changes in hydroid (Cnidaria Hydrozoa) assemblages: effect of Mediterranean warming? *Marine Ecology – An evolutionary perspective*, **30** (3), 313–326.

Pulliam, H.R. (1988) Sources, sinks, and population regulation. *American Naturalist*, **132** (5), 652–661.

Purcell, J.E., Milisenda, G., Rizzo, A. *et al.* (2015) Digestion and predation rates of zooplankton by the pleustonic hydroid *Velella velella* and widespread blooms in 2013 and 2014. *Journal of Plankton Research*, **37** (5), 1056–1067. doi:10.1093/plankt/fbv031

Ricklefs, R.E. (2012) Natural history, and the nature of biological diversity. *American Naturalist*, **179**, 423–435.

Rivetti, I., Fraschetti, S. and Lionello, P. *et al.* (2014) Global warming and mass mortalities of benthic invertebrates in the Mediterranean Sea. *PLoS ONE*, **9** (12), e115655. doi:10.1371/journal.pone.0115655

Shanks, A.L., Brian, A., Grantham, B.A. and Carr, M.H. (2003) Propagule dispersal distance and the size and spacing of marine reserves. *Ecological Applications*, **13** (1) Supplement, S159–S169.

Tewksbury, J.J., Anderson, J.G.T., Bakker, J.D. *et al.* (2014) Natural history's place in science and society. *BioScience*, **64** (4), 300–310. doi:10.1093/biosci/biu032

2

Ecological Effects and Benefits of Mediterranean Marine Protected Areas: Management Implications

Antoni Garcia-Rubies[1], Emma Cebrian[1,2], Patrick J. Schembri[3], Julian Evans[3] and Enrique Macpherson[1]

[1] *Centre d'Estudis Avançats de Blanes (CEAB-CSIC), Girona, Spain*
[2] *Department of Environmental Sciences, University of Montilivi, Girona, Spain*
[3] *Department of Biology, Faculty of Science, University of Malta, Msida, Malta*

Introduction

There is general consensus among scientists that marine life in the Mediterranean Sea, and in the world's oceans in general, is under considerable threat by human activities (Coll *et al.*, 2010; Micheli *et al.*, 2013). This strain on marine ecosystems worldwide has led to calls for new management approaches, especially for coastal areas (Botsford *et al.*, 1997). Such measures, for instance, include the regulation of fisheries towards more sustainable exploitation of resources and the establishment of networks of Marine Protected Areas (MPAs) (Olsen *et al.*, 2013). However, development of effective regulations for conservation must be based on sufficient knowledge and information about the protected systems. For example, it has been shown that the establishment of marine reserves that are too small or too scattered can have a reduced or nil effect on the protection of Mediterranean ecosystems (Abdulla *et al.*, 2008; Botsford *et al.*, 2009).

When appropriately designed, MPAs favour the recovery of harvested populations in the Mediterranean Sea and

elsewhere (Bell, 1983; Garcia-Rubies and Zabala, 1990; Harmelin *et al.*, 1995; Vacchi *et al.*, 1998; Claudet *et al.*, 2011; Fenberg *et al.*, 2012). The main reason for these MPA effects is the drastic reduction in overall mortality: when fishing mortality is removed or reduced, stock recovery is the most logical expected consequence (Bell, 1983). The more vulnerable to fishing a species is, the more it will respond to cessation of fishing mortality (Macpherson *et al.*, 2000). Therefore, the ecological benefits derived from these conservation units are essential for the sustainability of exploited ecosystems, and sagacious and effective management of MPAs is a key issue in an age of changing oceans and seas (Olsen *et al.*, 2013).

Ecological Benefits of MPAs

Effects on Fish Populations

Species vulnerability to fishing depends on the specific life history of the species involved (Molloy *et al.*, 2008). In general, large, long-lived, slow-growing, sedentary

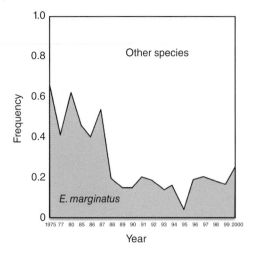

Year

Figure 2.1 Temporal pattern of the relative frequency of species among the largest specimens captured in the regional competitions of spear-fishing in the Balearic Islands. It can be seen that the dusky grouper *Epinephelus marginatus* lost its preponderance among the biggest specimens from the end of the 1980s. *Source*: Redrawn from Coll *et al.* (2004).

species with low natural mortality rates are more vulnerable than small, fast-growing species with high rates of natural mortality (Cheung *et al.*, 2005). Other features, such as a late sexual maturity, a delayed sex change or a limited reproductive capability, can also exacerbate the vulnerability of a given species. Furthermore, large predatory species tend to be more targeted by fisheries than smaller species, so their populations are much more likely to be depleted in fished areas (Figure 2.1) or by special fishing methods.

The recovery process of an exploited population inside an MPA is the reverse of the process of harvesting. Initial recovery rates can be relatively fast (Garcia-Charton *et al.*, 2008; Molloy *et al.*, 2009) (Figure 2.2), but total recovery can be extremely slow for the more vulnerable species. In many cases, attaining full recovery in an MPA may even be impossible within a human lifespan (e.g. the red coral *Corallium rubrum*; Garrabou

and Harmelin, 2002). Nonetheless, species targeted by fishing generally respond to protection in a positive way compared to non-targeted species (Micheli *et al.*, 2005; Claudet *et al.*, 2006; Guidetti and Sala, 2007), leading to a net increase in biomass. However, not all targeted species respond equally to protection, with response depending on their vulnerability and also on the carrying capacity of the system.

When a population reaches the carrying capacity (K), it can be considered to have fully recovered. Although this is one of the main objectives of MPAs, it has rarely been observed in practice and there is no consensus on successful K thresholds or on the time that is necessary to achieve full recovery. There are many differing descriptions in the literature, ranging from quick recoveries in less than five years (Côté *et al.*, 2001; Halpern and Warner, 2002) to estimated recovery times of 10–40 years for apex predator species (McClanahan *et al.*, 2007) in no-take zones.

Total recovery of harvested populations in MPAs has only recently been described in the Mediterranean Sea (Coll *et al.*, 2013; Garcia-Rubies *et al.*, 2013), in spite of the large number of Mediterranean MPAs. This is probably due to the relatively young age of most of these MPAs; however, it can also be attributed to the lack of long-term studies on the changes in protected populations in most MPAs.

Total recovery can vary greatly with time since protection was implemented. For instance, Coll *et al.* (2012, 2013) showed very fast recoveries of total target fish biomass in three Balearic MPAs. Garcia-Rubies *et al.* (2013), however, observed that reaching the carrying capacity could be a long process for highly vulnerable, long-lived species such as *Dicentrachus labrax* (20–25 years), *Diplodus cervinus* (13–16 years) and *Epinephelus marginatus* (21–24 years). Other species were still far from achieving total recovery, for example *Sciaena umbra*

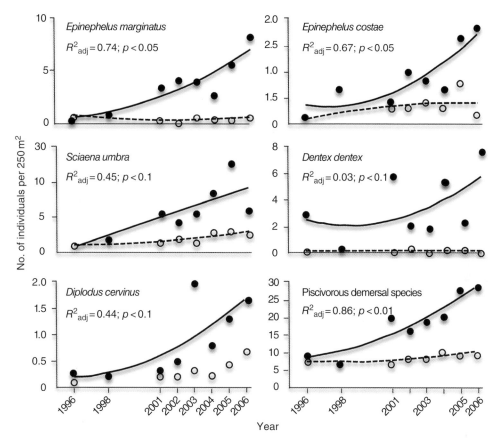

Figure 2.2 Temporal pattern of mean abundances of some large predatory species, and considering all piscivorous species together, in the Cabo de Palos – Islas Hormigas marine reserve, Cartagena, Spain (solid circles) and in an exploited control locality (open circles) after the establishment of protection measures in 1995. Although the first stages of recovery are fast in protected areas, total recovery is a long process. *Source*: Redrawn from Garcia-Charton *et al.* (2008).

(31–51 years), while *Dentex dentex* was still growing exponentially (see Figure 2.3).

Variations in carrying capacity values, and in the time it takes to reach them, can be explained by the effect of different environmental factors acting at small and medium scales. Achieving the maximum biomass, and the time to reach it, is a bottom-up regulated process influenced by environmental conditions (bottom features/substratum type, depth, slope and rugosity; Coll *et al.*, 2012, 2013) that favour or limit the development of the largest, long-lived species. This explains why the carrying capacity value is greater, and

the time to reach it is longer, in MPAs where environmental conditions are highly favourable (such as the Medes Islands MPA), compared to other MPAs sited in areas lacking these highly favourable conditions (Figure 2.4). Knowing the effect of these factors, one can predict how long it would take an ideal MPA to reach maximum values of K (Coll *et al.*, 2012, 2013). This ideal environment is very similar to that found in the Medes Islands marine reserve where, indeed, the value of K far exceeds that observed in the Balearic Islands MPAs, although the time required to achieve these values is much longer.

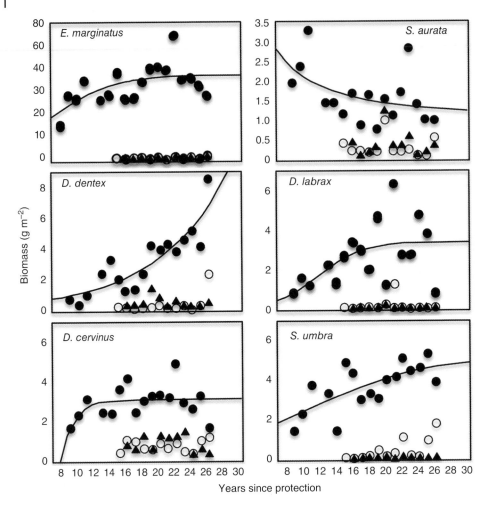

Figure 2.3 Differences in temporal patterns of biomass of six highly vulnerable fish species (*Epinephelus marginatus, Sparus aurata, Dentex dentex, Dicentrachus labrax, Diplodus cervinus* and *Sciaena umbra*) monitored vs. time of protection for marine reserve (solid circles), partially protected reserve (open circles) and non-reserve (solid triangles), in the Medes Islands marine reserve and neighbouring coast. *Source:* Redrawn from Garcia-Rubies *et al.* (2013).

Ecosystem Characteristics Affecting the Benefits of MPAs

Differences in carrying capacity are just one example of the wide range of results obtained from studies of different Mediterranean MPAs; ecological effects of Mediterranean MPAs have been found to vary in both magnitude and direction (Claudet *et al.*, 2011 and references therein). Although major differences between Mediterranean MPAs could

be attributed to the level of enforcement of, and compliance with, the protection measures, which is a significant socio-cultural factor (Guidetti *et al.*, 2008; Sala *et al.*, 2012), the results can be very different even between well-protected areas. As an example, Sala *et al.* (2012) used a large range of fish biomass (from 50 to 120 g m^{-2}) as the reference for a good conservation state for different well-enforced no-take areas. Such variation

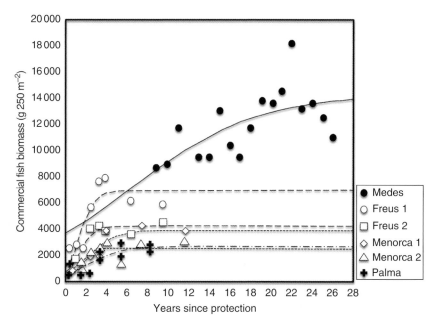

Figure 2.4 The differences in carrying capacity and recovery time between the Medes Islands MPA and three Balearic MPAs are evident despite the fact that the biomass of only six species were taken into account in the Medes Islands, whereas in the Balearic Islands MPAs total commercial fish biomass was included. *Source*: Redrawn from Coll *et al.* (2012) and Garcia-Rubies *et al.* (2013).

was attributed to the idiosyncrasies of the different MPAs, that is to say: there must be factors other than good enforcement that define fish biomass in any given well-protected no-take zone (Figure 2.5).

Some obvious factors such as age and size of MPAs have not been taken into account until recently (Guidetti and Sala, 2007; Claudet *et al.*, 2008; Molloy *et al.*, 2009). The effects of these factors have been mainly assessed indirectly by comparing MPAs of different ages through meta-analysis (Guidetti and Sala, 2007; Claudet *et al.*, 2008), whereas studies comparing a temporal evolution of single MPAs, or differences between different-sized coetaneous MPAs, are practically non-existent (but see Garcia-Rubies *et al.*, 2013 and Coll *et al.*, 2012, 2013). Age and size of the MPAs are among the main factors affecting the results of protection. Guidetti and Sala (2007) found that the

response of fish assemblages to protection was significantly related to reserve age only when evaluated at functional level, whereas reserve size did not appear to influence fish assemblages in terms of either species or functional level. In contrast, Claudet *et al.* (2008) found that the age of an MPA was less important than its size, and the size of the buffer zone, in determining commercial fish density in 12 Mediterranean MPAs, although commercial fish density increased at a rate of 8.3% per year in no-take protected zones.

Other factors such as depth range have rarely been taken into account in studies comparing protected and non-protected zones. In most studies, sampling depth is typically fixed within a narrow range, and no assessment of how protection effects vary with depth is carried out.

The role of environmental factors seems to be fundamental in explaining the effects

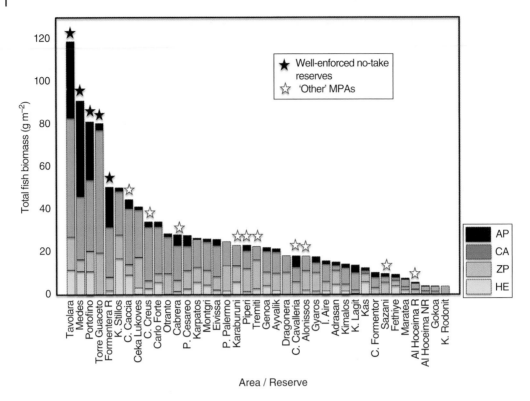

Figure 2.5 Total fish biomass in several MPAs and areas open to fishing in the Mediterranean Sea. The differences between well-enforced no-take reserves and other MPAs are obvious, but the differences between different well-enforced protected areas are also striking, indicating that factors other than effective protection have a strong influence on fish biomass. AP, apex predators; CA, carnivores; ZP, zooplanktivores; HE, herbivores and detritivores. *Source*: Redrawn from Sala *et al.* (2012), with some sites added to the original figure.

of MPAs (Garcia-Charton and Pérez-Ruzafa, 1999). The biomass of exploited populations in any no-take zone is clearly a result of a bottom-up process (Garcia-Rubies *et al.*, 2013). Knowing the key factors that regulate fish biomass is therefore of paramount importance in the design of future Mediterranean MPAs. However, one has to recognize that the combination of factors that must be present to lead to enhanced biomass at small and medium spatial scales can greatly reduce the number of potential candidate sites. In any case, rare privileged hotspots should be prioritized in any future conservation project in the Mediterranean Sea.

Partially Protected Areas

Most of the 677 Mediterranean MPAs (as included in Gabrié *et al.*, 2012) are barely protected against fishing. In fact, 507 of them are Natura 2000 areas with no specific management to avoid or limit extractive activities. Excluding the vast Pelagos Sanctuary (87 500 km^2), the area covered by coastal MPAs amounts to 18 965 km^2 (0.4% of the total surface of the Mediterranean Sea), while only 207 km^2 (0.012% of the total surface of the Mediterranean) can be considered as an actual fully protected no-take area. Out of the 170 true MPAs, 80 supplied management information indicating that only 31% (3 390 km^2) have some kind of special management, whether or not this includes fishing limitations. In short, only 0.14% of the Mediterranean Sea surface is known to enjoy some management so the vast majority of the Mediterranean MPAs are nothing more than partially protected areas (Gabrié *et al.*, 2012).

In spite of the fact that most MPAs are partially protected areas, studies on the benefits of partial protection are extremely scarce. Partially protected zones can have a broad range of protection regulations, going from well-protected zones, with limited fishing activities, to merely 'paper-parks' without any specific management or effective protection. This variability leads to a great diversity of results that, even when positive, are usually inferior to those obtained when full protection is in force. Many authors consider partially protected areas as inefficient (Denny and Babcock, 2004; Claudet *et al.*, 2008; Di Franco *et al.*, 2009; Lester *et al.*, 2009). In some cases, partially protected areas can even be counterproductive since they attract fishermen eager to fish near a no-take area, thus leading to an increase in fishing effort (Stelzenmuller *et al.*, 2007). This is possibly why large buffer zones can even have negative impacts on the overall ecological effectiveness of MPAs (Claudet *et al.*, 2006).

However, some buffer zones have shown positive trends although this mostly depends on habitat characteristics, as in the case of no-take zones, and current fisheries regulations (Coll *et al.*, 2012). Even the same regulation can lead to different results in two separate partially protected areas belonging to the same MPA. That was the case of the Freus of Eivissa and Formentera MPA, where one of the partially protected areas did not show any sign of improvement while in the other the commercial fish biomass increased by 330%. The differences were due to a combination of habitat features and fishing pressure. The seascape of the first

partially protected area had a low rugosity and few boulders, allowing fishing with trammel nets very close to the coast. In contrast, the second partially protected area had a highly complex rocky bottom with more large boulders making it very difficult to cast trammel nets there.

The buffer zone of the Medes Islands MPA showed a limited progression after 10 years of partial protection (only spear-fishing is absolutely banned there, while commercial and recreational fishing is regulated), and only three (*Epinephelus marginatus*, *Sciaena umbra* and *Dentex dentex*) out of six species studied by Garcia-Rubies *et al.* (2013) showed a positive trend in this zone. Moreover, total mean biomass of these species was 13 times lower than that observed in the no-take zone (see Figure 2.3). The modest recovery in the buffer zone could be explained by a limited spillover from the no-take zone (Garcia-Rubies *et al.*, 2013).

When protective measures are effective, and the partially protected areas are located where suitable environmental conditions prevail, the results can be quite surprising. For example, in the Nord de Menorca marine reserve one of the partially protected areas achieved a higher biomass of commercial fish than that attained in the no-take zone (Figure 2.6). Lack of spear-fishing and particularly suitable rocky bottoms in this partially protected area were the main factors allowing such a high fish biomass in spite of a moderate level of exploitation (Reñones *et al.*, 1999; Lloret *et al.*, 2008).

In general, one can conclude that the benefits of partially protected zones depend on the regulation of fishing activities, limiting the fishing effort and banning the most effective fishing methods, as well as the environmental conditions prevailing in the zone. The environmental factors that determine the success of no-take areas as fish biomass producers are exactly the same in the case of partially protected zones.

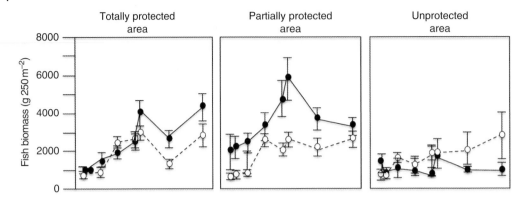

Figure 2.6 Mean commercial fish biomass at Nord de Menorca marine reserve at three protection levels; note that fish biomass at site 1 (closed circles) in the partially protected area is clearly superior to the biomass observed at site 2 (open circles) of the no-take area. *Source*: Redrawn from Coll *et al.* (2012).

Effects on Communities (Secondary Effects of Protection)

Fishing tends to remove a substantial part of exploited populations and this has been the pattern all around the Mediterranean Sea for millennia. New fishing methods such as spear-fishing have worsened the situation of large predatory fish from the second half of the 20th century onwards, especially in the Western Mediterranean (Spain, France and Italy). There are almost no historical quantitative data about the abundance of such species 70 years ago, but some books and movies from the 1940s show a very different picture from the current situation (Coll *et al.*, 2004; Guidetti and Micheli, 2011). One can conclude that these large predators have been seriously depleted from Mediterranean littoral rocky bottoms during the last 70 years or so.

The effects of such continued exploitation have led to a rarefaction of many species including many elasmobranchs (Ferretti *et al.*, 2008), monk seals *Monachus mona-chus* (Durant and Harwood, 1992), sea turtles, and blue fin tuna *Thunnus thynnus*. It could be concluded that there are no pristine sites left in the Mediterranean Sea (Sala *et al.*, 2012). Fishing can be considered as the main stressor of the littoral rocky reef communities, from which large predators have nearly disappeared following the widespread process of so-called fishing down the food webs (Pauly *et al.*, 1998).

Marine Protected Areas, whether including no-take zones or well-regulated partially protected zones, have demonstrated the recovery of large predatory and carnivorous fish, which reached much higher biomass values than in open areas (Sala *et al.*, 2012; Garcia-Rubies *et al.*, 2013). In fact, top predator biomass in well-protected MPAs can account for nearly half of the total fish biomass in the most successful Mediterranean MPAs (Sala *et al.*, 2012). For instance, apex predators represent up to 49% of the total fish biomass observed at the Medes Islands MPA. Although the dusky grouper has already reached the carrying capacity there (Garcia-Rubies *et al.*, 2013), the biomass of other large predators, such as *Dentex dentex*, is still increasing and has not yet reached an asymptote.

Effects of Fish on Other Fish

The pronounced increase of large predator biomass within the no-take protected zones (Russ and Alcala, 1996) must negatively affect prey fish populations, but there is little direct evidence for this effect within

Mediterranean MPAs, although the phenomenon has been reported elsewhere. Micheli *et al.* (2004) found, through a meta-analysis of the results of up to 20 studies and 31 sites, that 19% of fish species were negatively affected by the protection. The most affected fish were small benthic species such as Blennidae, Gobiidae, Pomacentridae, Atherinidae and Apogonidae.

In the Mediterranean, Macpherson (1994) showed a lower density of small gobies and blennies in the Medes Islands MPA in comparison with two unprotected areas. Sasal *et al.* (1996) observed an increase in size of male and female *Gobius bucchichi* within the no-take zone of Banyuls MPA. The authors explained this finding on the basis of increased predation on the smaller specimens, as well as the result of an increased competition for refugia. Garcia-Rubies (1999) also observed an increase in the number of mid- and large-sized *Diplodus sargus* inside the Medes Islands MPA but also a decrease in the number of young of a year (YOYs), which could be attributed to a rise in predation pressure on small individuals during the first year in the adults' habitat. However, the predation on settlers of *Diplodus* spp. was not related to protection level, since most of the predators of settlers were themselves small species not affected by protection (Macpherson *et al.*, 1997).

Effects of Fish on Invertebrates

Sea Urchins The first studies on the consequences of increasing fish abundance showed a direct effect on sea urchin populations in MPAs (Sala *et al.*, 1998). Thus, sea urchin density in the Medes Islands MPA was four times lower than in the unprotected area, and the results of a tethering experiment demonstrated that fish predation on sea urchins was five times higher in the no-take zone. Other effects of the increased predation inside the MPA were lower sea urchin mean size, changes in size–frequency distribution, and changes in the behaviour of the sea urchins. In a more recent study based on a long time series, carried out in the same Medes Islands MPA, Hereu *et al.* (2012) demonstrated that the relationship between fish and sea urchins is not as direct as it seemed in previous snapshot studies. In fact, these authors did not find significant long-term differences in the sea urchin density between the protected and unprotected areas. The most obvious result was that inter-annual variations in sea urchin abundance were less pronounced within the MPA than in open areas. But even the increase of predation pressure inside the MPA cannot cancel out the effects of episodic massive recruitment.

Although the effects of higher rates of urchin predation within MPAs were not as marked as suggested by short-term studies, there are more subtle results which demonstrate that fishes play an important role in regulating sea urchin populations in Mediterranean MPAs. For instance, there is a positive correlation between the abundance of juvenile and adult sea urchins inside the MPA, suggesting that the survival of juveniles is density-dependent and is facilitated by the presence of adults. Outside the MPA, no such relationship was observed since juvenile urchins can move more freely on open surfaces due to the reduced predation risk (Hereu *et al.*, 2005). One can conclude that although the increasing predation rate by fish does not have a determinant effect on total sea urchin abundance, it can buffer extreme variations in recruitment. Moreover, the total displacement and home range of sea urchins were significantly lower inside the MPA.

Spiny Lobsters As a rule, populations of exploited species recover inside MPAs, but there are some exceptions in which the trend is not so evident. In some cases, the population not only does not recover but actually tends to decline. This may be due to

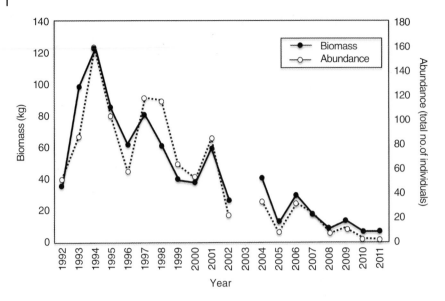

Figure 2.7 Temporal trends of abundance and biomass of the spiny lobster *Palinurus elephas* in the Medes Islands MPA over a period of 20 years. *Source*: Redrawn from Díaz (2013).

different factors: (i) protected areas are too small for those species with a large home range that regularly exceeds the boundaries of the protected area; (ii) an excessive fishing pressure outside the protected area can affect the populations inside; and (iii) the protection favours species that can damage others due to an increasing predation pressure. This seems to be the case for the spiny lobster *Palinurus elephas* in the Medes Islands MPA (Díaz *et al.*, 2005) where, according to a tethering experiment, juvenile mortality is much higher within the MPA (41%) than outside (17%). The increased juvenile mortality results in a progressive decline of the lobster population inside the Medes Islands MPA (Figure 2.7), contrasting with the trends of other highly vulnerable species (Garcia-Rubies *et al.*, 2013). In this particular case, increased predation pressure acts along with the home range of the lobster (Giacalone *et al.*, 2006) which is larger than the protected area of the no-take zone (93 ha), high fishing pressure in the area surrounding the MPA, and the high catchability of this species.

On the other hand, opposite trends can be observed in larger MPAs, such as Columbretes Island MPA, where the lobster population has shown a recovery trend according to what might be expected in a protected area (Díaz *et al.*, 2011) in spite of a high predatory fish biomass (Goñi *et al.*, 2006).

Trophic Cascades

A trophic cascade is an indirect effect of predators, not only upon prey populations, but also on the whole food web, involving more than two trophic levels. The simplest model of a trophic cascade would include a predator (e.g. carnivore), a prey (e.g. herbivore) and a primary producer. A trophic cascade is a top-down process in which variations in predator abundance can affect the structure of the whole community (Babcock *et al.*, 1999), that is, once predator populations exceed a certain threshold, predation can control prey populations and their effects on the community. However, when predation is weak, other factors become more important in structuring communities. The best documented trophic

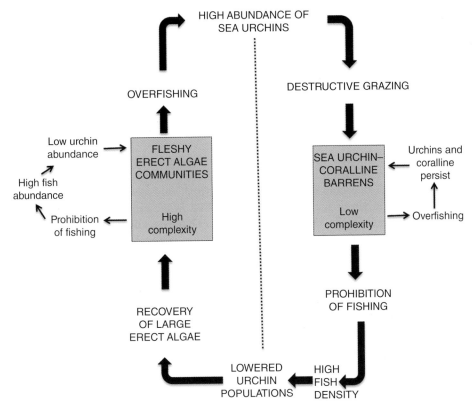

HIGH ABUNDANCE OF
SEA URCHINS

OVERFISHING

DESTRUCTIVE GRAZING

Low urchin
abundance

High fish
abundance

Prohibition
of fishing

FLESHY
ERECT ALGAE
COMMUNITIES

High
complexity

SEA URCHIN–
CORALLINE
BARRENS

Low
complexity

Urchins and
coralline
persist

Overfishing

RECOVERY
OF LARGE
ERECT ALGAE

PROHIBITION
OF FISHING

LOWERED
URCHIN
POPULATIONS

HIGH
FISH
DENSITY

Figure 2.8 Classical model of a trophic cascade due to overfishing in rocky infralittoral algal assemblages. Recent studies have demonstrated that factors other than fishing are important in regulating sea urchin densities in the Mediterranean Sea. *Source*: Redrawn from Sala *et al.* (1998).

cascade in the Mediterranean Sea is the relationship between fishes, sea urchins and macroalgae on sublittoral rocky bottoms (Sala *et al.*, 1998) (Figure 2.8).

It has been noted that in the Mediterranean Sea many localities suffered a progressive shift from macroalgal canopies to barrens as a result of overgrazing by sea urchins, with a concurrent large loss of species diversity and changes in community structure (Ling *et al.*, 2015). Lack of sea urchin predators due to overfishing was claimed to be the main factor responsible for barren formation, so it was expected that recovery of sea urchin predators inside the MPAs would ultimately prevent the overgrazing by urchins and thus arrest barren formation

through a trophic cascade, or even lead to the recovery of (former) macroalgal canopies. The first studies showed that the relationship between fishes, sea urchins and algae seemed to be quite straightforward: high predator abundance led to an increasing predation rate on sea urchins, keeping urchins in densities low enough to prevent barren formation or even to allow a recovery of macroalgal forests. Each step seemed to fit perfectly in the model: predatory fish were more abundant in the MPA than outside (Garcia-Rubies and Zabala, 1990), the predation rate increased five times inside the MPA, and the density of sea urchins was significantly lower within the MPA (Sala and Zabala, 1996).

This process is a clear paradigmatic example of a discontinuous catastrophic regime shift that meets all the requirements of such phenomena, namely: there is an abrupt change from one state to the other that occurs once the sea urchins exceed a certain threshold of density or biomass; the new state is very stable and persists over time as sea urchin biomass is maintained after the shift, preventing any reversion towards the previous state; hysteresis occurs since for a reversion to occur, the sea urchin biomass must be, at least, one order of magnitude less than that which caused the shift. An obvious temporal asymmetry is also introduced since the shift back to the original state takes much longer than the shift forward (Ling *et al.*, 2015).

Sea urchin biomass can be maintained after the establishment of a barren at the cost of sea urchins having to eat encrusting algae and sessile invertebrates. They also eat any newly settled macroalgae, thus preventing new re-colonization, although this has not yet been shown for the Mediterranean Sea. In addition, once the barren has been established there are many negative feedbacks tending to reinforce this alternative phase and preventing the shift backwards. It has been shown that overfishing of predators of sea urchins is a key factor (Guidetti, 2005) both in the development of barrens and in preventing reversion to the original state. In the case of the Mediterranean Sea, it has been argued that the overexploitation of fish that feed on sea urchins is the main cause of the formation of barrens, which can also be favoured by destructive fishing methods (Guidetti, 2011).

Marine Protected Areas are useful in maintaining high densities of sea urchin predators (mainly the sea bream *Diplodus sargus* and *D. vulgaris*; Guidetti, 2006). In a broad study Guidetti and Sala (2007) found that a minimum of 15 adult sea bream per $100\,m^2$ were necessary to reduce inter-annual variation in sea urchin densities by preventing peaks occurring due to abnormally high annual recruitments (Cardona *et al.*, 2007, 2013). Hereu *et al.* (2012) came to a similar conclusion after analysing the longest series of data on sea urchin density from the Mediterranean. While the average density of sea urchins in the Medes Islands marine reserve was similar to that found in the fished area near the MPA, annual variations within the MPA were much smaller. The peaks due to recruitment events leading to abnormally high sea urchin densities were suppressed in the marine reserve, reducing the risk of barren formation and maintaining the presence of communities of macroalgae (Figure 2.9).

The MPA as a Touchstone: Estimation of the Degree of Exploitation Outside MPAs

Even though the overall performance of no-take, well-enforced Mediterranean MPAs is positive in recovering exploited populations (Sala *et al.*, 2012), it must be pointed out that in 2012, only 0.012% of the total Mediterranean Sea surface was known to enjoy this highly protected status (Gabrié *et al.*, 2012). This very limited area prevents any significant effect at the scale of the whole Mediterranean, although local effects can be important, even in increasing fishing yields (Goñi *et al.*, 2006). In the current situation, protected areas are nothing but a small exception, so perhaps the best value of MPAs is as the best benchmarks available for assessing the resources exploited in the other 99.88% or so of the Mediterranean. Up to now only McClanahan *et al.* (2011), in the Indian Ocean, and Sala *et al.* (2012), in the Mediterranean, have established such baseline data.

Results obtained in well-protected Mediterranean MPAs are a good baseline reference for the state of exploitation of non-protected areas. According to Worm *et al.* (2009), an exploited stock whose

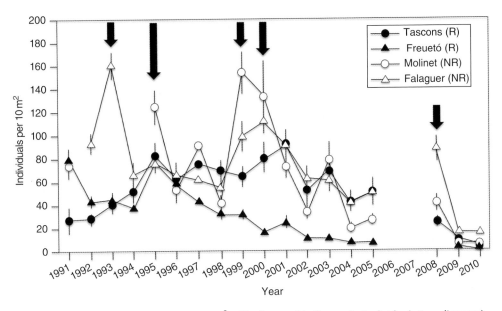

Figure 2.9 Mean density (individuals per $10\,m^2 \pm SE$) of sea urchin *Paracentrotus lividus* (>1 cm diameter) inside and outside the Medes Islands MPA on boulder substrates. Arrows indicate abnormally high peaks of annual recruitment. R, reserve; NR, non-reserve. *Source*: Redrawn from Hereu *et al.* (2012).

numbers do not reach 10% of the unexploited biomass (pristine biomass, according to McClanahan *et al.*, 2007) can be considered to be practically in ecological collapse; that, for instance, was the case of three species (*Epinephelus marginatus*, *Dicentrachus labrax* and *Sciaena umbra*) out of six analysed by Garcia-Rubies *et al.* (2013).

Although the differences between MPAs and fished areas are basically the result of a top-down process (fishing vs. no fishing) there are several environmental factors that can favour and enhance these differences. The Medes Islands MPA (Garcia-Rubies *et al.*, 2013) and other especially favoured or 'hotspot' Mediterranean MPAs (Sala *et al.*, 2012) approach what Coll *et al.* (2012) described as the ideal MPA, that is, a rocky outcrop in the open sea, totally exposed to winds and currents with a pronounced nearby slope and a highly complex (rugose) bottom composed of big rocky blocks. A surrounding boundary of sedimentary

bottoms, preventing spillover (Garcia-Rubies *et al.*, 2013), will further help to get an elevated rocky reef fish biomass.

Not all the Mediterranean rocky coast shows such favourable conditions, but Coll *et al.* (2012) proposed a model that allows prediction of maximum biomass of fish if the environmental conditions are known. This predictive biomass model was applied in 28 exploited sites ($N = 260$ transects) in the littoral of the Balearic archipelago. A mean value of expected total biomass (Bt_e) was obtained for each site, assuming that the sites had the maximum protection level. In short, Bt_e is the projected value closest to the potential carrying capacity (K) of each site, as a proxy of pristine biomass according to McClanahan *et al.* (2007). The difference between Bt_e and the mean biomass observed at each site (Bt_o) gives the degree of exploitation of that site as well as its potential for recovery. Moreover, one can establish a range of Bt_e values at each site to determine whether the resources are within the limits

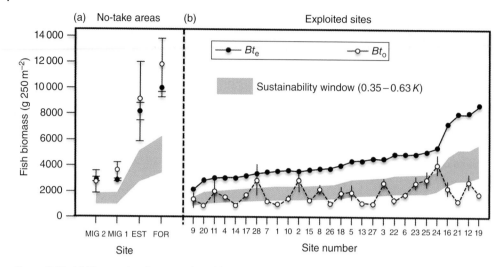

Figure 2.10 (a) Observed and expected total biomass values (Bt_o and Bt_e, respectively) in two no-take sites from two marine reserves where the model was tested: Mitgjorn Marine Reserve (MIG 1 and MIG 2) and Cabrera National Park (EST and FOR). (b) Bt_o in increasing order of Bt_e values from 28 exploited sites of the Balearic Islands; the grey band shows the biomass range window within the exploitation status of each site that could be considered sustainable ($0.35\,K < Bt < 0.63\,K$, K being the value of carrying capacity). *Source*: Redrawn from Coll *et al.* (2013).

of sustainability. The maximum and minimum of these limits were set to 63% and 35% of Bt_e taking into account both traditional target reference points (TRP) from maximum sustainable yield and the concept of multiple maximum sustainable yield (MMSY) for fish assemblages (Worm *et al.*, 2009), respectively.

Applying the model to exploited areas showed that 46% of them are within the limits of sustainability, while 43% are clearly overexploited and only 11% could be considered underexploited. The differences between no-take and open areas are evident (Figure 2.10). It is interesting to see that the areas that can potentially host the greatest biomasses are also the ones that show the highest differences from the expected biomass values. The fact that higher quality habitats that might support populations of highly vulnerable sedentary fish species with limited home ranges, high trophic status and long lifespan are especially affected by exploitation (i.e. *Epinephelus* spp., *Sciaena*

umbra, *Phycis phycis*), or are recruitment limited (such as *Diplodus* spp.), could be the cause.

The predictive biomass model described above has been developed for the littoral of the Balearic archipelago and it encompasses only the total biomass of 13 commercial species in a narrow range of depths (3–15 m). It would be desirable to know the K values for each fish species and also to improve the sampling methodology to determine the habitat requirements at a specific level. However, two main results show the applicability of the predictive biomass model. First the correlation between observed and predicted values for the 32 sites studied ($R = 0.60$; $p < 0.001$) suggests that the environmental variables included in the model are limiting factors for fish biomass, and that these factors are also good descriptors of the variation in fish biomass throughout the entire geographical area of the Balearic Islands, representing characteristic rocky habitats of the western Mediterranean.

The second key result of this study is the high predictive capacity of the model when no-take areas are tested as controls. The expected mean values are almost identical to the observed values and always fall within their margins of error.

Spillover and Larval Export

Adults and Juveniles

Marine Protected Areas can benefit neighbouring populations through 'spillover', that is, the net export of adult or juvenile individuals to non-protected areas (Russ *et al.*, 2004; Dudley and Hockings, this volume). There is some direct evidence of spillover in the Mediterranean MPAs. The study on spiny lobster in the Columbretes Islands MPA was the first to assess spillover from tag–recapture data (Goñi *et al.*, 2006, 2010). In this case it was estimated that 7% of the lobsters residing in the MPA emigrated every year to the adjacent fishing grounds, providing a net gain of over 10% of the catch in weight. A further study on spiny lobster using tag–recapture data was done by Follesa *et al.* (2011) to assess spillover in the Su Pallosu MPA in Sardinia. The authors found a clear gradient of catch per unit effort (CPUE) from the MPA boundaries towards fishing grounds, with a maximum located within about 6 km from the marine reserve boundary. The study by La Mesa *et al.* (2012) is the only one in the Mediterranean addressing, albeit indirectly, spillover using radio-tracking techniques. It showed that the home range of the parrotfish *Sparisoma cretense* extended beyond the boundaries of the no-take area of the Portofino MPA, and therefore that spillover was possible. Other studies have assessed density/biomass gradients across no-take area boundaries as a way to infer spillover, for example Guidetti (2007), Harmelin-Vivien *et al.* (2008), Forcada *et al.* (2008), La Mesa *et al.* (2011) and García-Rubies *et al.* (2013).

Eggs and Larvae

Spillover can be relevant in those overexploited species where reproduction occurs mostly in MPAs, for example the dusky grouper (Zabala *et al.*, 1997). Most marine organisms have complex life histories that include pelagic eggs and larval stages. These stages influence dispersal capabilities, affecting not only the geographical distribution of the species but also settlement rates and patterns of recruitment to the adult population, with the resulting effects on community structure. Recent studies suggest that the extent of dispersal between populations is more complex than previously assumed, with water flow dynamics and oceanographic discontinuities being important factors determining the population connectivity and settlement intensity (Cowen *et al.*, 2006; Galarza *et al.*, 2009; Schunter *et al.*, 2011a, 2011b, 2014).

In the Mediterranean Sea the large-scale circulation is superimposed on small-scale dynamics (see Boero, this volume). More than 500 canyons act as auxiliary engines to the main cold engines in the Western Mediterranean and the Adriatic Sea. The shape of the coast also generates gyres and eddies that concentrate nutrients and propagules. The portions of marine systems where production phenomena are generated by the interaction of physical, chemical, biological and ecological processes have been termed 'cells of ecosystem functioning' (Boero, 2015). Each one of these cells has its own specific characteristics and can be considered as relatively isolated from other cells due to physical boundaries (fronts) that reduce exchange of propagules. The cells are thus to be considered as the true biogeographical regions, each one with its own biological features. This is why Boero (this volume) suggested that each of these cells should contain, at least, an MPA network in order to preserve the main habitats of each cell of ecosystem functioning.

Populations that are not separated by evident oceanographic boundaries can also show a strong genetic isolation, usually related with a very low larval dispersal capability of the species and a sessile adult life (Duran *et al.*, 2004; Carreras-Carbonell *et al.*, 2006). This genetic isolation has been observed in many sessile organisms that constitute a fundamental part of the Mediterranean ecosystems (Uriz and Turon, 2012).

In addition, several studies have demonstrated that fish populations are not always open and that the proportion of larvae that may return to their natal population (self-recruitment) is very high (Galindo *et al.*, 2006; Almany *et al.*, 2007; Carreras-Carbonell *et al.*, 2007; Schunter *et al.*, 2014). These studies suggest that the extent of dispersal between populations is lower than currently assumed, as predicted by Cowen *et al.* (2006), affecting the connectivity among populations and having important implications for marine conservation policies.

Studies on the degree of self-recruitment in Mediterranean littoral fishes suggest that the self-recruited juveniles have lower probabilities of survivorship compared to juveniles from other localities (Carreras-Carbonell *et al.*, 2007; Planes *et al.*, 2009; Félix-Hackradt *et al.*, 2013). Therefore, the degree of connectivity among populations can also influence the spillover role of protected areas, the assessment of numerous fishery-exploited species and, in general, the management of marine ecosystems (Palumbi, 2004).

This scenario, however, is very different for species with very limited larval dispersal capabilities (e.g. sponges, ascidians, bryozoans and numerous algae). These species show a fine-scale genetic structure with genetic differences at distances in the range of metres (Duran *et al.*, 2004; Calderón *et al.*, 2007; Mokhtar-Jamaï *et al.*, 2011). This genetic structure may be common in invertebrates with lecitotrophic larvae.

Such invertebrates constitute an essential component of coastal rocky Mediterranean ecosystems, suggesting that the isolation in these species can have a strong effect in the dynamics of their populations. No spillover is expected for these organisms.

The approach of inferring population connectivity through genetic differentiation of locations is effective for the identification of major biogeographical or oceanographic barriers (Planes *et al.*, 2009) and allows for the measurement of gene flow across evolutionary timescales. Unfortunately, larval trajectories remain largely unknown and understanding present-day dispersal patterns is still a great challenge. Parentage analysis permits the direct estimation of connectivity, as the detection of parent–offspring pairs allows in many cases the movement of the offspring to be reconstructed, providing direct evidence of dispersal (e.g. Almany *et al.*, 2007; Planes *et al.*, 2009; Saenz-Agudelo *et al.*, 2011; Berumen *et al.*, 2012). At present, only one study (Schunter *et al.*, 2014) has used this direct measure of connectivity in the Mediterranean: it demonstrated a limited connectivity, with a decrease in dispersal success over 1 km distance and approximately 15% of the collected juveniles of *Tripterygion delaisi* identified as self-recruits; sibship reconstruction analysis found that full siblings in general did not recruit together to the same location, and that the distance between recruitment locations was more extensive (about 11.5 km).

The main conclusions from these results are that most coastal species in the Western Mediterranean have a reduced level of connectivity among populations, and it is likely that similar patterns of connectivity are present in the Eastern Mediterranean too, as well as the Black Sea (Öztürk *et al.*, this volume). As a result, islands (e.g. Balearic Islands, Ionian Islands, Aegean Islands) are mostly isolated from the continental coast and from each other, especially for littoral species, although a different situation may operate for species inhabiting deeper ecosystems. Therefore, the vulnerability of these island ecosystems is higher than for the continental coast. These results emphasize the necessity of establishment of MPA networks among islands and between the continent and the islands. Therefore the areas separated by oceanographic discontinuities (Schiavina *et al.*, 2014) should be considered as separate management units, requiring, at least, an MPA network for each unit (Boero, 2015). Finally, considering that many structural and/or key species (e.g. *Paramuricea clavata*, some sponges, some arborescent algae) have limited gene flow between populations, the distance between MPAs in the network should be established keeping these considerations in mind.

Management Implications

Anthropogenic factors that threaten marine ecosystems are many, complex and often act cumulatively or synergistically (Spalding *et al.*, 2013). Some are difficult to address, or virtually impossible to reverse in the short term (e.g. the effects of climate change), while others require management measures based on different approaches at different spatial scales. One of these management measures is the establishment of MPAs.

The IUCN defines an MPA as a 'clearly defined geographical space, recognized, dedicated and managed, through legal or other effective means, to achieve the long-term conservation of nature with associated ecosystem services and cultural values' (Laffoley, 2008). This is a very loose definition that can lead to many interpretations. The IUCN also defines up to seven categories of MPAs but only one includes a no-take zone and limited public access and can be considered a marine reserve.

There are various initiatives for the protection of Mediterranean ecosystems through MPAs. Many international organizations have proposed the creation of networks of MPAs for preserving Mediterranean ecosystems. Environmental NGOs (including WWF, ACCOBAMS, OCEANA and MedPAN) and scientific organizations (CIESM) have proposed various areas of special conservation interest including not only coastal habitats at a regional or national level, but also large transnational MPAs in offshore or deep-sea ecosystems (see Micheli *et al.*, 2013). The EU is also interested in the use of MPAs as a management measure in the Mediterranean Sea. Consequently, it has funded some international research projects, from the mid-1990s to the present (e.g. ECOMARE, BIOMEX, EMPAFISH, PERSEUS, CoCoNet), as well as others, such as LIFE projects. These projects have been instrumental in promoting international collaboration, including scientific publications, reports, websites, models and other tools, information sites for the general public, as well as training of personnel who can continue to contribute to increasing knowledge of the marine ecosystem and the role of MPAs. In general, these products have shown the utility of MPAs in increasing the biomass of natural resources, including increasing fisheries yields around MPAs, and conservation of littoral habitats.

However, the ecological benefits of protection contrast with the sad reality that most of the beneficial results shown by the Mediterranean MPAs come from no-take zones (i.e. true marine reserves) that represent only 0.012% of the total sea surface (Gabrié *et al.*, 2012). Although there is an increasing number of MPAs in the Mediterranean and around the world (Gabrié *et al.*, 2012; Costello and Ballantine, 2015), only a few can be considered as really effective. Most MPAs are nothing but a false image of protection. Currently, the main flaws of many Mediterranean MPAs include:

1) Few MPAs really work as protected areas due to poor or no management and lack of effective surveillance and enforcement, and, in some cases, no implementation of the management plans (assuming these exist).

2) Many MPAs are the result of political opportunism or spatial considerations that have little to do with ecology; that is, most MPAs have been established where and when it was opportunistically possible for mostly non-scientific reasons, such as in areas where there is likely to be least negative reaction from stakeholders.

3) Most MPAs seem to be 'cure-all' remedies aiming at the conservation of 'biodiversity' and at the same time favouring artisanal fisheries and sustainable use of resources. These all-purpose MPAs may sound good, but rarely incorporate adequate management measures and the ecological conditions to achieve all these high goals.

4) For many MPAs there is a lack of representation and of information on what is to be protected (no lists of species, no habitat mapping and no baselines which are necessary to test the effectiveness of protection).

5) Most MPAs lack long-term monitoring and adaptive management based on the monitoring results.

6) Some MPAs show serious deviations from the original objectives due to a bias towards economic interests (e.g. tourism).

7) Most MPAs are located in the western basin of the Mediterranean Sea; only a small number of MPAs have been established along the North African coast and in the eastern Mediterranean basin.

The number of EU-funded projects aimed at testing the effects of MPAs contrast with the caution shown by the EU when it comes to the effective protection of the Mediterranean through the use of MPAs. The Natura 2000 initiative can be considered only a hesitant approach to the issue of protecting the coastline and inshore waters; the guidelines for the establishment of the Natura 2000 network are based on the Birds Directive (2009/147/EC) and Habitats Directive (92/43/EEC). There is no question that birds are an important component of marine ecosystems, but they cannot be fundamental in the design of protected areas aiming to protect marine (mostly 'submarine') habitats. On the other hand, the Habitats Directive includes only 10 marine 'habitats' (actually mostly geomorphological units and habitat complexes), including *Posidonia oceanica* meadows and 'reefs', which are already protected by law (http://www.europa.eu.int./comm/environment/nature/hab-en.htm).

Considering Natura 2000 sites to be real MPAs may even prove counterproductive in the Mediterranean since they create a false impression of protection when, in fact, no actual protection is occurring (Agardy *et al.*, 2011). The management of these areas depends on national and regional governments and, in most cases, can be considered negligible or nil (Gabrié *et al.*, 2013). Only 25% of these spaces have some kind of management. In some countries (e.g. France, Spain), the marine Natura 2000 sites have been proposed as actual MPAs to meet the targets set out in the Aichi Biodiversity Targets of the Convention on Biological Diversity (Meinesz and Bienfune, 2015), which allows these countries to meet target No. 11 which states that '10 per cent of coastal and marine areas, especially areas of particular importance for biodiversity and ecosystem services, are conserved through effectively and equitably managed, ecologically representative and well connected systems of protected areas and other effective area-based conservation' by 2020 (see https://www.cbd.int/sp/targets/).

Spain, for instance, has not hesitated even to integrate some of these spaces in a brand new 'network' of MPAs (the so-called RAMPE: *Spanish Network of Marine Protected Areas*), grouping several areas with very different management levels (from Natura 2000 sites, to well-protected marine reserves). The possible connectivity among MPAs has almost certainly not been taken into account in this Spanish network. In short, this indicates that although the Aichi target No. 11 is officially met, in practice much less than 10% of the area is conserved and properly managed. The situation in other Mediterranean countries may be similar.

Among European countries, the most coherent policy towards MPAs seems to be that of Italy, which does not include the Natura 2000 sites in the Aichi target, but is based on the establishment of 29 well-managed MPAs (Meinesz and Bienfune, 2015), all of them with the same zoning, effective protection (including at least one no-take area) and a similar management regime. Hence the Italian MPAs form effective conservation and geographical networks in which no-take areas are included.

Although the number of Mediterranean MPAs that function as effective protected areas are few, several international organizations are presently promoting the establishment of networks of MPAs. These networks have to meet the minimum requirements of representativeness, effectiveness, replicability and connectivity (IUCN-WCPA, 2008). The first step in creating a conservation MPA network (Beal *et al.*, this volume) is that the component sites have effective management to ensure good protection of threatened habitats and species. In this sense, the initiative to establish a list of *Specially Protected Areas of Mediterranean Importance* (SPAMI) seems much more realistic than to merely consider marine Natura 2000 sites as effective MPAs. At least, the candidate areas to be included in SPAMI meet certain minimum requirements,

including having a legal status and protection, planning and management measures. However, laudable initiatives such as the detailed Mediterranean MPAs Roadmap (http://www.medmpaforum2012.org/sites/default/files/mediterranean_mpa_roadmap.pdf) do not impose the minimum requirements of management and effective enforcement for the MPAs to be integrated into the networks, and include Natura 2000 zones as MPAs. Assembling 'paper MPAs' into a nominal network will not improve the situation.

It seems that the EU lacks the political will to enforce the minimum requirements that any European MPA should have to be considered a real MPA. In this regard the EU should evaluate the existing and future MPAs based on some criteria such as serious management and means to ensure the effective protection of the ecological and biological features that the MPAs were set up to conserve. The minimum requirements for any coastal MPA to be approved by the EU would be similar to those proposed by Meinesz and Bienfune (2015):

1) All the MPAs must include a representative no-take area in which fishing should be absolutely banned and a buffer zone in which a limited amount of fishing could be allowed, excluding the more harmful methods (e.g. spear-fishing) in order to preserve the local artisanal fishery.
2) There must be effective enforcement of protection measures with a sufficient number of wardens and the means that allow an efficient surveillance of the protected area.
3) There must be clear protection objectives, and avoidance of the misinterpretation and wrong implementation of such objectives that could lead to misuse of protected areas.

Such effective MPAs may serve to protect biologically rich habitats, restore overexploited stocks of target species, resolve user conflicts and ameliorate degraded areas. Therefore the establishment of MPAs taking into account the above minimum criteria will eventually lead to better management and protection of marine species and habitats, particularly if MPAs form an ecologically coherent network linking cells of ecosystem functioning. However, routine monitoring of reserves is far from common and, in general, the elements conferring effectiveness have in many cases not yet been established. Management of future Mediterranean MPA networks is also challenged by lack of information on habitat distribution, and on how populations are connected between habitats and MPAs through dispersal of pelagic larvae or propagules.

The present overview of the ecological effects and benefits of MPAs gives rise to several issues that have implications for management. In particular, good management of MPAs and MPA networks should take the following into account:

1) The carrying capacity (K) for exploited species is a key factor that is indicative of management effectiveness; management actions should focus on attaining K values.

2) The magnitude of K values and time to attain them depend on various factors, including the life history of the species, extent of protection and environmental features of the area. Predictive biomass modelling can be used to determine the K values under a particular set of circumstances and hence to monitor the effectiveness of MPAs; therefore collection of the required environmental data for biomass modelling should be built into management plans.

3) Since environmental features play a role in determining the extent of population recovery, the physical environment itself must also be managed. In addition, when designing MPA zoning schemes, those areas having the 'best' habitat for the most vulnerable species, or habitats which are more susceptible to adverse effects, should be chosen as no-take zones.

4) Activities that are of particular concern (e.g. spear-fishing in the case of large long-lived predatory fish species) may need to be banned even outside no-take zones, while other activities will need to be regulated. The extent of regulation for different activities will depend on the objectives of the MPA.

5) Buffer zones can only serve as 'buffers' if they are adequately managed. In the absence of management measures, these zones would be similar to non-protected areas and may even be counterproductive, leading to excessive fishing pressure outside the fully protected areas that can even affect the populations inside. Management plans should include carefully designed zoning schemes to reduce such impacts, for instance through having a set of nested buffer zones, each with a different regulatory regime.

6) MPAs should not be managed solely for recovery of top predators since prey species are equally important for maintaining a functioning ecosystem and some prey species are themselves of conservation concern. In addition, prey abundance is itself a factor that can influence the carrying capacity of the system for top predators. Management must therefore take prey species into account and actions that favour a balance between predator and prey populations may be needed; for example, management of the physical habitat should not focus solely on habitats that yield maximum biomass of predators but also habitats that offer shelter to their prey. This should ultimately lead to MPAs having an ecosystem structure and function that is similar to that found in pristine environments.

7) MPAs should be large enough to protect substantial portions of populations of sedentary species since MPAs that are much smaller than the home range of a species offer inadequate protection for such species. In such circumstances, management may therefore involve extending the areas of MPAs or linking adjacent MPAs through connected corridors where activities are regulated.

8) MPAs should be linked into an ecologically coherent network; 'connectivity' should therefore be built into any management plan. This can be taken into consideration when designing MPA networks to allow connectivity through eggs, larvae and propagules; such connectivity cannot be assumed and must be ascertained through appropriate research. For already existing MPAs, connectivity may be enhanced through establishing 'protected corridors' between the MPAs or having large buffer zones linking neighbouring MPAs, where activities are regulated.

9) MPA effectiveness is also linked to the level of enforcement of, and compliance with, the protection measures; setting protection levels on paper may be easy but enforcement is difficult. Compliance is a function of both enforcement and stakeholder education; therefore, management actions should also focus on educating stakeholders.

We hope that these considerations will aid the formulation and implementation of appropriate management plans for Mediterranean MPAs in the near future, which, together with the establishment of a functioning network of MPAs whose design is based on sound scientific data on the distribution of species, habitats and their connectivity patterns, will ultimately serve to improve the health of one of the world's hotspots of marine biodiversity.

References

Abdulla, A., Gomei, M., Maison, E. and Piante, C. (2008) *Status of Marine Protected Areas in the Mediterranean Sea*. IUCN, Malaga and WWF, France. 152 pp.

Agardy, T., Notarbartolo Di Sciara, G. and Christie, P. (2011) Mind the gap: addressing the shortcomings of marine protected areas through large scale marine spatial planning. *Marine Policy*, **35** (2), 226–232.

Almany, G.R., Berumen, M.L., Thorrold, S.R. *et al.* (2007) Local replenishment of coral reef fish populations in a marine reserve. *Science*, **316**, 742–744.

Babcock, R.C., Kelly, S., Shears, N.T. *et al.* (1999) Changes in community structure in temperate marine reserves. *Marine Ecology Progress Series*, **189**, 125–134.

Bell, J.D. (1983) Effects of depth and marine reserve fishing restrictions on the structure of a rocky reef fish assemblage in the northwestern Mediterranean Sea. *Journal of Applied Ecology*, **20** (2), 357–369.

Berumen, M.L., Almany, G.R., Planes, S. *et al.* (2012) Persistence of self recruitment and patterns of larval connectivity in a marine protected area network. *Ecology and Evolution*, **2**, 444–452.

Boero, F. (2015) The future of the Mediterranean Sea ecosystem: towards a different tomorrow. *Rendiconti Lincei*, **26**, 3–12.

Botsford, L.W., Castilla, J.C. and Peterson, C.H. (1997) The management of fisheries and marine ecosystems. *Science*, **277**, 509–515.

Botsford, L.W., Brumbaugh, D.R., Grimes, C. *et al.* (2009) Connectivity, sustainability, and yield: bridging the gap between conventional fisheries management and marine protected areas. *Reviews in Fish Biology and Fisheries*, **19** (1), 69–95.

Calderón, I., Ortega, N., Duran, S. *et al.* (2007) Finding the relevant scale: clonality and genetic structure in a marine invertebrate (*Crambe crambe*, Porifera). *Molecular Ecology*, **16**, 1799–1810.

Cardona, L., Sales, M. and Lopez, D. (2007) Changes in fish abundance do not cascade to sea urchins and erect algae in one of the most oligotrophic parts of the Mediterranean. *Estuarine, Coastal and Shelf Science*, **72** (1–2), 273–282.

Cardona, L., Moranta, J., Reñones, O. and Hereu, B. (2013) Pulses of phytoplanktonic productivity may enhance sea urchin abundance and induce state shifts in Mediterranean rocky reefs. *Estuarine, Coastal and Shelf Science*, **133**, 88–96.

Carreras-Carbonell, J., Macpherson, E. and Pascual, M. (2006) Population structure within and between subspecies of the Mediterranean triplefin fish *Tripterygion delaisi* revealed by highly polymorphic microsatellite loci. *Molecular Ecology*, **15**, 3527–3539.

Carreras-Carbonell, J., Macpherson, E. and Pascual, M. (2007) High self-recruitment levels in a Mediterranean littoral fish population revealed by microsatellite markers. *Marine Biology*, **151**, 719–727.

Cheung, W.L., Pitcher, T.J. and Pauly, D. (2005) A fuzzy logic expert system to estimate intrinsic extinction vulnerabilities of marine fishes to fishing. *Biological Conservation*, **124** (1), 97–111.

Claudet, J., Pelletier, D., Jouvenel, J.Y. *et al.* (2006) Assessing the effects of a marine protected area (MPA) on a reef fish assemblage in a northwestern Mediterranean marine reserve: identifying community-based indicators. *Biological Conservation*, **130** (3), 349–369.

Claudet, J., Osenberg, C.W., Benedetti-Cecchi, L. *et al.* (2008) Marine reserves: size and age do matter. *Ecology Letters*, **11** (5), 481–489.

Claudet, J., Garcia-Charton, J.A. and Lenfant, P. (2011) Combined effects of levels of protection and environmental variables at different spatial resolutions on fish assemblages in a marine protected area. *Conservation Biology*, **25** (1), 105–114.

Coll, J., Linde, M., Garcia-Rubies, A. *et al.* (2004) Spear fishing in the Balearic Islands (west central Mediterranean): species affected and catch evolution during the period 1975–2001. *Fisheries Research*, **70**, 97–111.

Coll, J., Garcia-Rubies, A., Morey, G. and Grau, A.M. (2012) The carrying capacity and the effects of protection level in three marine protected areas in the Balearic Islands (NW Mediterranean). *Scientia Marina*, **76** (4), 809–826.

Coll, J., Garcia-Rubies, A., Morey, G. *et al.* (2013) Using no-take marine reserves as a tool for evaluating rocky-reef fish resources in the western Mediterranean. *ICES Journal of Marine Science*, **70** (3), 578–590.

Coll, M., Piroddi, C., Steenbeek, J. *et al.* (2010) The biodiversity of the Mediterranean Sea: estimates, patterns, and threats. *PLoS ONE*, **5** (8), e11842. doi:10.1371/journal.pone.0011842

Costello, M.J. and Ballantine, B. (2015) Biodiversity conservation should focus on no-take Marine Reserves: 94% of Marine Protected Areas allow fishing. *Trends in Ecology and Evolution*, **30** (9), 507–509.

Côté, I.M., Mosqueira, G. and Reynolds, J.D. (2001) Effects of marine reserves characteristics on the protection of fish populations: a meta-analysis. *Journal of Fish Biology*, **59** (Supplement A), 178–189.

Cowen, R.K., Paris, C.B. and Srinivasan, A. (2006) Scaling of connectivity in marine populations. *Science*, **311**, 522–527.

Denny, C. and Babcock, R. (2004) Do partial marine reserves protect reef fish assemblages? *Biological Conservation*, **116** (1), 119–129.

Di Franco, A., Milazzo, M., Baiata, P. *et al.* (2009) Scuba diver behaviour and its effects on the biota of a Mediterranean marine protected area. *Environmental Conservation*, **36** (1), 32–40.

Díaz, D. (2013) Els grans decàpodes de la costa del Montgrí i les Illes Medes, in 'Els fons marins de les Illes Medes i el Montgrí: quatre dècades de recerca' (eds B. Hereu and X. Quintana). *Recerca i Territori*, **4**, 139–154.

Díaz, D., Zabala, M., Linares, C. *et al.* (2005) Increased predation of juvenile European spiny lobster (*Palinurus elephas*) in a marine protected area. *New Zealand Journal of Marine and Freshwater Research*, **39** (2), 447–453.

Díaz, D., Mallol, S., Parma, A.M. and Goñi, R. (2011) Decadal trend in lobster reproductive output from a temperate marine protected area. *Marine Ecology Progress Series*, **433**, 149–157.

Duran, S., Pascual, M., Estoup, A. and Turon, X. (2004) Strong population structure in the marine sponge *Crambe crambe* (Poecilosclerida) as revealed by microsatellite markers. *Molecular Ecology*, **13**, 511–522.

Durant, S.M. and Harwood, J. (1992) Assessment of monitoring and management strategies for local populations of the Mediterranean monk seal *Monachus monachus*. *Biological Conservation*, **61** (2), 81–92.

Félix-Hackradt, F.C., Hackradt, C.W., Pérez-Ruzafa, A. and García-Charton, J.A. (2013) Discordant patterns of genetic connectivity between two sympatric species, *Mullus barbatus* (Linnaeus, 1758) and *Mullus surmuletus* (Linnaeus, 1758), in south-western Mediterranean Sea. *Marine Environmental Research*, **92**, 23–34.

Fenberg, P.B., Caselle, J.E., Claudet, J. *et al.* (2012) The science of European marine reserves: status, efficacy, and future needs. *Marine Policy*, **36** (5), 1012–1021.

Ferretti, F., Myers, R.A., Serena, F. and Lotze, H.K. (2008) Loss of large predatory sharks from the Mediterranean Sea. *Conservation Biology*, **22** (4), 952–964.

Follesa, M.C., Cannas, R., Cau, A. *et al.* (2011) Spillover effects of a Mediterranean marine protected area on the European spiny lobster *Palinurus elephas* (Fabricius, 1787) resource. *Aquatic Conservation of Marine and Freshwater Ecosystems*, **21** (6), 564–572.

Forcada, A., Bayle Sempere, J., Valle, C. and Sánchez-Pérez, P. (2008) Habitat continuity effects on gradients of fish biomass across marine protected area boundaries. *Marine Environmental Research*, **66** (5), 536–547.

Gabrié, C., Lagabrielle, E., Bissery, C. *et al.* (2012) *The Status of Marine Protected Areas in the Mediterranean Sea*. MedPAN, Marseilles and RAC/SPA, Tunis. MedPAN Collection. 256 pp.

Galarza, J., Carreras-Carbonell, J., Macpherson, E. *et al.* (2009) The influence of oceanographic fronts and early-life history traits on connectivity among fish populations: a multi-species approach. *Proceedings of the National Academy of Sciences*, **106**, 1473–1478.

Galindo, H.M., Olson, D.B. and Palumbi, S.R. (2006) Seascape genetics: a coupled oceanographic–genetic model predicts population structure of Caribbean corals. *Current Biology*, **16**, 1622–1626.

Garcia-Charton, J.A.A. and Pérez-Ruzafa, Á. (1999) Ecological heterogeneity and the evaluation of the effects of marine reserves. *Fisheries Research*, **42** (1), 1–20.

Garcia-Charton, J.A., Pérez-Ruzafa, Á., Marcos, C. *et al.* (2008) Effectiveness of European Atlanto-Mediterranean MPAs: do they accomplish the expected effects on populations, communities and ecosystems? *Journal for Nature Conservation*, **16** (4), 193–221.

Garcia-Rubies, A. (1999) Effects of fishing on community structure and on selected populations of Mediterranean coastal reef fish. *Naturalista Siciliano*, **23** (Supplement), 59–81.

Garcia-Rubies, A. and Zabala, M. (1990) Effects of total fishing prohibition on the rocky fish assemblages of Medes Islands marine reserve (NW Mediterranean). *Scientia Marina*, **54**, 317–328.

Garcia-Rubies, A., Hereu, B. and Zabala, M. (2013) Long-term recovery patterns and limited spillover of large predatory fish in a Mediterranean MPA. *PLoS ONE*, **8** (9), e73922. doi:10.1371/journal.pone.0073922

Garrabou, J. and Harmelin, J.G. (2002) A 20-year study on life-history traits of a harvested long-lived temperate coral in the NW Mediterranean: insights into conservation and management needs. *Journal of Animal Ecology*, **71** (6), 966–978.

Giacalone, V.M., D'Anna, G., Pipitone, C. and Badalamenti, F. (2006) Movements and residence time of spiny lobsters *Palinurus elephas* released in a marine protected area: an investigation by ultrasonic telemetry. *Journal of the Marine Biological Association*, **86** (5), 1101–1106.

Goñi, R., Quetglas, A. and Reñones, O. (2006) Spillover of spiny lobsters *Palinurus elephas* from a marine reserve to an adjoining fishery. *Marine Ecology Progress Series*, **308**, 207–219.

Goñi, R., Hilborn, R., Díaz, D. *et al.* (2010) Net contribution of spillover from a marine reserve to fishery catches. *Marine Ecology Progress Series*, **400**, 233–243.

Guidetti, P. (2005) Reserve effect and trophic cascades in Mediterranean sublittoral rocky habitats: a case study at the Marine Protected Area of Torre Guaceto (southern Adriatic Sea). *Biologia Marina Mediterranea*, **12** (1), 99–105.

Guidetti, P. (2006) Marine reserves re-establish lost predatory interactions and cause community changes in rocky reefs. *Ecological Applications*, **16**, 963–976.

Guidetti, P. (2007) Potential of marine reserves to cause community-wide changes beyond their boundaries. *Conservation Biology*, **21** (2), 540–545.

Guidetti, P. (2011) The destructive date-mussel fishery and the persistence of barrens in Mediterranean rocky reefs. *Marine Pollution Bulletin*, **62** (4), 691–695.

Guidetti, P. and Micheli, F. (2011) Ancient art serving marine conservation. *Frontiers in Ecology and Environment*, **9**, 374–375.

Guidetti, P. and Sala, E. (2007) Community-wide effects of marine reserves in the Mediterranean Sea. *Marine Ecology Progress Series*, **335**, 43–56.

Guidetti, P., Milazzo, M., Bussotti, S. *et al.* (2008) Italian Marine Reserve effectiveness: does enforcement matter? *Biological Conservation*, **141**, 699–709.

Halpern, B.S. and Warner, R.R. (2002) Marine reserves have rapid and lasting effects. *Ecology Letters*, **5** (3), 361–366.

Harmelin, J-G., Bachet, F. and Garcia, F. (1995) Mediterranean marine reserves: fish indices as tests of protection efficiencies. *Marine Ecology*, **16** (3), 233–250.

Harmelin-Vivien, M., Le Direach, L., Bayle-Sempere, J. *et al.* (2008) Gradients of abundance across reserve boundaries in six Mediterranean marine protected areas: evidence of spillover? *Biological Conservation*, **141** (7), 1829–1839.

Hereu, B., Zabala, M., Linares, C. and Sala, E. (2005) The effects of predator abundance and habitat structural complexity on survival of juvenile sea urchins. *Marine Biology*, **146**, 293–299.

Hereu, B., Linares, C., Sala, E. *et al.* (2012) Multiple processes regulate long-term population dynamics of sea urchins on Mediterranean rocky reefs. *PloS ONE*, **7** (5), e36901. doi:10.1371/journal.pone.0036901

International Union for Conservation of Nature – World Commission on Protected Areas (IUCN-WCPA) (2008) *Establishing Marine Protected Area Networks: Making It Happen*. IUCN-WCPA, National Oceanic and Atmospheric Administration and The Nature Conservancy, Washington, DC. 118 pp.

La Mesa, G., Molinari, A., Gambaccini, S. and Tunesi, L. (2011) Spatial pattern of coastal fish assemblages in different habitats in North-western Mediterranean. *Marine Ecology*, **32** (1), 104–114.

La Mesa, G., Consalvo, I., Annunziatelis, A. and Canese, S. (2012) Movement patterns of the parrotfish *Sparisoma cretense* in a Mediterranean marine protected area. *Marine Environmental Research*, **82**, 59–68.

Laffoley, D. (2008) *Towards Networks of Marine Protected Areas: The MPA Plan of Action for IUCN's World Commission on Protected Areas.* IUCN-WCPA, Gland, Switzerland. 28 pp.

Lester, S.E., Halpern, B.S., Grorud-Colvert, K. *et al.* (2009) Biological effects within no-take marine reserves: a global synthesis. *Marine Ecology Progress Series*, **384** (2), 33–46.

Ling, S., Scheibling, R., Rassweiler, A., Johnson, C. *et al.* (2015) Global regime shift dynamics of catastrophic sea urchin overgrazing. *Philosophical Transactions of the Royal Society B: Biological Sciences*, **370** (1659), 20130269.

Lloret, J., Zaragoza, N., Caballero, D. and Riera, V. (2008) Biological and socioeconomic implications of recreational boat fishing for the management of fishery resources in the marine reserve of Cap de Creus (NW Mediterranean). *Fisheries Research*, **91** (2–3), 252–259.

Macpherson, E. (1994) Substrate utilization in a Mediterranean littoral fish community. *Marine Ecology Progress Series*, **114**, 211–218.

Macpherson, E., Biagi, F., Francour, P. *et al.* (1997) Mortality of juvenile fishes of the genus *Diplodus* in protected and unprotected areas in the western Mediterranean Sea. *Marine Ecology Progress Series*, **160**, 135–147.

Macpherson, E., García-Rubies, A. and Gordoa, A. (2000) Direct estimation of natural mortality rates for littoral marine fishes using populational data from a marine reserve. *Marine Biology*, **137** (5–6), 1067–1076.

McClanahan, T.R., Graham, N.A., Calnan, J.M. and MacNeil, M.A. (2007) Toward pristine biomass: reef fish recovery in coral reef marine protected areas in Kenya. *Ecological Applications*, **17** (4), 1055–1067.

McClanahan, T.R., Graham, N.A., MacNeil, M.A. *et al.* (2011) Critical thresholds and tangible targets for ecosystem-based management of coral reef fisheries. *Proceedings of the National Academy of Sciences*, **108** (41), 17230–17233.

Meinesz, A. and Bienfune, A. (2015) 1983–2013: Development of marine protected areas along the French Mediterranean coasts and perspectives for achievement of the Aichi target. *Marine Policy*, **54**, 10–16.

Micheli, F., Halpern, B.S., Botsford, L.W. and Warner, R.R. (2004) Trajectories and correlates of community change in no-take marine reserves. *Ecological Applications*, **14** (6), 1709–1723.

Micheli, F., Benedetti-Cecchi, L., Gambaccini, S. *et al.* (2005) Cascading human impacts, marine protected areas, and the structure of Mediterranean reef assemblages. *Ecological Monographs*, **75** (1), 81–102.

Micheli, F., Levin, N., Giakoumi, S. *et al.* (2013) Setting priorities for regional conservation planning in the Mediterranean Sea. *PLoS ONE*, **8** (4), e259038. doi:10.1371/journal.pone.00259038

Mokhtar-Jamaï, K., Pascua, M., Ledoux, J.B. *et al.* (2011) From global to genetic structuring in the red gorgonian *Paramuricea clavata*: the interplay between oceanographic conditions and limited larval dispersal. *Molecular Ecology*, **20** (16), 3291–3305.

Molloy, P.P., Reynolds, J.D., Gage, M.J.G. *et al.* (2008) Links between sex change and fish densities in marine protected areas. *Biological Conservation*, **141** (1), 187–197.

Molloy, P.P., McLean, I.B. and Côté, I.M. (2009) Effects of marine reserve age on fish populations: a global meta-analysis. *Journal of Applied Ecology*, **46** (4), 743–751.

Olsen, E.M., Johnson, D., Weaver, P. *et al.* (2013) *Achieving Ecologically Coherent MPA Networks in Europe: Science Needs and Priorities* (eds K.E. Larkin and N. McDonough). Marine Board Position Paper 18. European Marine Board, Ostend, Belgium. 88 pp.

Palumbi, S.R. (2004) Marine reserves and ocean neighborhoods: the spatial scale of marine populations and their management. *Annual Review of Environment and Resources*, **29**, 31–68.

Pauly, D., Christensen, V., Dalscaard, J. *et al.* (1998) Fishing down marine food webs. *Science*, **279**, 860–863.

Planes, S., Jones, G.P. and Thorrold, S.R. (2009) Larval dispersal connects fish populations in a network of marine protected areas. *Proceedings of the National Academy of Sciences*, **106**, 5693–5697.

Reñones, O., Goñi, R., Pozo, M. *et al.* (1999) Effects of protection on the demographic structure and abundance of *Epinephelus marginatus* (Lowe, 1834): evidence from Cabrera Archipelago National Park (West-central Mediterranean). *Marine Life*, **9** (2), 45–53.

Russ, G.R. and Alcala, A.C. (1996) Marine reserves: rates and patterns of recovery and decline of large predatory fish. *Ecological Applications*, **6** (3), 947–961.

Russ, G.R., Alcala, A.C., Maypa, A.P. *et al.* (2004) Marine reserve benefits local fisheries. *Ecological Applications*, **14**, 597–606.

Saenz-Agudelo, P., Jones, G.P., Thorrold, S.P. and Planes, S. (2011) Connectivity dominates larval replenishment in a coastal reef fish metapopulation. *Proceedings of the Royal Society of London B, Biological Sciences*, rspb20102780.

Sala, E. and Zabala, M. (1996) Fish predation and the structure of the sea urchin *Paracentrotus lividus* populations in the NW Mediterranean. *Marine Ecology Progress Series*, **140** (1), 71–81.

Sala, E., Ribes, M., Hereu, B. *et al.* (1998) Temporal variability in abundance of the sea urchins *Paracentrotus lividus* and *Arbacia lixula* in the northwestern Mediterranean: comparison between a marine reserve and an unprotected area. *Marine Ecology Progress Series*, **168**, 135–145.

Sala, E., Ballesteros, E., Dendrinos, P. *et al.* (2012) The structure of Mediterranean rocky reef ecosystems across environmental and human gradients, and conservation implications. *PLoS ONE*, **7** (2), e32742.

Sasal, P., Faliex, E. and Morand, S. (1996) Population structure of *Gobius bucchichii* in a Mediterranean marine reserve and in an unprotected area. *Journal of Fish Biology*, **49** (2), 352–356.

Schiavina, M., Marino, I.A.M., Zane, L. and Melià, P. (2014) Matching oceanography and genetics at the basin scale: seascape connectivity of the Mediterranean shore crab in the Adriatic Sea. *Molecular Ecology*, **23**, 5496–5507.

Schunter, C., Carreras-Carbonell, J., Planes, S. *et al.* (2011a) Genetic connectivity patterns in an endangered species: the dusky grouper (*Epinephelus marginatus*). *Journal of Experimental Marine Biology and Ecology*, **401**, 126–133.

Schunter, C., Carreras-Carbonell, J., Macpherson, E. *et al.* (2011b) Matching genetics with oceanography: directional gene flow in a Mediterranean fish species. *Molecular Ecology*, **20**, 5167–5181.

Schunter, C., Pascual, M., Garza, J.C. *et al.* (2014) Retention and fish larval dispersal potential on a highly connected open coast line. *Proceedings of the Royal Society of London B*, **281** (1785), 20140556.

Spalding, M.D., Meliane, I., Milam, A. *et al.* (2013) Protecting marine spaces: global targets and changing approaches. *Ocean Yearbook*, **27** (1), 213–248.

Stelzenmuller, V., Maynou, F. and Martin, P. (2007) Spatial assessment of benefits of a coastal Mediterranean Marine Protected Area. *Biological Conservation*, **136** (4), 571–583.

Uriz, M.J. and Turon, X. (2012) Sponge ecology in the molecular era. *Advances in Marine Biology*, **61**, 345–410.

Vacchi, M., Bussotti, S., Guidetti, P. and La Mesa, G. (1998) Study of the coastal fish assemblage in the marine reserve of the Ustica Island (southern Tyrrhenian Sea). *Italian Journal of Zoology*, **65**, 281–286.

Worm, B., Hilborn, R., Baum, J.K. *et al.* (2009) Rebuilding global fisheries. *Science*, **325**, 578–585.

Zabala, M., Garcia-Rubies, A., Louisy, P. and Sala, E. (1997) Spawning behaviour of the Mediterranean dusky grouper (*Epinephelus marginatus* (Lowe, 1834) (Pisces, Serranidae) in the Medes Islands Marine Reserve (NW Mediterranean, Spain). *Scientia Marina*, **61** (1), 65–77.

3

Typology, Management and Monitoring of Marine Protected Area Networks

Stephen Beal, Paul D. Goriup and Thomas Haynes

NatureBureau, Newbury, UK

Introduction

Marine Protected Area (MPA) networks are a means of amplifying the ecological, social and economic benefits of single MPAs. The International Union for Conservation of Nature (IUCN-WCPA, 2008) defines MPA networks as

> a collection of individual marine protected areas (MPAs) or reserves operating co-operatively and synergistically, at various spatial scales and with a range of protection levels, that are designed to meet objectives that a single reserve cannot achieve.

Well-planned networks should provide connectivity and ensure that ecosystem processes are maintained (NRC, 2000; Garcia-Rubies *et al.*, this volume). Networks can also spread risk by protecting threatened species, habitats and ecosystems across a wide area, securing their survival in the event of local extirpations.

The importance of developing MPA networks to halt biodiversity loss and protect marine ecosystems has been widely endorsed (SPA/BD Protocol, 1995; IUCN-WCPA, 2008; UNEP-WCMC, 2008; JNCC and Natural England, 2010), and supported by various guidelines for designing and establishing MPA networks (e.g. IUCN-WCPA, 2008; Laffoley, 2014). However, in spite of repeated calls to establish networks of MPAs, little attention has been paid to the specific characteristics of, and management approaches for, such networks. Indeed, as the very broad scope of the IUCN definition makes clear, 'MPA networks' can be a very flexible concept which may defy attempts to devise standardized approaches. For example, while various guidance on managing and assessing the effectiveness of individual MPAs is available (see Dudley and Hockings, this volume), the basic principles for network management and indicators to measure the success of achieving network objectives have not been clearly identified (UNEP-WCMC, 2008).

Based on a review of a wide variety of MPAs in the Mediterranean and Black Seas, this chapter sets out to show how networks are formed, and propose a typology of networks based on the MPA management objectives. As such, it seeks to elucidate from the empirical evidence what the IUCN definition encompasses in practice. The conclusions drawn are that there are at least seven categories of network, broadly

Management of Marine Protected Areas: A Network Perspective, First Edition. Edited by Paul D. Goriup.
© 2017 John Wiley & Sons Ltd. Published 2017 by John Wiley & Sons Ltd.

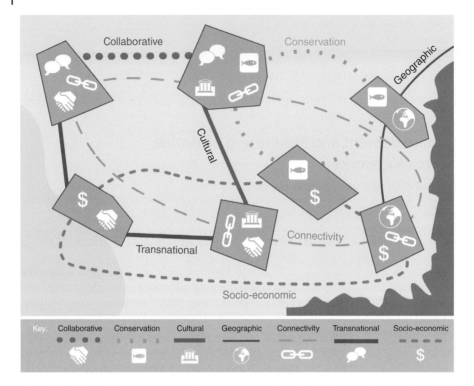

Figure 3.1 Schematic representation of nested and overlapping MPA network types. Artwork: Barbara Creed.

understood in legislation, literature and by MPA managers and stakeholders; that these are formed through various mechanisms ranging from ad hoc to more or less systematic; and that they can overlap both spatially and temporally.

Network Formation Characteristics

There are several general characteristics shown across all MPA network types. Some, such as planned or unplanned networks, arise from the network formation process. Others, such as spatial and temporal overlaps, are an emergent property of network formation itself. These characteristics interact at the site level so that no MPA will belong to a single network type, but will represent a node in different network types according to its own properties and

functions. This multiplicity of MPA networks is shown schematically in Figure 3.1.

Unplanned (Ad Hoc) Networks

Unplanned or ad hoc (Grorud-Colvert *et al.*, 2011) networks typically consist of MPAs which have been designated on a case by case basis to protect particular charismatic features (species and habitats) which are well known to scientists and/or attract many visitors (especially divers), and which were under threat. Most early MPAs, particularly in the Mediterranean which has both high biodiversity and intense human use, were established on these grounds and were not deliberately intended to form a network (Grorud-Colvert *et al.*, 2014). In due course, as the number of MPAs grew, they were first grouped in national geographic networks (also called MPA systems), and then incorporated in various functional networks that are more or less recognized today (see next section). As a result, unplanned networks have become rare, but their vestiges can still be discerned (sometime as anomalies) in planned networks.

Planned and Semi-Planned Networks

(Semi-)planned networks are formed by the application of predetermined criteria, usually enshrined in legislation or internationally accepted principles (e.g. Convention on Biological Diversity, 2004), in order to reach explicit conservation targets rooted in the concept of systematic conservation planning (Margules and Pressey, 2000). In planned networks, for example the proposed UK Marine Conservation Zone network (JNCC and Natural England, 2010, 2012), the locations of the MPAs are largely defined prior to their designation.

Semi-planned networks also have targets and criteria, but the exact locations of MPAs are not predicated on them. For example, the European Natura 2000 network (established from implementation of the EU Birds Directive, 2009/147/EC, and Habitats Directive, 92/43/EEC) is designed around unifying conservation features (habitats and species of European importance) to ensure that representative and replicate examples of them are included in the network (Evans, 2006; Braun, this volume). However, the locations of specific sites are not defined in advance (e.g. from models) but from an ongoing process of information gathering and negotiation among stakeholders. Moreover, the criteria (the lists of habitats and species of European concern) are amended each time a new member state joins the EU.

Overlapping Networks

It is axiomatic that all network types will overlap spatially and temporally, and that individual sites will serve as nodes in several network types. This means that sites and networks can be mutually supporting and synergistic in achieving their aims.

At the site level, for example, the Telašćica Nature Park in Croatia has management objectives relevant to conservation, geographic, socio-economic, collaborative and cultural network types (Puhov *et al.*, 2012). The conservation network is supported because the site is designated based on the presence of important habitats and species, in particular *Posidonia* meadows. The MPA forms part of the Croatian, Adriatic and Mediterranean geographic networks. It has aims relating to sustainable fisheries management and improving the quality of tourist experiences of the park, so it could be inside a socio-economic network. As a member of MedPAN, the park administration contributes towards the Mediterranean collaborative network. Finally, Telašćica promotes the preservation, protection and promotion of cultural and historical heritage which qualifies it for membership of a cultural network.

Networks as a whole will overlap in many different combinations, each strengthening the others. For example, the Spanish Marine Reserves Network is primarily a socio-economic network (Goñi *et al.*, 2015). Its main aim is to protect the fish resources of artisanal fishermen by creating no-take zones which prohibit all forms of fishing. It also contributes towards a conservation network by protecting biodiversity through preventing harmful fishing activities (Goñi *et al.*, 2015). Furthermore, it supports a cultural network by preserving traditional fishing for local people. Another example of how networks can assist each other is the Mediterranean collaborative network formed by MedPAN. This network conducts workshops which promote good governance, socio-economic development and cultural preservation (Gabrié *et al.*, 2012).

Nested Networks

Nested networks (or network subtypes) occur when one or more networks are situated within one of a higher level and therefore have a more direct relationship than sharing some degree of overlap. The most obvious examples are geographic networks which can start as a quite local network (a stretch of coast or a group of islands), within a national network, which in turn is part of a regional network (e.g. sea basin) and ultimately all geographic networks are part of the global MPA network. Other networks can also be nested according to particular themes. Thus a conservation network could include subsidiary networks for marine mammals, fish, habitat types and so on each with its own associated interest groups, stakeholders and activities.

MPA Network Typology

Building on previous work by Grorud-Colvert *et al.* (2011, 2014), a review was conducted of MPA management plans and networks described in the Mediterranean and Black Seas (including existing and proposed MPA networks, guidance documents and the wider MPA network literature). In total, 15 management plans were reviewed: six from the Black Sea, six from the Mediterranean Sea, plus for comparison two from the North Sea (Germany) and one from the Caspian Sea (Russia) (Table 3.1). In addition, 36 documents relating to MPA networks were reviewed. As a result, an MPA network typology was developed (Table 3.2), comprising seven main categories:

- Conservation
- Connectivity
- Socio-economic
- Geographic
- Collaborative
- Cultural
- Transnational.

Each of the main categories has several subtypes that further refine the respective network characteristics and are described further below.

Conservation Networks

Conservation networks are those in which a system of sites aim at protecting features of conservation importance through the principles of representation, replication and adequacy. Networks which utilize these principles are intended to protect the fullest possible range of biodiversity and preserve viable populations (Margules and Pressey, 2000).

Conservation networks are typically established according to predetermined criteria, typically identified in legislation or guidelines. Conservation features normally refer to species and habitats, but geological and geomorphological formations are often included. There is also an increasing trend to protect ecological processes or ecosystem functions (Pereira and Navarro, 2015; Boero, this volume). Sites within a particular conservation network should be linked by a common list of conservation features, allowing coherent management strategies to be developed across all sites.

Connectivity Networks

Connectivity networks are specifically designed to preserve the movement and dispersal of propagules, larvae, juveniles or adults, including habitat-forming species. Ecological coherence is achieved by protecting sites where genetic exchanges are known to occur. Connectivity is often a stated aim of conservation networks (IUCN-WCPA, 2008; JNCC and Natural England, 2010) and is assumed to be attained by an appropriate spacing of MPAs within the network. However, due to a lack of knowledge of connectivity patterns, and/or because of the complexity and plethora of species-specific connectivity patterns, it is difficult if not impossible for one network to ensure connectivity for all species of interest (except perhaps at very broad scales). Moreover, the degree of connectivity between sites can vary according to the approach taken

Table 3.1 MPA management plans included in the MPA network review process.

Site	Country	Region	Reference
Strandzha–Igneada	Bulgaria/ Turkey	Black Sea	Muresan *et al.* (2014)
Kholketi National Park	Georgia	Black Sea	Gabunia *et al.* (2004)
Vama Veche – 2 Mai	Romania	Black Sea	Magda Nenciu, *in litt.*, 17 March 2015
Black Sea Biosphere Reserve	Ukraine	Black Sea	Galina Minicheva, *in litt.*, 31 March 2015
Small Phyllophora Field	Ukraine	Black Sea	Goriup (2009)
Zernov's Phyllophora Field	Ukraine	Black Sea	Galina Minicheva, *in litt.*, 31 March 2015
Telašćica Nature Park	Croatia	Mediterranean Sea	Puhov *et al.* (2012)
Pelagos Sanctuary	France/ Monaco/ Italy	Mediterranean Sea	Notarbartolo di Sciara and Hyrenbach (2007)
Rosh HaNiqra – Akhziv Nature Reserve	Israel	Mediterranean Sea	López *et al.* (2004)
Qawra/Dwejra Heritage Park	Malta	Mediterranean Sea	MEPA (2010)
Rdum Majjiesa to Ras ir-Raheb Cave	Malta	Mediterranean Sea	Tunesi *et al.* (2003)
Zembra and Zembretta Marine National Park	Tunisia	Mediterranean Sea	Orueta and Limam (n.d.)
Borkum Riffgrund Nature Conservation Area	Germany	North Sea	Eva Schachtner, *in litt.*, 1 April 2015
Vorpommersche Boddenlandschaft National Park	Germany	North Sea	Eva Schachtner, *in litt.*, 1 April 2015
Astrakhanskii State Nature Biosphere Reserve	Russia	Caspian Sea	UNESCO-MBP (2007)

(e.g. physical oceanography, propagule dispersion, beta diversity index or genetics). Due to these factors, connectivity networks should have clearly stated aims, probably limited to a small number of key features (e.g. habitat-formers or species with fragmented distributions).

Socio-economic Networks

The ecosystem approach recognizes that humans are an integral part of the marine environment and therefore human uses and impacts need to be taken into account and managed in a sustainable way; thus the IUCN guidelines on establishing MPA networks (IUCN-WCPA, 2008) state that socio-economic development should be an objective of all networks and in fact many existing MPAs and MPA networks fit within the socio-economic network category. Socio-economic networks provide a discrete platform for managing marine resources and uses with positive environmental, social and economic benefits. Conservation is not

Table 3.2 MPA network typology for the Mediterranean and Black Seas.

Network type	Subtypes	Description	Purpose	References
Conservation	Habitats, Species, Habitats and Species, Geological, Geomorphological, Legislation	A system of sites aimed at protecting features (species and habitats) of conservation importance established according to predetermined criteria. Features to be protected are typically identified in legislation	Generally designed to protect features showing the full range of their variation, by representation, replication and adequacy of features, across a range of sites	Grorud-Colvert et al. (2011); Catchpole (2012)
Connectivity	Genetic, Species/ Populations, Habitats/ Ecosystems	A system of sites connected by, and designed to maintain, the movement and dispersal of propagules, larvae, juveniles or adults (including habitat-forming species)	To ensure ecological coherence of marine biodiversity by providing protection to sites important for genetic exchange	Grorud-Colvert et al. (2014)
Socio-economic	Fisheries, Aquaculture, Renewable energy, Ecotourism, Education	A system of sites where marine resources are managed for social and economic benefits	To protect and manage the use of marine resources in a sustainable manner	Grorud-Colvert et al. (2011); Goñi et al. (2015)
Geographic	International, Regional Sea, Ecoregion	A system of sites in a given area contributing to protected area coverage and conservation targets within that area	To achieve conservation and protected area coverage targets within a defined geographical area	Grorud-Colvert et al. (2011); Laffoley (2014)
Collaborative	Network managers, MPA managers, Stakeholders	A system of sites or networks whose managers, practitioners, stakeholders, decision-makers, scientists, and others interact to develop and share best practice	To promote and improve area-based planning, management or monitoring of marine resources	Grorud-Colvert et al. (2011); MedPAN (2015)
Cultural	Seascapes, Archaeological – World Heritage Sites	A system of sites aimed at protecting features and seascapes of historical and cultural importance	To protect sites and areas where significant historical, cultural and seascape features are present, by preserving and promoting traditional management practices and preventing harmful activities	UNESCO 1972; Council of Europe (2000); Natural England (2012)
Transnational	Peace parks, Transboundary MPAs	A system of sites which are managed in common by two or more countries, usually with supranational coordination	To establish coordinated management of natural resources between and beyond national jurisdictions	EU Habitats Directive 92/43/EEC; Sandwith et al. (2001)

always a primary objective but is always addressed. While some socio-economic networks are based purely on managing marine resources for exploitation (e.g. the Spanish Marine Reserves Network; see box) the sustainable management methods often align with conservation methods.

Case Study: Spanish Marine Reserves Network

Ten reserves were established between 1986 and 2007, by the General Secretary for Fisheries (GSF), with the primary goal of sustaining fisheries for artisanal purposes. No fishing activities are allowed in the reserves, which range in size from 457 ha to 70 439 ha and include sites in the Mediterranean Sea and Canary Islands. The results have been positive for both fishermen and biodiversity, with noticeable increases in biomass (both inside and outside the reserves) for the dusky grouper *Epinephelus marginatus* and European spiny lobster *Palinurus elephas*. The reserves also act as study sites for assessing the effects of invasive species and climate change (Goñi *et al.*, 2015).

Geographic Networks

Most existing MPA networks are built on geographic principles within jurisdictional boundaries based on the UN Convention on the Law of the Sea (see Braun, this volume). They range from relatively local (provincial) to national systems and ultimately international, for example the Mediterranean SPAMI network (SPA/BD Protocol, 1995), the HELCOM MPA network in the Baltic Sea (Baltic Marine Environment Protection Commission [HELCOM], 2013) and the OSPAR north-east Atlantic MPA network (Commission for the Protection of the Marine Environment of the North-East Atlantic [OSPAR], 2013). The EU Natura

2000 network of MPAs is a unique example of an international geographic network established by common legislation (the EU Birds and Habitats Directives). By building conservation networks from the bottom up, through a geographic approach, the foundations for larger networks can be laid. For instance, national networks form the basis of ecoregion networks, which form the basis of regional networks, which form a global network. It follows that all other network types will fall within a geographic network at some level, even if not formally declared. Geographic networks enhance the other networks by nesting different levels of governance and features within appropriate spatial boundaries. The establishment and/or expansion of geographic MPA networks is currently driven by the 10% marine area coverage target set by the Parties to the Convention on Biological Diversity (UNEP/CBD/COP/10/27/Add.1).

Collaborative Networks

The objectives of collaborative networks are to promote interaction between members in order to improve area-based planning, management, implementation or monitoring of marine resources and associated uses. Essentially, these are social networks where members share knowledge and experiences.

Examples of such networks exist in the Mediterranean (MedPAN; see Webster, this volume), the Caribbean (CaMPAM) and Pacific (LMMA) (UNEP-WCMC, 2008; CaMPAM, 2009; LMMA, 2015). They are often important facilitators for creating ecologically coherent networks (UNEP-WCMC, 2008), especially where the existing networks are unplanned and lack a coherent governance structure.

Cultural Networks

Cultural networks comprise sites protected for their historical or cultural importance (e.g. shipwrecks or sunken settlements)

and/or where special seascapes are present. They can be seen in the various UNESCO networks (UNESCO, 1972; Dizon *et al.*, 2013) and implementation of the European Landscape Convention (Council of Europe, 2000). Protection measures typically involve preventing activities that would damage the features and preserving cultural heritage. Cultural networks, like socio-economic networks, include and encourage responsible human activities, and have important roles in supporting nature conservation and promoting public awareness (IUCN-WCPA, 2008; Laffoley, 2014).

Transnational Networks

The objective of transnational networks is to promote the co-management of natural resources across national boundaries, requiring special cooperation with common goals to ensure effective management (e.g. Natura 2000 network, Pelagos Marine Mammal Sanctuary). Transnational networks of sites may overlap heavily with conservation networks, but they require specific governance approaches which support their recognition as an individual network type. They are recognized by IUCN as Transboundary Protected Areas and Peace Parks (Vasilijević *et al.*, 2015) and may comprise planned or semi-planned networks. For instance, the EU Natura 2000 network is a semi-planned transnational network co-managed by all Member States with the aim of achieving favourable conservation status for habitats and species of European importance.

MPA Network Management and Monitoring

It is evident from the foregoing analysis and discussion of MPA network types that management and monitoring of individual sites at a network level must take explicit account of which types of network that site lies within and contributes to. Otherwise, synergistic opportunities may be missed, or worse, important network features eroded because they were overlooked. The effective management and monitoring of MPA networks therefore requires both top-down and bottom-up approaches. In the former, individual sites should be categorized according to the network types relevant to them; and in the latter, the management plans should include specific objectives and activities that ensure they sustain their contributions to those networks (Table 3.3). This section provides a starting point for developing site-based, network-aware management approaches for MPAs in the Mediterranean and Black Seas.

Conservation Network Sites

General Description
A system of sites aimed at protecting features (species, habitats, geological and geomorphological formations) of conservation importance established according to predetermined criteria. Features to be protected are typically identified in legislation.

Purpose
Generally designed to protect certain specific features, showing the full range of their variation, in a viable condition, by representation, replication and adequacy of features, across a range of sites.

Features/Uses
A conservation network is based on including enough sites to ensure representation, replication, adequacy, viability, connectivity and protection for specified features using best available evidence (JNCC and Natural England, 2010; Laffoley, 2014). Correspondingly, at a network level, there should be a common list of features of conservation interest (Laffoley, 2014). The required management activities are focused on protecting the designated features and so

Table 3.3 MPA networks and appropriate site-level management and monitoring actions.

Network type	Network purpose	Management actions required at MPA site level	Monitoring actions required at MPA site level
Conservation	Generally designed to protect features showing the full range of their variation, by representation, replication and adequacy of features, across a range of sites	1) Identify legal protection measures for features 2) Identify activities harmful to features 3) Identify appropriate restrictive activities	1) Identify features of conservation importance 2) Instigate monitoring programme of features in line with Marine Strategy Framework Directive (MSFD) indicators
Connectivity	To ensure ecological coherence by providing protection to sites between which genetic exchanges are known to occur	No specific measures are likely to be effective at site level	1) Instigate connectivity monitoring programme
Socio-economic	To protect and manage marine resources in a sustainable manner, while optimizing coastal uses and avoiding conflicts	1) Identify and engage key stakeholders 2) Identify appropriate economic instruments 3) Assess likely success of economic instruments	1) Identify socio-economic activities 2) Instigate socio-economic monitoring programme
Geographic	To achieve conservation and protected area coverage targets within a defined geographical area	1) Identify MPAs sharing features at a range of geographic scales 2) Create links with identified MPAs	1) Identify conservation status of features at appropriate geographic scales 2) Establish monitoring indicator thresholds for features based on geographic scales
Collaborative	To promote interaction among members to effectively plan, manage, implement or monitor area-based management of marine resources and associated uses	1) Identify appropriate collaborative networks 2) Become an active member of collaborative network	1) Identify features of conservation importance 2) Instigate monitoring programme of features in line with MSFD indicators
Cultural	To protect sites and areas where significant historical and cultural features and seascapes are present, by preserving and promoting traditional management practices and preventing harmful activities	1) Identify legal protection measures for features 2) Identify activities harmful to features 3) Identify appropriate restrictive activities	1) Identify cultural features 2) Instigate monitoring programme of cultural features
Transnational	Co-management of natural resources beyond existing political boundaries	1) Identify other sites within transnational network 2) Instigate cooperation programmes with other network members	1) Identify features of conservation importance 2) Instigate monitoring programme of features in line with MSFD indicators 3) Establish appropriate monitoring indicator thresholds

are dependent on the specific features present, but they usually rely on zonation (from strictly protected to regulated use zones; Ballantine, 2014), and reduction of external impacts. A further important aspect of management is enforcing rules and regulations designed to protect designated features. Network members should make use of remote surveillance systems for detecting illegal activities, such as the Virtual Watch Room (Pew Charitable Trusts, 2015) and Global Fishing Watch (Global Fishing Watch, 2014) which assist MPA managers to identify illegal fishing activities. Where necessary, surveillance should be supported by coercive enforcement at site level, by conducting physical patrols to discourage all illegal activities.

Monitoring and Assessing Effectiveness

By definition, a conservation network is deemed effective if all the specified features show a full range of variation and are present in a viable condition (in terms of for example spatial distribution, abundance and population dynamics). This can be determined by implementing appropriate monitoring protocols for each feature that are consistent across each site in the network. Ideally, the network should deliver a spillover or source effect that helps to maintain features outside the MPA network. Clearly, external effects such as climate change, pollution, altered hydrological patterns and invasion by alien species can undermine the effectiveness of a conservation network. Including non-network sites in the monitoring programme can help to determine whether the network is performing better than having no network.

Governance

The governance structures of conservation networks will generally flow from the site administration up to a multinational body,

with each level having responsibility for the design, coordination and assessment of management and monitoring delegated to the lowest appropriate level. The overall structure should be concerned with issues such as:

- Establishing the common conservation features of the network
- Identifying gaps and adding/designating new sites in the network
- Developing standards for monitoring conservation features
- Ensuring data quality and control, and sharing information within and beyond the network (applying the guidelines for the INSPIRE Directive, 2007/2/EC)
- Developing 'best practice' codes for site and network managers, including emergency response procedures
- Sharing resources between network members for enforcement and monitoring activities
- Developing national and multilateral legislation to strengthen the network.

Connectivity Network Sites

General Description

A system of sites connected by, and designed to maintain, the unimpeded movement and dispersal of propagules, larvae, juveniles or adults (including habitat-forming species).

Purpose

To ensure ecological coherence of marine biodiversity by providing protection to sites important for genetic exchange.

Features/Uses

The key feature of a connectivity network is its deliberate design based on evidence. Some legislation (e.g. CBD resolutions, EU Habitats Directive – see Braun, this volume) refers to connectivity as a criterion for establishing MPA networks. However, connectivity processes vary greatly among species (depending on larval lifespans and recruitment techniques), regions (depending on current patterns) and over time (due to variations in fecundity and currents) (Ballantine, 2014). Moreover, connectivity can be assessed in relation to oceanography (how currents spread propagules), beta diversity (the degree of similarity of site communities), and genetics (closeness of genotypes between sites). Therefore, a connectivity network has to define which habitats and/or species it is addressing so that enough sites, at a relevant scale and scope (i.e. including the high seas), can be included to maintain a sufficient degree of genetic exchange between them (whether as sources or sinks or both).

Monitoring and Assessing Effectiveness
Assessing and confirming connectivity within a network is a significant challenge (OSPAR, 2013; Grorud-Colvert *et al.*, 2014): considerable inter-species variation means it is difficult to assess connectivity in a general sense. Large-scale movements of propagules and larvae dispersed by established current systems are fairly well understood and connectivity between sites can be inferred through analysis of them. However, more localized current patterns are less well known but could have equally important effects on connectivity through local dispersal of propagules. Conducting systematic sampling and DNA analysis of the target organisms within existing MPAs can establish the degree of genetic similarity and isolation between sites, helping to identify which sites are well connected to each other, and where gaps in the network are apparent, with respect to the target organisms. However, this is a time-consuming and expensive process and is likely to be beyond the resources of network and site managers at the current time. Ironically, monitoring the spread of invasive alien species can help to elucidate connectivity between sites

(Otero *et al.*, 2013) while improving understanding of how invasive species colonize new areas. Meanwhile, the OSPAR network uses the proxies of replication, representation and adequacy to assess connectivity (OSPAR, 2013).

Governance
Connectivity networks are most likely to be nested within conservation and geographic network types, so the governance structures of those networks should be responsible for monitoring connectivity. This can be done by including expert knowledge (e.g. setting up a connectivity working group) within the governance system to ensure the topic is properly addressed.

Socio-economic Network Sites

General Description
A system of sites where marine resources are managed for social and economic benefits.

Purpose
To protect and manage the use of marine resources in a sustainable manner.

Features/Uses
Socio-economic networks are defined by the presence of activities which exploit marine resources (whether consumptive or non-consumptive). The primary aim is not always conservation but it is always a by-product at both network and site levels. Typical activities include fisheries, mineral extraction, tourism and renewable energy generation. Where fisheries are managed, no-take areas should be established which provide benefits (increased fish catches) outside the no-take zone due to spillover effects (FAO, 2011). Tourism generates economic benefits by providing jobs and additional income to local people, and social benefits for the visitors (recreation, health improvement). Renewable energy schemes like offshore

wind farms (OWFs) and tidal turbines deliver socio-economic benefits through the creation of jobs and clean energy production, and there can be conservation gains from increased habitat heterogeneity (Ashley, 2014). Offshore wind farms also relieve pressures on fisheries by acting as de-facto no-take zones.

Monitoring and Assessing Effectiveness

The monitoring programme for the sites and network has to cover direct and indirect parameters. Direct parameters include evaluating the status of biodiversity, especially of any wild species that are harvested, and for aquaculture, the impact of feeding and chemicals used. Indirect parameters include level of harvests outside the network sites, amount of power generated, employment changes, visitor numbers, entry charges and so on. MedPAN has developed guidance on assessing recreational and artisanal fishing (Font *et al.*, 2012) and on monitoring tourist/visitor numbers at a site level (Le Berre *et al.*, 2013) which should be adopted across the network.

Governance

The governance structure of a socio-economic network depends largely on whether the network is planned (e.g. a series of OWFs, or fishery sanctuaries) or nested within another type (e.g. sustainable tourism within conservation networks), or even a combination of both. Planned networks will normally come under the supervision of a national, or in the high seas an international, authority. A network nested in another type is usually not under a dedicated authority, and is more informally managed through cooperation between the network members. Socio-economic networks are most likely to have some form of dedicated enforcement measures to protect their attributes, whether coercive (patrols

and physical presence of guards) and/or through the use of economic instruments (EIs). As EIs comprise market-based fiscal incentives they are often more effective than legislation-based tools (Ojea *et al.*, this volume). In any case, the governance structure should be responsible for the following aspects of network management:

- Setting sustainable use thresholds (e.g. fishing quotas, visitor numbers, suitable tourism activities, etc.)
- Establishing methods for assessing effectiveness
- Setting standards for direct and indirect monitoring schemes
- Sharing resources between network members for enforcement and monitoring activities
- Ensuring data quality and control, and sharing information within and beyond the network (applying the guidelines for the INSPIRE Directive, 2007/2/EC)
- Developing 'best practice' codes for site and network users.

Geographic Network Sites

General Description

A system of sites in a given area contributing to protected area coverage and conservation targets within that area.

Purpose

To achieve conservation and protected area coverage targets within a defined geographical area.

Features/Uses

Geographic networks are typically scaled versions of one or more networks, and ultimately all MPAs form part of a global geographic network. They have explicit spatial boundaries (such as regional, national, sea basin) and always contribute towards conservation targets within that area. Links between geographic network sites can be multiple and include locality, conservation features or management objectives. Examples of geographic networks include:

- Natura 2000 network (protecting marine conservation features within the EU region)
- Specially Protected Areas of Mediterranean Importance under the Barcelona Convention
- MPA network of Ukraine.

Monitoring and Assessing Effectiveness

The effectiveness of national MPA geographic networks is related to the degree to which they meet coverage targets of the marine environment. The effectiveness of other geographic networks can be assessed according to the strength of collaboration and joint activities between its constituent MPAs.

Governance

Governance structures will vary depending on the scale and general purpose(s) of the network, as well as the extent to which the network is embedded in legislation. Small-scale geographic networks will normally operate more coherently than large, multinational ones as decisions can be made more quickly the fewer the parties involved. This suggests that geographic networks should be composed of relatively small discrete regional units nested within increasingly larger scale areas. The overall aim of any governance structure should be to ensure that network members contribute to the success of the network by:

- Identifying gaps and adding/designating new sites in the network
- Developing national and multilateral legislation to strengthen the network
- Identifying and jointly acting on common concerns across the network

- Ensuring data quality and control, and sharing information within and beyond the network (applying the guidelines for the INSPIRE Directive, 2007/2/EC)
- Developing 'best practice' codes for site and network managers, including emergency response procedures
- Sharing resources between network members for enforcement and monitoring activities.

Collaborative Network Sites

General Description

A system of sites or networks whose managers, practitioners, stakeholders, decision-makers, scientists, and others interact to develop and share best practice.

Purpose

To promote and improve area-based planning, management or monitoring of marine resources.

Features/Uses

Collaborative networks are formalized platforms (such as legally registered non-government entities) that provide opportunities for site managers, staff members, network managers, government officials and other relevant stakeholders to share knowledge and develop best practices, as well as raise awareness and promote the values and protection of MPAs (Olsen *et al.*, 2013). Collaborative networks often coincide with other networks in relation to topics and/or geographic coverage. The chief example of a collaborative network in Europe is the Mediterranean network of MPA managers (MedPAN, 2015).

Monitoring and Assessing Effectiveness

Assessing the effectiveness of collaborative networks should take place on two levels. The first level should evaluate the scale and functionality of the network using parameters such as:

- Number of members
- Types of members (e.g. MPA management bodies, government institutions, NGOs, etc.)
- Number of workshops organized
- Website visitor rates
- Publications released
- Membership satisfaction (e.g. conducting questionnaire surveys among workshop participants).

The second level, which ought to follow from the first, should assess whether sites within the network are more effectively managed for their intended purposes (conservation, socio-economics, geographic coverage) than those not in the network.

Governance

The governance of a collaborative network is based on a legal entity (usually with some support structures such as a board, secretariat, and working groups) which encourages membership from a wide variety of sites and representatives of their stakeholders. Although an element of top-down management is required for organizational and logistical purposes, there should be a strong focus on bottom-up participation to ensure the network serves the needs of its members.

Cultural Network Sites

General Description

A system of sites aimed at protecting features and seascapes of historical and cultural importance.

Purpose

To protect sites and areas where significant historical, cultural and seascape features are present, by preserving and promoting traditional management practices and preventing harmful activities.

Features/Uses

Cultural networks are usually unplanned, developing gradually over space and time (e.g. UNESCO World Heritage network). The main characteristic of a cultural network is that it contains sites which are designated and managed for their historical, aesthetic or cultural importance, including features such as:

- Coastal landmarks, whether man-made (e.g. sea defences, harbours) or natural (e.g. sea caves, bays, inlets)
- Surface water objects, both man-made (e.g. lighthouses) and natural (e.g. rocky outcrops)
- Current uses (e.g. fishing, diving, sailing)
- Historic uses (e.g. traditional fishing methods)
- Cultural associations (e.g. site of naval battles)
- Sunken objects (e.g. submerged wrecks, archaeological sites, palaeo-landscape features).

Monitoring and Assessing Effectiveness

Cultural features are often spatially unique, although some cultural practices may be spread over many sites in the network. The most appropriate way to evaluate their effectiveness is to assess the status of individual designated features at site level, to check that they retain their character (not impaired by intrusions affecting intangible qualities) and wholeness (for physical objects). The status of cultural activities can be assessed by surveying the number of practitioners and their attitudes towards the continuation of the activity (e.g. using the UNESCO Framework for Cultural Statistics Handbook – UNESCO, 2009). Natural physical features are typically geological or geomorphological formations, assessed by the visibility of the feature, its physical integrity and extent (JNCC, 2004). Archaeological features often only survive underwater due to a delicate balance between the surrounding chemical and physical environment, so intrusive monitoring should be minimized (Dizon *et al.*, 2013), relying on photographs and video.

Governance

The governance structure for cultural networks can often be incorporated within conservation and/or geographic networks, ensuring that cultural feature experts are included within the structure. Where sites and networks contain cultural features of international importance, external governance organs may also be involved (e.g. UNESCO for sites in the World Heritage List or covered by the Convention on Protection of Underwater Cultural Heritage). Because cultural sites are normally subject to use or access, enforcement of regulations will be a significant issue for governance, especially to prevent damage to physical features (e.g. trawling and diving around archaeological sites), and surveillance systems (e.g. Global Fishing Watch, 2014; Pew Charitable Trusts, 2015) should be put in place.

Transnational Network Sites

General Description

A system of sites which are managed in common by two or more countries, usually with supranational coordination.

Purpose

To establish coordinated management of natural resources between and beyond national jurisdictions.

Features/Uses

Transnational networks are characterized by an overt form of collaboration and co-management of resources between and beyond national jurisdictions (e.g. EU Natura 2000 network). They include transboundary sites where MPAs are contiguous across borders, and can overlap with conservation, socio-economic, geographic and cultural

network types but normally enhance them by having supranational governance coordination (Vasilijević *et al.*, 2015).

Monitoring and Assessing Effectiveness

The effectiveness of the network is based on monitoring and assessing the co-managed resource(s), such as commercial fish stocks or marine mammal populations, especially those in the high seas. The effectiveness of management within the network can be assessed at a site level using management effectiveness evaluation (MEE) techniques (e.g. Pomeroy *et al.*, 2004; Tempesta and Otero, 2013).

Governance

Transnational networks (and their sites) are likely to have a supranational coordinating body, usually established in the framework of a bi- or multilateral treaty (Lausche, 2011), that supervises overall network strategy and engages with members of the network to enhance effectiveness at the site level, as well as sharing resources for network-wide enforcement measures (including possible sanctions for breaches occurring in the high seas). However, the governance body will face significant challenges due to differing legal systems, conservation features, socio-economics, cultural elements and political aspects.

Conclusions

This chapter has described seven types of MPA networks, based on a review of literature and analysis of a representative sample of MPA management plans from the Mediterranean and Black Seas. It has been shown that although each MPA network type can be considered individually (and will have its own set of priorities and supporters), none of them are mutually exclusive. Indeed, it is only when the coordinators of the networks proceed in concert that the most effective and holistic results can be obtained.

For instance, by referring to the purposes of each network type, MPA managers can identify to which of them their site contributes, and identify any gaps that could be addressed in their management plans. Similarly, protected area authorities can review their MPA estate from a network perspective and determine whether to set up more overt MPA networks where they do not exist. As a result, the roles of individual MPAs within specific networks can be more clearly identified (along with gaps in network coverage); network-level objectives and indicators designed; and appropriate governance, management and monitoring practices put in place. Moreover, the process of designating MPAs should take account of which network types they can support while not compromising their primary intended functions. This will ensure that each MPA contributes effectively to relevant networks and that an ecosystem-based approach is taken towards marine conservation.

Acknowledgements

This chapter is based on research carried out with funding from the European Community's Seventh Framework Programme (FP7/2007–2013) under Grant Agreement No. 287844 for the project 'Towards COast to COast NETworks of marine protected areas (from the shore to the high and deep sea), coupled with sea-based wind energy potential' (CoCoNet).

References

Ashley, M. (2014) *The implications of co-locating marine protected areas around offshore wind farms.* PhD thesis, University of Plymouth.

Ballantine, B. (2014) Fifty years on: lessons from marine reserves in New Zealand and principles for a worldwide network. *Biological Conservation,* **176**, 297–307. 10.1016/j.biocon.2014.01.014

Baltic Marine Environment Protection Commission (HELCOM) (2013) *HELCOM PROTECT: Overview of the Status of the Network of Baltic Sea Marine Protected Areas.* Helsinki Commission, Helsinki. 31 pp.

CaMPAM (2009) *The Caribbean Marine Protected Area Management Network and Forum: Factsheet.* http://campam.gcfi.org/ campam.php. Accessed 7 July 2015.

Catchpole, R. (2012) *Ecological Coherence Definitions in Policy and Practice: Final Report.* Scottish Natural Heritage, Edinburgh. http://www.snh.gov.uk/docs/ B1028804.pdf.Accessed 10 November 2014.

Commission for the Protection of the Marine Environment of the North-East Atlantic (OSPAR) (2013) *2012 Status Report on the OSPAR Network of Marine Protected Areas.* OSPAR Biodiversity Series. http:// www.ospar.org/ospar-data/p00618_2012_ mpa_status%20report.pdf. Accessed 30 July 2015.

Convention on Biological Diversity (2004) *Technical Advice on the Establishment and Management of a National System of Marine and Coastal Protected Areas.* CBD Technical Series No. 13. 40 pp.

Council of Europe (2000) *European Landscape Convention.* CETS No. 176. http://www.coe.int/en/web/conventions/ full-list/-/conventions/treaty/176. Accessed 9 July 2015.

Dizon, E., Egger, B., Elkin, D. *et al.* (2013) *Manual for Activities directed at Underwater Cultural Heritage: Guidelines to the Annex of the UNESCO 2001 Convention.* UNESCO, Paris.

Evans, D. (2006) The habitats of the European Union Habitats Directive. *Biology and Environment: Proceedings of the Royal Irish Academy,* **106B**(3), 167–173.

FAO (2011) *Fisheries Management: Marine Protected Areas and Fisheries.* FAO Technical Guidelines for Responsible Fisheries No. 4, Supplement 4. FAO, Rome.

Font, T., Lloret, J. and Piante, C. (2012) *Recreational Fishing within Marine Protected Areas in the Mediterranean.* MedPAN North Project. WWF France. 168 pp.

Gabrié, C., Lagabrielle, E., Bissery, C. *et al.* (2012) *The Status of Marine Protected Areas in the Mediterranean Sea.* MedPAN, Marseilles and RAC/SPA, Tunis. MedPAN Collection. 256 pp.

Gabunia, M., Janelidze, C., Potskhishvili, K. *et al.* (2004) *Kholketi National Park Management Plan.* Unpublished document for the State Department of Protected Areas, Nature Reserves and Hunting of Georgia.

Global Fishing Watch (2014) *Global Fishing Watch.* http://www.globalfishingwatch.org. Accessed 11 June 2015.

Goñi, R., Revenga, S., Laborda, C. *et al.* (2015) *Spain's Network of Marine Reserves: More than 25 years of protecting our waters.* Ministerio de Agricultura, Alimentacion y Medio Ambiente. http://publicacionesoficiales.boe.es/detail.php?id=614228015-0001. Accessed 17 April 2016.

Goriup, P. (2009) *Preliminary Management Plan for the Small Phyllophora Field Marine Protected Area Karkinitsky Bay, Black Sea, Ukraine.* Environmental Collaboration for the Black Sea Project (ECBSea). EuropeAid/120117/C/SV/Multi.

Grorud-Colvert, K., Claudet, J., Carr, M.H. *et al.* (2011) The assessment of marine reserve networks: guidelines for ecological evaluation, in *Marine Protected Areas: A Multidisciplinary Approach* (ed. J. Claudet). Cambridge University Press, Cambridge.

Grorud-Colvert, K., Claudet, J., Tissot, B.N. *et al.* (2014) Marine protected area networks: assessing whether the whole is greater than the sum of its parts. *PloS ONE,*

9 (8), e102298. doi:10.1371/journal.pone.0102298

International Union for Conservation of Nature – World Commission on Protected Areas (IUCN-WCPA) (2008) *Establishing Marine Protected Area Networks: Making It Happen.* IUCN-WCPA, National Oceanic and Atmospheric Administration and The Nature Conservancy, Washington, DC. 118 pp.

Joint Nature Conservation Committee (JNCC) (2004) *Common Standards Monitoring for Earth Science Sites.* http://jncc.defra.gov.uk/pdf/CSM_earth_science.pdf.Accessed 10 July 2015.

Joint Nature Conservation Committee (JNCC) and Natural England (2010) *The Marine Conservation Zone Project: Ecological Network Guidance.* Sheffield and Peterborough, UK.

Joint Nature Conservation Committee (JNCC) and Natural England (2012) *Marine Conservation Zone Project: JNCC and Natural England's Advice to Defra on Recommended Marine Conservation Zones.* http://jncc.defra.gov.uk/PDF/MCZProjectSNCBAdviceBookmarked.pdf. Accessed 10 July 2015.

Laffoley, D. (ed.) (2014) *Guidelines to Improve the Implementation of the Mediterranean Specially Protected Areas Network and Connectivity between Specially Protected Areas.* RAC/SPA, Tunis. 32 pp.

Lausche, B. (2011) *Guidelines for Protected Areas Legislation.* IUCN, Gland, Switzerland. xxvi + 370 pp.

Le Berre, S., Peuziat, I., Le Corre, N. and Brigand, L. (2013) *Visitor Use Observation and Monitoring in Mediterranean Marine Protected Areas.* MedPAN North Project. WWF-France and Parc National de Port-Cros.

LMMA (2015) *The Locally-Managed Marine Areas (LMMA) Network.* http://www.lmmanetwork.org. Accessed 30 July 2015.

López, A., Vilar, G. and Frieyro, J.E. (2004) *Management Plan for the Marine Area of Rosh HaNiqra – Akhziv Nature Reserve.* UNEP-RAC/SPA, Tunis.

Malta Environment and Planning Authority (MEPA) (2010) *Qawra/Dwejra Heritage Park Action Plan.* https://www.mepa.org.mt. Accessed 19 April 2015.

Margules, C.R. and Pressey, R.L. (2000) Systematic conservation planning. *Nature,* **405** (6783), 243–253.

MedPAN (2015) *Network of Marine Protected Area Managers in the Mediterranean.* http://www.medpan.org. Accessed 30 July 2015.

Muresan, M., Begun, T., Teaca, A. *et al.* (2014) *Draft Management Plan of the Strandzha–Igneada Area.* EC DG Env. MISIS Project Deliverables. 159 pp.

National Research Council (NRC) (2000) *Marine Protected Areas: Tools for Sustaining Ocean Ecosystems.* National Research Council, Washington, DC.

Natural England (2012) *An Approach to Seascape Character Assessment.* https://www.gov.uk/government/uploads/system/uploads/attachment_data/file/396177/seascape-character-assessment.pdf. Accessed 17 April 2016.

Notarbartolo di Sciara, G. and Hyrenbach, D. (2007). *The Pelagos Sanctuary for Mediterranean Marine Mammals: Case Study.* http://www.cetaceanalliance.org/download/literature/NotarbartolodiSciara_etal_2007.pdf. Accessed 17 April 2016.

Olsen, E.M., Johnson, D., Weaver, P. *et al.* (2013) *Achieving Ecologically Coherent MPA Networks in Europe: Science Needs and Priorities* (eds K.E. Larkin and N. McDonough). Marine Board Position Paper 18. European Marine Board, Ostend, Belgium. 88 pp.

Orueta, J.F. and Limam, A. (n.d.) *Plan de gestion de la partie marine du parc national de Zembra et Zembretta.* Project Med MPA. UNEP-RAC/SPA, Tunis.

Otero, M., Cebrian, E., Francour, P. *et al.* (2013) *Monitoring Marine Invasive Species in Mediterranean Marine Protected Areas (MPAs): A Strategy and Practical Guide for Managers.* MedPAN North Project. IUCN, Malaga, Spain. 136 pp.

Pereira, H.M. and Navarro, L.M. (eds) (2015) *Rewilding European Landscapes.* Springer Open.

Pew Charitable Trusts (2015) *Virtual Watch Room.* http://www.pewtrusts.org/en/research-and-analysis/fact-sheets/2015/01/virtual-watch-room. Accessed 11 June 2015.

Pomeroy, R.S., Parks, J.E. and Watson, L.M. (2004) *How is Your MPA Doing? A Guidebook of Natural and Social Indicators for Evaluating Marine Protected Area Effectiveness.* IUCN, Gland, Switzerland. 216 pp.

Puhov, B., Baković, N., Ramov, M. *et al.* (2012) *Telašćica Nature Park, Croatia: Extract of Management Plan.* MedPAN, Marseilles.

Sandwith, T., Shine, C., Hamilton, L. and Sheppard, D. (2001) *Transboundary Protected Areas for Peace and Co-operation.* IUCN, Gland, Switzerland and Cambridge, UK. 111 pp.

SPA/BD Protocol (1995) *Protocol Concerning Specially Protected Areas and Biological Diversity in the Mediterranean.* Barcelona. UNEP-RAC/SPA, Tunis.

Tempesta, M. and Otero, M. (2013) *Guide for Quick Evaluation of Management in Mediterranean MPAs.* WWF Italy.

Tunesi, L. *et al.* (2003) *Management Plan for the Marine Protected Area from Rdum Majjiesa to Ras ir-Raheb Cave.* Project Med MPA. UNEP-RAC/SPA, Tunis.

United Nations Educational, Scientific and Cultural Organization (UNESCO) (1972) *Convention Concerning the Protection of the World Cultural and Natural Heritage.* http://whc.unesco.org/en/conventiontext/. Accessed 7 October 2016.

United Nations Educational, Scientific and Cultural Organization (UNESCO) (2009) *Measuring Cultural Participation.* http://www.uis.unesco.org/culture/Documents/fcs-handbook–2-cultural-participation-en.pdf. Accessed 30 July 2015.

United Nations Educational, Scientific and Cultural Organization Man and the Biosphere Programme (UNESCO-MBP) (2007) *Astrakhanskii Biosphere Reserve Information.* http://www.unesco.org/mabdb/br/brdir/directory/biores.asp?mode=genandcode=RUS+05. Accessed 16 April 2015.

United Nations Environment Programme-World Conservation Monitoring Centre (UNEP-WCMC) (2008) *National and Regional Networks of Marine Protected Areas: A Review of Progress.* UNEP-WCMC, Cambridge, UK.

Vasilijević, M., Zunckel, K., McKinney, M. *et al.* (2015) *Transboundary Conservation: A Systematic and Integrated Approach.* Best Practice Protected Area Guidelines Series No. 23. IUCN, Gland, Switzerland. xii + 107 pp.

4

Marine Protected Area Governance and Effectiveness Across Networks

Nigel Dudley[1,2] and Marc Hockings[2]

[1] *Equilibrium Research, Bristol, UK*
[2] *School of Geography, Planning and Environmental Management at the University of Queensland, Brisbane, Australia*

Introduction

Marine Protected Areas (MPAs) are almost certainly the fastest growing protected area type in the world. But they are also one of the most complicated forms of protected area, displaying a sometimes bewildering array of types, management approaches and governance structures and still beset with definitional arguments about what should and should not fall under the title of an 'MPA'. At present, there are over 12 000 MPAs listed on the World Database on Protected Areas. Some 8.4% of all marine areas within national jurisdiction (0–200 nautical miles) are covered by protected areas while only 0.25% of marine areas beyond national jurisdiction are similarly protected, demonstrating the particular challenges involved in agreeing protection in international waters. Together this adds up to 3.4% of the marine biome under protection (Juffe-Bignoli *et al.*, 2014).

Marine Protected Areas exhibit enormous variety of size, location and management. They include long-term, traditional management systems that are now being incorporated into national protected area networks, along with many newly designated areas imposed by governments and sometimes, tentatively, by the international community. International law of the sea has an effect; protected areas usually have to allow ships passage for example.

Management of marine areas for a specific conservation purpose has lagged far behind similar efforts in the terrestrial environment (Watson *et al.*, 2014). There are many factors contributing to this delayed attention to the marine realm, including the status of marine areas as a 'commons'; the lack of visibility of marine species and long-standing cultural traditions that regard fish as cold, slimy and slippery; and deep-seated beliefs that the resources of the sea are effectively limitless (Kenchington, 2014).

Additionally, many MPAs are not wholly marine: large protected areas can sometimes contain terrestrial, freshwater and marine components. Some MPAs only protect part of the marine ecosystem – for example open water but not the seabed, or vice versa – and sometimes protection only covers certain parts of the water column (Grober-Dunsmore *et al.*, 2008). The readiness with which local or resident human communities accept MPAs also differs around the world. Some coastal cultures feel very comfortable with the concepts of protection, while others do not. In consequence, some MPAs

Management of Marine Protected Areas: A Network Perspective, First Edition. Edited by Paul D. Goriup.
© 2017 John Wiley & Sons Ltd. Published 2017 by John Wiley & Sons Ltd.

are welcomed by the people who live there while others are resented, opposed and undermined (Mascia and Claus, 2008).

In 2010, signatories of the Convention on Biological Diversity (CBD) agreed a global conservation target of protection for 10% of marine and coastal area by 2020. If this is to be met, an additional 2 million square kilometres of marine area within national jurisdiction will need to be designated as MPAs over the next few years (Figure 4.1).

All the signs are that MPAs will continue to be designated at a rapid pace for the next few years. The real target might even be greater if governments aim to protect a full range of coastal and marine habitats. Aichi Biodiversity Target 11 on increasing protected area networks, agreed by the CBD in Nagoya, Japan in 2010, is: 'By 2020, at least 17% of terrestrial and inland water, and 10% of coastal and marine areas, *especially areas of particular importance for biodiversity and ecosystem services*, are conserved through *effectively and equitably managed, ecologically representative* and well connected systems of protected areas and other effective area-based conservation measures, and integrated into the wider landscapes and

seascapes' (emphasis added). The current statistics mask a major disparity in the geographical spread of protection, with almost half the total consisting of huge areas designated recently in Australia, around New Caledonia in the Pacific and around the UK's South Georgia and South Sandwich islands. Full ecological representation will require protection in many other parts of the ocean as well and would necessarily exceed the numerical target of 10%, especially given what has already been designated.

The political momentum towards creation of MPAs means that many are very new, so that we are still learning how they are best managed and if such management is achieving the desired results in terms of biodiversity conservation, ecosystem services and socio-economic values. And because an increasing proportion of humanity lives within a few miles of the coast, all but a very few MPAs in the high seas or around remote islands have resident or local human communities that generally have long-term links with the sea and its resources. The massive MPA around the Chagos Islands is a rare example of an uninhabited MPA, although even here the islanders (deposed to make a

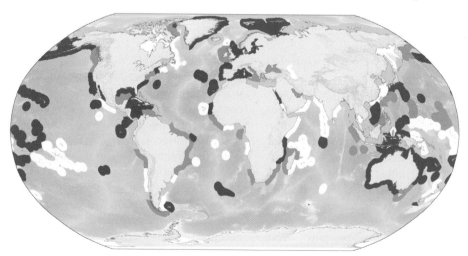

Figure 4.1 Protected area coverage in percentage for the 232 marine ecoregions of the world.
Source: Protected Planet Report 2014 (Juffe-Bignoli *et al.,* 2014). Reproduced with permission of United Nations Environment Programme-World Conservation Monitoring Centre (UNEP-WCMC).

US–UK air base) are lobbying to return. Aldabra Atoll in the Indian Ocean is one of the few examples of an MPA that is largely uninhabited apart from a small research station and management presence, although even in this remote site there was periodic but very limited habitation between the 1700s and the 1950s (Stoddart, 1971).

The considerable amount of time and money usually needed to set up an MPA is only justified if the results match the investment. Finding out what this means, and whether MPAs are actually proving to be effective conservation mechanisms, is the subject of this chapter. Three influences are important:

- How the MPA is managed
- Who makes the decisions
- How effective the MPA is in conserving biodiversity and ecosystem services.

None of these issues are particularly straightforward. But the combination of all three tells us a great deal about the overall performance of an individual MPA and, in combination, of MPA networks.

How the MPA is Managed: Management Structures within MPAs

A 'protected area' is not a monolithic, changeless entity, but rather a collective term and overarching framework for many different approaches to space-based conservation. Two overarching definitions of protected areas exist: one from the CBD and another from the International Union for Conservation of Nature (IUCN). The CBD definition is binding in international law while the IUCN definition has been adopted by many countries and written into national laws; there is tacit agreement between the two institutions that the definitions are equivalent (Lopoukhine and Dias, 2012).

IUCN is an international conservation body with both state and non-governmental members. Its policy is set by resolution and voting at periodic assemblies of the World Conservation Congress. In the last few decades, the 89 state members have voluntarily adopted a range of resolutions on MPAs. IUCN has attempted to set some philosophical boundaries on the protected area concept by agreeing a definition and, under this, six different management categories. Together these provide one way of arriving at consensus on what MPA priorities are likely to be and how multiple objectives can be accommodated. The categories, which classify protected areas according to their management objectives, are recognized as the global standard for defining and recording protected areas. They can help in the planning of protected areas and protected area systems; manage information about protected areas; and regulate activities in protected areas. They also provide the basis for legislation in many countries; help set budgets; interpret or clarify land tenure and governance; and are a tool for advocacy. To some extent, debate about the management categories has provided the forum for philosophical enquiries about the nature of protection and the relationship between protected areas and other forms of land and water use (Dudley *et al.*, 2010).

In applying the management categories, the first step according to IUCN is to determine whether the site meets the definition of a protected area and then to identify the most suitable category for the site in question. According to IUCN, a protected area is: 'A clearly defined geographical space, recognised, dedicated and managed, through legal or other effective means, to achieve the long-term conservation of nature with associated ecosystem services and cultural values' (Dudley, 2008). This gives primacy to nature conservation – a distinction emphasized by an accompanying principle – but also recognizes the importance of associated ecosystem and cultural values. The CBD definition is by contrast a 'geographically defined area which is designated or regulated and managed to achieve specific conservation objectives'. The CBD also recognizes the IUCN management categories, for example in its Programme of Work on Protected Areas (CBD, n.d.).

In 2012, IUCN issued additional guidance on the definition and management categories for MPAs (Day *et al.*, 2012), which sought to distinguish them from other ways of managing coastal and marine areas that also deliver conservation benefits but where these are incidental to other management priorities. The guidance stated that spatial areas which may incidentally appear to deliver nature conservation but do not have stated nature conservation objectives should not automatically be classified as MPAs, as defined by IUCN, in large part because they have far less security (if management objectives change, the conservation values may be lost). Examples given of such areas include:

- Fishery management areas, such as temporary set asides, with no wider stated conservation aims
- Community areas managed primarily for sustainable extraction of marine products (e.g. coral, fish, shells, etc.)
- Marine and coastal management systems managed primarily for tourism, which also include areas of conservation interest
- Wind farms and oil platforms that incidentally help to build up biodiversity around underwater structures and by excluding fishing and other vessels
- Marine and coastal areas set aside for other purposes but which also have conservation benefit: military training areas or their buffer areas (e.g. exclusion zones); disaster mitigation management (e.g. coastal defences that also harbour significant biodiversity); communications cable or pipeline protection areas; shipping lanes, etc.

- Large areas (e.g. regions, provinces, countries) where certain species are protected by law across the entire region, such as protection of whales or protection of depleted populations of commercially important fish species.

Any of the above management approaches could be classified as an MPA if instead they had a primary stated aim to deliver nature conservation; this would generally also have some additional management implications. In the future, many such areas may be categorized as 'other effective area-based conservation measures (OECMs)', the new designation that emerged in Aichi Target 11 from the CBD Conference of Parties in 2010 aimed at describing some of the places that provide conservation benefits outside protected areas. At the time of writing OECMs are still being defined, but will likely include several marine and coastal management systems that currently fall outside the IUCN definition of a protected area.

Importantly in the current context, the IUCN protected area definition is expanded by recognition of six categories (one with a sub-division), defined by the management model, summarized in Table 4.1. Together they demonstrate the wide range of approaches taken to management within an MPA.

The category should be based around the primary management objective(s), which should apply to at least three-quarters of the protected area – the so-called 75% rule. Many MPAs may have specific zones within them where other uses such as tourist lodges or villages are situated, or areas where fishing is permitted in what is otherwise a strictly protected MPA. In some cases, the 25% may be movable: for example, designated zones where local fisherfolk have access may be moved occasionally to prevent over-exploitation or to allow the rebuilding of stocks. Distinct protected areas nested within larger protected areas

Table 4.1 IUCN protected area categories.

No.	Name	Description
Ia	Strict nature reserve	Strictly protected for biodiversity and also possibly geological/geomorphological features, where human visitation, use and impacts are controlled and limited to ensure protection of the conservation values.
Ib	Wilderness area	Usually large unmodified or slightly modified areas, retaining their natural character and influence, without permanent or significant human habitation, protected and managed to preserve their natural condition.
II	National park	Large natural or near-natural areas protecting large-scale ecological processes with characteristic species and ecosystems, which also provide environmentally and culturally compatible spiritual, scientific, educational, recreational and visitor opportunities.
III	Natural monument or feature	Areas set aside to protect a specific natural monument, which can be a landform, sea mount, marine cavern, geological feature such as a cave, or living feature such as an ancient grove.
IV	Habitat/species management area	Areas to protect particular species or habitats, where management reflects this priority. Many will need regular, active interventions to meet the needs of particular species or habitats, but this is not a requirement of the category.
V	Protected landscape or seascape	Where the interaction of people and nature over time has produced a distinct character with significant ecological, biological, cultural and scenic value; and where safeguarding the integrity of this interaction is vital to protecting and sustaining the area and its associated nature conservation and other values.
VI	Protected areas with sustainable use of natural resources	Areas which conserve ecosystems, together with associated cultural values and traditional natural resource management systems. Generally large, mainly in a natural condition, with a proportion under sustainable natural resource management and where low-level non-industrial natural resource use compatible with nature conservation is seen as one of the main aims.

Source: Dudley (2008).

can have their own category if they are managed differently, such as a small strictly protected area inside a protected seascape. In addition, different *zones* in protected areas can have their own category, if the zones are described and fixed in law – a factor that is important in some of the larger MPAs such as the Australian Great Barrier Reef Marine Park (Emslie *et al.*, 2015). Different protected areas making up a transboundary protected area between two countries may also have different categories.

The categories do not imply a hierarchy, either in quality and or in other ways – such as the degree of intervention allowed or the implied 'naturalness' of the area. But nor are all categories equally useful in any situation. A well-balanced protected area system should consider using all categories, although not all are practical everywhere. In most situations, at least a proportion of protected areas should be in the more strictly protected categories, namely I–IV.

What does this mean in practice? While IUCN has quite a precise definition of protected area, and thus of an MPA, within this a whole range of different management approaches are possible, from strictly protected no-go zones to managed seascapes where conservation is integrated with other forms of use. Given that in practice the precise hierarchy of management priorities is often hard to pin down, the definition of an MPA is probably less exact than simple guidelines suggest and ultimately the international community relies on governments to determine what does and does not fall within their own protected areas system. The strictness with which the IUCN definition is applied varies between countries. The situation is further complicated because the CBD (CBD, COP 7, Decision VII/5 [note 11]) and the UN Food and Agricultural Organization (http://www.fao.org/fishery/topic/4400/en) each have slightly different definitions of MPAs, which sometimes creates confusion about what sites are included within national MPA networks. The management categories are a constant however, being recognized by all parties, and an understanding of these provides a clear picture of the types of MPAs present in a country and how they are being managed.

That said, information on MPA categories remains partial; by 2014 only half of all global MPAs had a category assigned by governments. Of these, the commonest MPA categories were IV (habitat or species management areas), followed by Ia (strictly protected) and VI (extractive reserves) (Juffe-Bignoli *et al.*, 2014; see Table 4.1). This would suggest that a significant proportion of MPAs are quite strictly protected. But MPAs can be found in the full range of IUCN categories and governments are increasingly understanding and utilizing the different management approaches available. Some examples of MPAs in different categories are given in Table 4.2 (examples drawn from Day *et al.*, 2012). Furthermore, research shows that very few MPAs ban all kinds of fishing, and no-take reserves are a tiny proportion of the total: data published in 2015 found 94% of MPAs allow fishing (Costello and Ballantine, 2015), suggesting that the IUCN categories are in some cases being misapplied.

Who Makes Decisions: Different Governance Types within MPAs

Closely allied with management approach is the related question of who makes decisions about management: on the degree of protection, the boundaries of the MPA and the way in which it is managed on a day-to-day basis; in other words, what stakeholder or group of stakeholders has a role in both deciding if and where an MPA might be situated and in determining management plans, regulations, enforcement, and so on. The question of governance is increasingly seen as critical in determining the eventual effectiveness of an MPA (Jones *et al.*, 2011); if the governance structure is not appropriate for the social and cultural context of the site then the chances of success are dramatically reduced.

As in the case of management, IUCN has agreed a typology of governance types to describe who holds authority and responsibility in protected areas (Table 4.3).

Examples of these are given in the text below. It should be noted that any category can be applied with any governance type as shown in the matrix in Table 4.4. Protected area networks would usually be expected to contain examples of many different management approaches and governance types. Even individual protected areas are often zoned for differing levels of use (say a strictly protected core zone surrounded by zones where controlled access and fishing are allowed). Different zones within protected areas can even be given their own category if these zones are fixed by law (Day *et al.*, 2012) although in practice this is only likely to be worth doing for the largest MPAs, such as the Great Barrier Reef in Australia.

This flexibility becomes increasingly important in the context of MPA networks. Effective marine conservation will usually involve a seascape-wide approach that

Table 4.2 IUCN protected area categories: examples from MPAs.

Category	Main aims	Example
Ia	Strictly protected for biodiversity and also possibly geological/geomorphological features	The 11 Marine Reserves within the Channel Islands National Marine Sanctuary, California, USA are assigned to category Ia within the category IV National Park, to protect both biological and cultural resources, including deep water corals. The Marine Reserves are established for scientific purposes and to preserve biodiversity.
Ib	Wilderness areas without permanent or significant human habitation, protected and managed to preserve their natural condition	The Chassahowitzka Wilderness (category Ib) covers 95 km^2 or 77% of the Chassahowitzka National Wildlife Refuge (category IV) in the USA. It comprises saltwater bays, estuaries and brackish marshes at the mouth of the Chassahowitzka River, and provides critical habitat to a diversity of wildlife, including endangered species such as the West Indian manatee and whooping crane.
II	Large natural or near-natural areas managed for ecosystem protection and recreational opportunities	In South Korea, Hallyeohaesang National Park (76% of which is marine) and most of Dadohaehaesang National Park (80% marine) are assigned to category II (Shadie *et al.*, 2012).
III	Natural monuments such as a sea mount, marine cavern, or geological feature such as a cave	Blue Hole Natural Monument, Belize is an almost perfectly circular underwater sinkhole, surrounded by coral, and at around 300 metres across is the largest marine blue hole in the world. It is managed with the goal of protecting and preserving natural resources and nationally significant natural features of special interest or unique characteristics.
IV	Habitat/species management area to protect particular species or habitats, where management reflects this priority	Montague Island Habitat Protection Zone is category IV in Bateman's Marine Park in New South Wales, Australia and is designed to protect critical habitat of the grey nurse shark (*Carcharias taurus*); the island itself is an important seabird colony.
V	Protected seascape where the interaction of people and nature over time has produced a distinct character with significant ecological, biological, cultural and scenic value	Apo Island, in the Philippines, mixes traditional use of marine resources with ecotourism, generating revenue for communities. It is one of the best known community-managed marine reserves in the world and has in addition become an important recreational dive site. Visitors pay a fee that helps to maintain the sanctuary.
VI	Protected areas with sustainable use of natural resources	Misali Island Marine Conservation Area, Zanzibar, Tanzania was set up to protect important marine corals and other biodiversity whilst allowing sustainable use for over 11000 fishermen. A strictly protected zone helps to maintain fish populations.

Source: Day *et al.* (2012).

combines many different types of management: both protected areas and sustainable use areas. Protected areas will often vary from strict no-go sites to those with a strong emphasis on cultural management systems and sustainable use. Other areas that do not fit into the definition of a protected area may nonetheless have important conservation

Table 4.3 Typology of governance types in protected areas.

Type	Name	Description
A	Governance by government	• Federal or national ministry/agency in charge • Sub-national ministry/agency in charge • Government-delegated management (e.g. to NGO)
B	Shared governance	• Collaborative management (various degrees of influence) • Joint management (pluralist management board) • Transboundary management (various levels over frontiers)
C	Private governance	• By individual owner • By non-profit organizations (NGOs, universities, cooperatives) • By for-profit organizations (individuals or corporate)
D	Governance by indigenous peoples and local communities	• Indigenous peoples' conserved areas and territories • Community conserved areas – declared and run by local communities

Source: Dudley (2008).

functions (some of these may eventually be categorized as the new 'other effective area-based conservation measures'), and there will also likely be areas of more general, even intensive, use.

In the marine environment, national governments control coastal waters, while control of the high seas falls under international jurisdiction, which might be a subset of governance by governments (Type A) or perhaps of shared governance (Type B). The limit of a nation's control varies; generally territorial waters stretch up to 12 nautical miles from shore while Exclusive Economic Zones reach 200 nautical miles from the shore, where the state has special rights regarding the use of natural resources, including energy resources. This means that states can and do make the decision to set up MPAs in relatively near-shore waters, while high-seas MPAs require agreement from many states and are in consequence much more difficult to establish (and also more difficult to enforce).

Ownership and rights over coastal waters and onshore coastal habitats remain complex. Coastal land can be owned by governments, companies or private individuals depending on national laws, while coastal waters are in some cultures under traditional management agreements that stretch back beyond any written records. In many countries, governments control all coastal waters, but for instance in the Pacific customary rights have primacy (Govan and Jupiter, 2013). Almost everywhere people have rights of access and in many cases also extraction of resources and other forms of use. Determining the real governance systems in MPAs can be tricky but is an essential part of understanding how they function.

In practice, *governments* remain the commonest decision-makers in MPAs, particularly in offshore sites. Governance can in these cases be within the remit of national governments, local state, county or provincial governments and sometimes right down to individual municipalities or coastal communities. In a few cases, while the government retains ultimate decision-making power, it will delegate responsibility to another body such as an NGO. In the Seychelles Islands, for example, the remote Aldabra Atoll is a government MPA but day-to-day responsibility is in the hands of the charitable Seychelles Island Foundation (Hockings *et al.*, 2008).

Table 4.4 The IUCN protected area matrix of management categories and governance types.

Governance types / Protected area categories	A. Governance by government			B. Shared governance			C. Private governance			D. Governance by indigenous peoples and local communities	
	Federal or national ministry or agency in charge	Sub-national ministry or agency in charge	Government-delegated management (e.g. to an NGO)	Transboundary management	Collaborative management (various forms of pluralist influence)	Joint management (pluralist management board)	Declared and run by individual landowner	…by non-profit organizations (e.g. NGOs, universities, cooperatives)	…by for-profit organizations (e.g. individual or corporate landowners)	Indigenous peoples' and conserved areas and territories – established and run by indigenous peoples	Community conserved areas – declared and run by local communities
Ia) Strict nature reserve											
Ib) Wilderness area											
II) National park											
III) Natural monument											
IV) Habitat/species management											
V) Protected landscape/ seascape											
VI) Managed resource protected area											

Source: Borrini-Feyerabend *et al.* (2012).

Privately protected areas are relatively rare in the marine environment, although not unknown. The Nature Conservancy (TNC) in the USA has been involved in private protection of important marine eco-systems since the early years of the century. The umbrella term *Marine Conservation Agreement* (MCA) has been adopted by TNC to mean *any formal or informal con-tractual arrangement that aims to achieve ocean or coastal conservation goals in which one or more parties (usually right-holders) voluntarily commit to taking certain actions, refraining from certain actions, or transfer-ring certain rights and responsibilities in exchange for one or more other parties (usually conservation-oriented entities) vol-untarily committing to deliver explicit (direct or indirect) economic incentives.* Not all the ocean and coastal projects that TNC oper-ates involving MCAs result in the establish-ment and management of private MPAs, but many do. By 2014, some 167 MCAs were under development that had or were likely to result in privately protected areas in the marine environment, with examples ranging from West Papua to Long Island, New York (de Groot and Bush, 2010; Savage *et al.*, 2013; Udelhoven, 2014).

The situation regarding community or indigenous management is more complex. IUCN recognizes a governance type it has labelled *Indigenous Peoples' and Community Conserved Territories and Areas* (ICCAs), defined as: 'natural and/or modified ecosys-tems, containing significant biodiversity val-ues, ecological benefits and cultural values, voluntarily conserved by indigenous peoples and local communities, both sedentary and mobile, through customary laws or other effective means' (Borrini-Feyerabend *et al.*, 2012). The relationship between ICCAs and protected areas is complicated; some ICCAs are also protected areas, but some are not either because the community or indigenous people involved do not want their territory to be identified as a protected area, or because the management aims are not the same as those of a protected area as defined by IUCN. As a result many, but not all, ICCAs are also protected areas.

In the marine realm, the situation with ICCAs is even more complicated. In some parts of the world, particularly but not exclusively in parts of the Pacific, long-established fish management systems include many elements that resemble those of MPAs, including particularly the tempo-rary or permanent setting aside of areas to allow fish breeding. The distribution of marine ICCAs is highly skewed, with for example well over 500 established in the Philippines (Haribon Foundation, 2005), but being virtually unknown in other areas. There has also been a long and sometimes acrimonious debate about whether such fishery set asides are the equivalent of pro-tected areas (Govan and Jupiter, 2013). The Locally Managed Marine Areas (LMMA) movement has promoted such bottom-up initiatives, some of which are rooted in centuries of traditional management. Proponents argue that LMAAs and other community-driven initiatives (which are roughly equivalent to ICCAs) have stronger social legitimacy and support, and thus greater effectiveness, than top-down MPAs imposed by the government (Govan and Jupiter, 2013). Sceptics argue that while LMMAs may be effective ways of maintain-ing stocks of fish species that have com-mercial or subsistence value, they may offer little else in terms of overall ecological pro-tection. As with ICCAs, it may well turn out that many LMMAs will be eventually recognized as MPAs while others will fall under the new definition of OECMs. While these acronyms are confusing and to a large extent irrelevant to many onlookers, for the people managing such sites the way in which marine areas are designated often has legal and thus immediately practical implications (Fitzsimons, 2011).

Finally, within its governance typology IUCN identifies *shared governance* as a dis-tinct governance type, defined as formal or informal sharing authority or responsibility amongst several actors. In MPAs, where governments generally have such legal control over coastal waters, involving local communities or other stakeholders in management is generally a form of shared governance.

Perhaps even more important than the form of governance is the quality of governance, or as has been suggested the *vitality* of governance (Borrini-Feyerabend *et al.*, 2014). Increasing attention has been paid to the governance quality in MPAs and particularly to identifying what does and does not work (Gjertsen, 2004; Christie and White, 2007; Mascia *et al.*, 2010). The consensus, as with terrestrial and freshwater protected areas, is that the time taken to reach participatory, consensus-driven approaches (Lundquist and Granek, 2005) to protection is worth taking in terms of long-term effectiveness, and that successful MPAs need to be actively supported by local communities, albeit within a strong framework of government laws and policies.

The Effectiveness of MPAs

Finally and most significantly, MPAs are judged on whether or not they work. As larger areas of land and water are put under protection, and larger sums of money invested into conservation, pressure has grown to develop a better understanding of whether or not a protected area is delivering the benefits it promises, and if not what aspects of management need to be changed. Understanding of management effectiveness has been built both through the development of specific approaches to assessment and also through individual research projects that have collected information about the performance of MPAs in protecting particular species, groups or habitats. The IUCN World Commission on Protected Areas provided an overall framework for assessment of management effectiveness,

along with technical guidance about how this might be applied (Hockings *et al.*, 2000, 2006) (see Figure 4.2).

The WCPA framework is based on the idea that good protected area management follows a process that has six distinct stages, or elements:

- It begins with understanding the *context* of existing values and threats, understanding what is most important about the protected area both in ecological terms and also associated values such as those relating to ecosystem services, livelihoods, tourism and economic development along with the immediate and longer term pressures on the site
- progresses through *planning*, covering such issues as the original location and design of the site and the quality of both management planning and day-to-day planning of activities and
- allocation of resources (*inputs*) including information, money, equipment, infrastructure and staff
- then as a result of management actions (*processes*) such as ecological management, tourism, community relations, monitoring and evaluation, enforcement, etc.
- eventually produces products and services (*outputs*), which broadly relate to achievement of management plans
- that result in impacts or *outcomes*, the most important thing of all being whether or not the site's values are being maintained, which is also often the most difficult to measure over time.

The WCPA framework is a skeleton around which to build an assessment system. Since it was agreed in 2000, some 95 different protected area assessment systems have been developed throughout the world (Coad *et al.*, 2015). Some have been designed for analysis of just a few sites or for a particular research project while the most commonly used have now been applied many thousand times around the world, and the results are increasingly being used to drive management

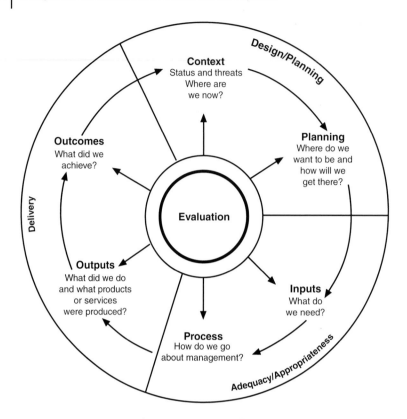

Figure 4.2 WCPA protected area management effectiveness framework. *Source*: Hockings *et al.* (2006).

policies (Leverington *et al.*, 2010). While many systems can be used in any protected area, terrestrial or marine, a number of assessments aimed particularly at MPAs were also developed, ranging from quite detailed (Pomeroy *et al.*, 2004) to simple question-naire approaches (Staub and Hatziolos, 2004). Assessment systems have been developed for particular conservation issues, such as migra-tory species (Hinch and de Santo, 2011). Additionally, a growing number of tech-niques aimed at measuring biological status, such as the health of coral communities (e.g. Kaufman *et al.*, 2011) provide tools for use in assessment.

So we have some tools, but what do they tell us? Unsurprisingly, the greatest effort has been put into looking at whether or not MPAs boost fish numbers – the question most immediately relevant to most people

in coastal communities and often the issue that clinches whether or not an MPA is supported and enforced, or opposed and undermined.

In a review carried out for WWF, Roberts and Hawkins (2000) identified five main benefits that MPAs supply to fisheries.

1) Enhancing the production of offspring to re-stock fishing grounds
2) Allowing spillover of adults and juveniles into fishing grounds
3) Preventing habitat damage
4) Promoting development of natural bio-logical communities
5) Facilitating recovery from catastrophic human disturbance.

Marine Protected Areas create sheltered conditions that increase the success of fish breeding and the survival of young fish, thus boosting overall populations. Three factors are important: (i) maintaining larger fish within the overall population; (ii) providing suitable habitat; and (iii) ensuring protection at vulnerable stages of the life cycle. Overfishing often removes most of the older and thus largest fish, but bigger fish generally produce disproportionately many more eggs than smaller ones and are thus particularly important for maintaining population. For example, one 10-kg red snapper (*Lutjanus campechanus*) produces over 20 times more eggs at a spawning than ten 1-kilogram snappers. Habitat protection means that whole populations are able to survive instead of being fragmented. Size of population itself is important for the breeding success of some marine species: many species that are attached to substrate or have limited powers of movement (e.g. oysters, clams or abalones) often only reproduce successfully at high population densities. Marine Protected Areas can ensure that species are protected at vulnerable stages of their life cycle, in particular through protection of fish nurseries and spawning grounds, thus increasing the chances that fish reproduce before being caught (Roberts and Hawkins, 2000; see also Gell and Roberts, 2003).

Researchers conclude that in most cases, fish density inside MPAs is higher than in the surrounding ocean, particularly in places where there are high levels of fishing (Pérez-Ruzafa *et al.*, 2008). This not particularly surprising finding has been backed by a series of large meta-studies. A review of research in 80 different MPAs found strikingly higher fish populations, compared with both the surrounding area and the same area before an MPA was established: population densities were 91% higher, biomass 192% higher, and average organism size and diversity 20–30% higher in MPAs. Changes on this scale were measured quite quickly, often between one and three years after establishment, and occurred even in small MPAs (Halpern, 2003). Similarly, a

global meta-analysis of 124 temperate and tropical MPAs in 29 countries identified dramatic increases in biomass (+446%) and densities (+166%) of organisms inside protected areas, along with moderate increases of individual size (+28%) and species richness (+21%) (Lester *et al.*, 2009).

While these factors are important from a conservation perspective, they will not be sufficient to persuade fishing communities that setting aside areas of marine habitat is a good thing. A much more important question from a social perspective is whether the higher populations within MPAs also help to boost populations in surrounding areas – the so-called 'spillover effect'. Since most fish have free-floating larvae or eggs, and because many fish move, animals can swim or drift out of reserves, thus re-stocking nearby fishing grounds. Research shows that spillover does occur and that the net effect for fish catch can be positive, although here there is more variation between results (see Table 4.5). For example, research from the Great Barrier Reef found that for two commercially exploited coral species, MPAs accounting for 28% of local reef area contributed half of all juvenile recruitment through the area within a distance of 30 km (Harrison *et al.*, 2012). Similarly, in the Mediterranean, effective management was linked closely with increased fish numbers (Guidetti *et al.*, 2014).

Six factors have been suggested to affect spillover: (i) the success of protection; (ii) the time that the MPA has been established; (iii) intensity of fishing outside the MPA; (iv) mobility of species; (v) boundary length of the reserve (greater edge to area ratio giving increased spillover); and (vi) boundary porosity, with out-migration encouraged if there is continuous habitat type (Roberts and Hawkins, 2000).

A study of 87 MPAs worldwide identified five critical factors for success: (i) the presence of no-take policies; (ii) effective enforcement; (iii) age of the MPA (those over

Table 4.5 Impact of MPAs on fisheries: some recent research examples from around the world.[a]

MPA	Increased fish numbers	Spillover
Channel Island network of MPAs, California, USA (Caselle *et al.*, 2015)	✓	✓
Network of deep water MPAs near Hawai'i, USA (Sackett *et al.*, 2014)	✓	
Goukamma MPA, South Africa (Kerwath *et al.*, 2013)	✓	✓
Columbretes Islands Marine Reserve, Spain (Stobart *et al.*, 2009)	✓	✓
Saldanha Bay, Langebaan Lagoon, South Africa (Kerwath *et al.*, 2009)	✓	✓
Cerbere-Banyuls and Carry-le-Rouet MPAs in France, and Medes, Cabrera, Tabarca, and Cabo de Palos MPAs in Spain (Goñi *et al.*, 2008)		✓
Medes Islands MPA, Spain (Stelzenmüller *et al.*, 2008)	✓	✓
Abrolhos National Marine Park, Brazil (Francini-Filho and Leão de Moura, 2008)	✓	
Rottnest Island, Western Australia (Babcock *et al.*, 2007)	✓	
Wakatobi Marine National Park, Indonesia (Unsworth *et al.*, 2007)	✓	
Côte Bleue MPA, France (Claudet *et al.*, 2006)		✓
Apo Island, Philippines (Abesamis and Russ, 2005)	✓	✓
Nabq Managed Resource Protected Area, Egypt (Ashworth and Ormond, 2005)	✓	✓
Malindi and Watamu Marine National Parks, Kenya (Kaunda-Arara and Rose, 2004)	✓	✓
Soufrière Marine Management Area, St Lucia (Roberts *et al.*, 2001)	✓	
Mombasa MPA, Kenya (McClanahan and Mangi, 2000)		✓
Monterey Bay National Marine Sanctuary; Hopkins Marine Life Refuge; Point Lobos State & Ecological Reserve; Big Greek Marine Ecological Reserve, USA (Paddack and Estes, 2000)	✓	

Source: Adapted from Stolton and Dudley (2010).

a)   Not all studies above looked at both fish numbers and spillover.

10 years old being more effective); (iv) size, with areas in excess of $100\,\mathrm{km}^2$ being preferred; and (v) isolation from other areas by deep water and sand (Edgar *et al.*, 2014). But while there is growing consensus that well-managed protected areas can boost fish numbers, they do not offer a panacea in all cases. For example, MPAs offer far less to fisheries that target highly mobile, single species with little by-catch, being better suited to fisheries that target multiple, more sedentary species (Hilborn *et al.*, 2004). Researchers also point out that MPAs still lack rigorous evaluation against counterfactuals (e.g. Ahmadia *et al.*, 2015) to pinpoint more precisely what management approaches work, what don't work and in which particular conditions.

As MPAs become better established, information is starting to accumulate about other aspects of effectiveness. For example, Selig and Bruno (2010) found evidence that MPAs were effective in maintaining coral cover, with benefits appearing to increase with the age of the MPA. Research in the Bahamas also found MPAs facilitated recovery of corals after bleaching events, possibly because higher fish numbers reduced algal cover (Mumby and Harborne, 2010). However, success is not inevitable; analysis of 78 coral reef-based MPAs found a quarter failing to meet aims in increasing coral cover over time and that in some regions MPAs were simply mirroring outside changes, in other words that it was sometimes hard to prove additionality (Hargreaves-Allen *et al.*, 2011). In Australia, the Great Barrier Reef Marine Park, one of the largest MPAs in the world with significant management resources, extensive science input and with one-third of the MPA being in no-take zones, has nevertheless seen coral cover reduce by 50% over a 27-year period (De'ath *et al.*, 2012).Perhaps unsurprisingly, most of this loss has been in the southern inshore section of the Great Barrier Reef, adjacent to areas of coastal development with significantly degraded water quality from nutrient, sediment and pesticide input to the coastal waters (Great Barrier Reef Marine Park Authority, 2014).

There has been a long debate about the effectiveness of MPAs in mitigating extreme weather events and geological movements, such as typhoons, hurricanes and tsunamis, and the longer term impacts of sea-level rise. The consensus seems to be that healthy natural ecosystems such as coral reefs, coastal mangroves and coastal lagoons and marshes can provide important protection but that there will be limits in the extent to which these work – the largest events such as huge tsunamis will overwhelm any kind of defence, natural or made by humans (Stolton *et al.*, 2008). Steps to increase the use of MPAs as natural defence mechanisms are being promoted (Dudley *et al.*, 2015).

So we have mounting evidence that MPAs can work, *if* they are well managed. But do they *actually* work? Are researchers cherry-picking well-managed reserves in their research, giving a false impression of overall effectiveness? Some researchers definitely think so. Writing in 2006, and therefore before the latest flurry of MPA creation, Mora and colleagues (Mora *et al.*, 2006) calculated that although 18.7% of the world's corals were in some kind of MPA, most were in extractive or multipurpose MPAs where corals remain under pressure, or in MPAs with no effective management, and that less than 0.1% were in no-take MPAs with effective management.

Conclusion: Priorities for MPA Management

There seem to be four urgent priorities with MPA management. First is to build the arguments, the constituency and the political and social support for an increase in no-take zones within MPAs (Costello and Ballantine, 2015) – not for all MPAs but to address those ecosystems that are most damaged or most fragile. Secondly, the commitment to creating MPAs needs to be matched with a commitment to managing those MPAs that have been designated: too many MPAs are currently being created to satisfy international obligations or a domestic conservation movement without apparently any serious intention to make them anything but paper parks (Jameson *et al.*, 2002). Next, while the evidence for the benefits of MPAs is growing all the time (Halpern and Warner, 2002), as has been pointed out we still need more and more rigorous studies, and just as important the results of these studies need better publicity. And lastly, the lessons learned from successful MPAs deserve far more attention: a global network of working MPAs could provide coordination, learning capacity and demonstration sites, and also serve as ambassadors for other places wrestling with the question of how to maintain their marine resources. None of these things are impossible; all require political will and willing hands to make them a reality.

References

Abesamis, R.A. and Russ, G.R. (2005) Density-dependent spillover from a marine reserve: long-term evidence. *Ecological Applications*, **15**, 1798–1812.

Ahmadia, G.N., Glew, L., Provost, M. *et al.* (2015) Integrating impact evaluation in the design and implementation of monitoring marine protected areas. *Philosophical Transactions of the Royal Society B*, **370** (1681). http://dx.doi.org/10.1098/rstb.2014.0275

Ashworth, J.S. and Ormond, R.F.G. (2005) Effects of fishing pressure and trophic group on abundance and spillover across boundaries of a no-take zone. *Biological Conservation*, **121** (3), 333–344.

Babcock, R.C., Phillips, J.C., Lourey, M. and Clapin, G. (2007) Increased density, biomass and egg production in an unfished population of Western Rock Lobster (*Panulirus cygnus*) at Rottnest Island, Western Australia. *Marine and Freshwater Research*, **58**, 286–292.

Borrini-Feyerabend, G., Dudley, N., Lassen, B. *et al.* (2012) *Governance of Protected Areas: From Understanding to Action*. IUCN, GIZ and ICCA Consortium, Gland, Switzerland.

Borrini-Feyerabend, G., Bueno, P., Hay-Edie, T. *et al.* (2014) *A Primer on Governance for Protected and Conserved Areas*. Stream on Enhancing Diversity and Quality of Governance, 2014 IUCN World Parks Congress. IUCN, Gland, Switzerland.

Caselle, J.E., Rassweiler, A., Hamilton, S.L. and Warner, R.R. (2015) *Recovery Trajectories of Kelp Forest Animals are Rapid yet Spatially Variable Across a Network of Temperate Marine Protected Areas*. Nature Scientific Reports. doi:10.1038/srep14102

Christie, P. and White, A.T. (2007) Best practices for improved governance of coral reef marine protected areas. *Coral Reefs*, **26**, 1047–1056.

Claudet, J., Pelletier, D., Jouvenel, J.Y. *et al.* (2006) Assessing the effects of a marine protected area (MPA) on a reef fish assemblage in a northwestern Mediterranean marine reserve: identifying community-based indicators. *Biological Conservation*, **130** (3), 349–369.

Coad, L., Leverington, F., Knights, K. *et al.* (2015) Measuring the impact of protected area management interventions: current and future use of the global database of protected area management effectiveness. *Philosophical Transactions of the Royal Society B*, http://dx.doi.org/10.1098/rstb.2014.0281

Convention on Biological Diversity (CBD) (n.d.) http://www.cbd.int/programmes/cross-cutting/ecosystem/default.shtml

Costello, M.J. and Ballantine, B. (2015) Biodiversity conservation should focus on no-take Marine Reserves: 94% of Marine Protected Areas allow fishing. *Trends in Ecology and Evolution*, **30** (9), 507–509.

Day, J., Dudley, N., Hockings, M. *et al.* (2012) *Guidelines for Applying the IUCN Protected Area Management Categories to Marine Protected Areas*. IUCN, Gland, Switzerland.

De'ath, G., Fabricius, K.E., Sweatman, H. and Puotinen, M. (2012) The 27-year decline of coral cover on the Great Barrier Reef and its causes. *Proceedings of the National Academy of Sciences*, **109** (44), 17995–17999.

Dudley, N. (ed.) (2008) *Guidelines for Applying Protected Area Management Categories*. IUCN, Gland, Switzerland. WITH Stolton, S., Shadie, P. and Dudley, N. (2013) *IUCN WCPA Best Practice Guidance on Recognising Protected Areas and Assigning Management Categories and Governance Types*. Best Practice Protected Area Guidelines Series No. 21. IUCN, Gland, Switzerland.

Dudley, N., Parrish, J.D., Redford, K.R. and Stolton S. (2010) The revised IUCN protected area management categories: the debate and ways forward. *Oryx*, **44** (4), 485–490. doi:10.1017/S0030605310000566

Dudley, N., Buyck, C., Furuta, N. *et al.* (2015) *Protected Areas as Tools for Disaster Risk Reduction: A Handbook for Practitioners.* MOEJ, Tokyo and IUCN, Gland, Switzerland. 44 pp.

Edgar, G.J., Stuart-Smith, R.D., Willis, T.J. *et al.* (2014) Global conservation outcomes depend on marine protected areas with five key features. *Nature,* **506,** 216–220. doi:10.1038/nature13022

Emslie, M.J., Logan, M., Williamson, D.H. *et al.* (2015) Expectations and outcomes of reserve network performance following re-zoning of the Great Barrier Reef Marine Park. *Current Biology,* **25,** 1–10.

Fitzsimons, J. (2011) Mislabelling marine protected areas and why it matters: a case study of Australia. *Conservation Letters,* **4,** 340–345.

Francini-Filho, R.B. and Leão de Moura, R. (2008) Dynamics of fish assemblages on coral reefs subjected to different management regimes in the Abrolhos Bank, eastern Brazil. *Aquatic Conservation in Marine and Freshwater Ecosystems,* **18,** 1166–1179.

Gell, F.R. and Roberts, C.M. (2003) *The Fishery Effects of Marine Reserves and Fishery Closures.* WWF-US, Washington, DC.

Gjertsen, H. (2004) Can habitat protection lead to improvement in human well-being? Evidence from marine protected areas in the Philippines. *World Development,* **33,** 199–217.

Goñi, R., Adlerstein, S., Alvarez-Berastegui, D. *et al.* (2008) Spillover from six western Mediterranean marine protected areas: evidence from artisanal fisheries. *Marine Ecology Progress Series,* **366,** 159–174.

Govan, H. and Jupiter, S. (2013) Can the IUCN 2008 protected areas management categories support Pacific Island approaches to conservation? *PARKS,* **19,** 73–80.

Great Barrier Reef Marine Park Authority (2014) *Great Barrier Reef Outlook Report 2014.* GBRMPA, Townsville. 311 pp.

Grober-Dunsmore, R., Wooninck, L., Field, J. *et al.* (2008) Vertical zoning in marine protected areas: ecological considerations for balancing pelagic fishing with conservation of benthic communities. *Fisheries,* **33,** 598–610.

de Groot, J. and Bush, S.R. (2010) The potential for dive tourism led entrepreneurial marine protected areas in Curacao. *Marine Policy,* **34,** 1051–1059.

Guidetti, P., Baiata, P., Ballesteros, E. *et al.* (2014) Large-scale assessment of Mediterranean Marine Protected Areas effects on fish assemblages. *PLoS ONE,* **9** (4), e91841. doi:10.1371/journal.pone.0091841

Halpern, B.S. (2003) The impact of marine reserves: do reserves work and does reserve size matter? *Ecological Applications,* **13,** 117–137.

Halpern, B.S. and Warner, R.R. (2002) Marine reserves have rapid and lasting effects. *Ecology Letters,* **5** (3), 361–366.

Hargreaves-Allen, V., Mourato, S. and Milner-Gulland, E.J. (2011) A global evaluation of coral reef management performance: are MPAs producing conservation and socio-economic improvements? *Environmental Management,* **47,** 684–700.

Haribon Foundation (2005) *Atlas of Community-based Marine Protected Areas in the Philippines.* Manila.

Harrison, H.B., Williamson, D.H., Evans, R.D. *et al.* (2012) Larval export from marine reserves and the recruitment benefit for fish and fisheries. *Current Biology,* **22,** 1023–1028. doi:10.1016/j.cub.2012.04.008

Hilborn, R., Stokes, K.R., Maguire, J.J. *et al.* (2004) When can marine reserves improve fisheries management? *Ocean and Coastal Management,* **47,** 197–205.

Hinch, P.R. and de Santo, E.M. (2011) Factors to consider in evaluating the management and conservation effectiveness of a whale sanctuary to protect and conserve the North Atlantic right whale (*Eubalaena glacialis*). *Marine Policy,* **35,** 163–180.

Hockings, M., Stolton, S. and Dudley, N. (2000) *Evaluating Effectiveness: A Framework for Assessing Management Effectiveness of Protected Areas*. IUCN, Gland, Switzerland.

Hockings, M., Stolton, S., Leverington, F. *et al.* (2006) *Evaluating Effectiveness: A Framework for Assessing Management Effectiveness of Protected Areas* (2nd edition). IUCN, Gland, Switzerland.

Hockings, M., Stolton, S., James, R. *et al.* (2008) *Enhancing our Heritage Toolkit: Assessing Management Effectiveness of Natural World Heritage Sites*. World Heritage Papers 23. UNESCO, Paris.

Jameson, S.C., Tupper, M.H. and Ridley, J.M. (2002) The three screen doors: can marine 'protected' areas be effective? *Marine Pollution Bulletin*, **44**, 1177–1183.

Jones, P.J.S., Qiu, W. and De Santo, E.M. (2011) *Governing Marine Protected Areas: Getting the Balance Right*. Technical Report, United Nations Environment Programme.

Juffe-Bignoli, D., Burgess, N.D., Bingham, H. *et al.* (2014) *Protected Planet Report 2014*. UNEP-WCMC, Cambridge, UK.

Kaufman, L., Obdura, D., Rohwer, F. *et al.* (2011) *Coral Health Index: Measuring Coral Community Health*. Conservation International, Arlington, VA, USA.

Kaunda-Arara, B. and Rose, G.A. (2004) Effects of marine reef National Parks on fishery CPUE in coastal Kenya. *Biological Conservation*, **118**, 1–13.

Kenchington, R.A. (2003) Managing marine environments: an introduction to issues of sustainability, conservation, planning and implementation, in *Conserving Marine Environments: Out of Sight, Out of Mind* (eds P. Hutchings and D. Lunney). Royal Zoological Society of New South Wales, Mosman, NSW, Australia.

Kerwath, S.E., Thorstad, E.B., Næsje, T.F. *et al.* (2009) Crossing invisible boundaries: the effectiveness of the Langebaan Lagoon Marine Protected Area as a harvest refuge for a migratory fish species in South Africa. *Conservation Biology*, **23**, 653–661.

Kerwath, S.E., Winker, H., Götz, A. and Attwood, C.G. (2013) Marine protected area improves yield without disadvantaging fishers. *Nature Communications*, **4**, 2347. doi:10.1038/ncomms3347

Lester, S.E., Halpern, B.S., Grorud-Colvert, K. *et al.* (2009) Biological effects within no-take marine reserves: a global synthesis. *Marine Ecology Progress Series*, **384** (2), 33–46.

Leverington, F., Lemos Costa, K., Pavese, H. *et al.* (2010) A global analysis of protected area management effectiveness. *Environmental Management*, **46**, 685–698. doi:10.1007/s00267-010-9564-5

Lopoukhine, N. and Dias, B. Ferreira de Souza (2012) What does target 11 really mean? *PARKS*, **18** (1), 5–8.

Lundquist, C.J. and Granek, E.F. (2005) Strategies for successful marine conservation: integrating socioeconomic, political and scientific factors. *Conservation Biology*, **19**, 1771–1778.

Mascia, M.B. and Claus, C.A. (2008) A property rights approach to understanding human displacement from protected areas: the case of marine protected areas. *Conservation Biology*, **23** (1), 16–23.

Mascia, M.B., Claus, C.A. and Naidoo, R. (2010) Impacts of marine protected areas on fishing communities. *Conservation Biology*, **24**, 1424–1429.

McClanahan, T.R. and Mangi, S. (2000) Spillover of exploitable fishes from a marine park and its effect on the adjacent fishery. *Ecological Applications*, **10** (6), 1792–1805.

Mora, C., Andréfouët, S., Costello, M.J. *et al.* (2006) Coral reefs and the global network of marine protected areas. *Science*, **312**, 1750–1751.

Mumby, P.J. and Harborne, A.R. (2010) Marine reserves enhance the recovery of corals on Caribbean reefs. *PLoS ONE*, **5** (1), e8657. doi:10.1371/journal.pone.0008657

Paddack, M.J. and Estes, J.A. (2000) Kelp forest fish populations in marine reserves and adjacent exploited areas of central California. *Ecological Applications*, **10**, 855–870.

Pérez-Ruzafa, A., Martín, E., Marcos, C. *et al.* (2008) Modelling spatial and temporal scales for spill-over and biomass exportation from MPAs and their potential for fisheries enhancement. *Journal for Nature Conservation*, **16** (4), 234–255.

Pomeroy, R.S., Parks, J.E. and Watson, L.M. (2004) *How is Your MPA Doing? A Guidebook of Natural and Social Indicators for Evaluating Marine Protected Area Effectiveness.* IUCN, Gland, Switzerland. 216 pp.

Roberts, C.M. and Hawkins, J.P. (2000) *Fully-protected Marine Reserves: A Guide.* WWF Endangered Seas Campaign, Washington, DC, USA and Environment Department, University of York, UK.

Roberts, C.M., Bohnsack, J.A., Gell, F. *et al.* (2001) Effects of marine reserves on adjacent fisheries. *Science*, **294**, 1920–1923.

Sackett, D.K., Drazen, J.C., Moriwake, V.N. *et al.* (2014) Marine protected areas for deepwater fish populations: an evaluation of their effects in Hawai'i. *Marine Biology*, **161**, 411–425.

Savage, J., Osborne, P. and Hudson, M. (2013) Abundance and diversity of marine flora and fauna of protected and unprotected reefs of the Koh Rong Archipelago, Cambodia. *Cambodian Journal of Natural History*, **2013** (2), 83–94.

Selig, E.R. and Bruno, J.F. (2010) A global analysis of the effectiveness of marine protected areas in preventing coral loss. *PLoS ONE*, **5** (2), e9278. doi:10.1371/journal.pone.0009278

Shadie, P., Young Heo, H., Stolton, S. and Dudley, N. (2012) *Protected Area Management Categories and Korea: Experience to Date and Future Directions.* IUCN, Gland, Switzerland and KNPS, Seoul, Republic of Korea.

Staub, F. and Hatziolos, M.E. (2004) *Score Card to Assess Progress in Achieving Management Effectiveness Goals for Marine Protected Areas* (revised version). The World Bank, Washington, DC.

Stelzenmüller, V., Maynou, F. and Martín, P. (2008) Patterns of species and functional diversity around a coastal marine reserve: a fisheries perspective. *Aquatic Conservation: Marine and Freshwater Ecosystem*, **19** (5), 554–565.

Stobart, B., Warwick, R., Gonzalez, C. *et al.* (2009) Long-term and spillover effects of a marine protected area on an exploited fish community. *Marine Ecology Progress Series*, **384**, 47–60.

Stoddart, D.W. (1971) Settlement, development and conservation of Aldabra. *Philosophical Transactions of the Royal Society B*, **260**, 611–628. doi:10.1098/rstb.1971.0028

Stolton, S. and Dudley, N. (eds) (2010) *Arguments for Protected Areas: Multiple Benefits for Conservation and Use.* Earthscan, London.

Stolton, S., Dudley, N. and Randall, J. (2008) *Natural Security: Protected Areas and Hazard Mitigation.* WWF, Gland, Switzerland.

Udelhoven, J. (2014) Marine PPAs: mythical sea creature or ocean of opportunity? in *The Futures of Privately Protected Areas* (eds S. Stolton, K.H. Redford and N. Dudley). IUCN, Gland, Switzerland.

Unsworth, R.K.F., Powell, A., Hukom, F. and Smith, D.J. (2007) The ecology of Indo-Pacific grouper (Serranidae) species and the effects of a small scale no take area on grouper assemblage, abundance and size frequency distribution. *Marine Biology*, **152**, 243–254.

Watson, J.E., Dudley, N., Segan, D.B. and Hockings, M. (2014) The performance and potential of protected areas. *Nature*, **515** (7525), 67–73.

5

Marine Protected Areas as Spatial Protection Measures under the Marine Strategy Framework Directive

Daniel Braun

Research Group of Prof. Dr. Detlef Czybulka, Faculty of Law, University of Rostock, Germany

Introduction

In 2002 the European Union (EU) developed a thematic strategy for the protection and conservation of the marine environment with the overall aim of promoting sustainable use of the seas and conserving marine ecosystems in line with the Sixth Community Environment Action Programme.[1] In 2008 a legally binding framework was adopted as the environmental pillar of the future maritime policy for the EU[2]: the Marine Strategy Framework Directive (MSFD, 2008/56/EC). This Directive now applies to 28 Member States – of which 22 have a marine zone[3] – and obliges them to achieve or maintain a Good Environmental Status (GES) in the marine environment by 2020.

To this end, the Member States shall identify and afterwards take the necessary measures. These include spatial protection measures as an element of the so-called 'programmes of measures'. The MSFD pays particular attention to these measures, as it contains specific provisions about the possible types and criteria for networks of Marine Protected Areas (MPAs). Within this chapter, the requirements of spatial protection measures and their relevance with respect to the goals of the MSFD are discussed.

Area of Application of the MSFD

According to Art. 2(1), the MSFD is applicable to all marine waters. This term is defined in Art. 3(1)(a) MSFD for the purpose of the Directive as 'waters, the seabed and subsoil on the seaward side of the baseline from which the extent of territorial waters is measured extending to the outmost reach of the area where a Member State has and/or exercises jurisdictional rights, in accordance with the UNCLOS [...]'. The abbreviation 'UNCLOS' within this definition stands for the United Nations Convention on the Law of the Sea of 1982, in force since 1994. This convention, often referred to as 'constitution for the oceans'[4], defines different marine zones; within some of them coastal states may exercise sovereignty or jurisdiction. The (full) sovereignty of the coastal state extends to a sea area described as the territorial sea which extends up to a distance of 12 nautical miles seawards from the baselines (Art. 2(1) and 3 UNCLOS). The baselines are defined in accordance with Art. 5 to 7 UNCLOS. As the concept of sovereignty reaches further than jurisdictional rights in the meaning of Art. 3(1)(a) MSFD, territorial seas of the Member States belong to marine waters under the Directive.

Management of Marine Protected Areas: A Network Perspective, First Edition. Edited by Paul D. Goriup.
© 2017 John Wiley & Sons Ltd. Published 2017 by John Wiley & Sons Ltd.

Moreover, UNCLOS confers jurisdictional rights to the coastal state on the continental shelf and within the exclusive economic zone (EEZ). Article 76(1) UNCLOS defines the continental shelf as the 'seabed and sub-soil of the submarine areas that extend beyond its territorial sea throughout the natural prolongation of its land territory to the outer edge of the continental margin, or to a distance of 200 nautical miles from the baselines from which the breadth of the territorial sea is measured where the outer edge of the continental margin does not extend up to that distance'. The coastal state exercises exclusive rights with regard to the exploration and exploitation of the natural resources of the continental shelf (Art. 77 UNCLOS). The EEZ comprises the waters beyond and adjacent to the territorial sea up to a distance of 200 nautical miles from the baselines. This follows from the regulations in Art. 55, 57 UNCLOS. Within this zone, the coastal state exercises the sovereign rights and jurisdiction mentioned in Art. 56 UNCLOS and related detailed regulations. While the exclusive rights on the continental shelf exist *ipso iure* (Cacaud, 2005), the establishment of an EEZ requires a proclamation, because a provision equivalent to Art. 77(3) UNCLOS does not exist. In cases where a coastal state has proclaimed an EEZ, the underlying continental shelf is integrated in the regime of the EEZ by Art. 56(3) UNCLOS. It is also noteworthy that the MSFD is applicable in marine areas where Member States in their role as coastal states only proclaim some of the exclusive rights encompassed by the full regime of the EEZ provided by UNCLOS. Such (exclusive) fishing zones and ecological protection zones currently exist in the Mediterranean Sea. Depending on the individual case it remains unclear to what extent it is possible to make a contribution to the GES by exercising the exclusive rights in these zones. Due to its applicability in marine waters, the MSFD represents the marine counterpart to the Water Framework Directive (WFD, 2000/60/EC) which was adopted in 2000. The area of application of the WFD covers waters on the landward side of the baseline. These include transitional waters and the coastal waters. As this Directive aims at the protection and improvement of the aquatic environment *inter alia* against discharges, emissions and losses of hazardous substances (see Art. 1 lit. (c) WFD), it provides an indirect contribution to the GES of the marine environment which must not be neglected.

Implementation Process

The implementation of the MSFD is divided into two successive stages (Art. 5(3) MSFD). It began with a preparation stage comprising an initial assessment, a determination of the GES and the establishment of environmental targets. This stage should have been completed with the establishment and implementation of a monitoring programme by July 2014. The second stage concerns the programmes of measures: these had to be developed by 2015 at the latest, and implementation had to start by 2016 at the latest. According to this schedule, monitoring programmes should also have already been established and implemented. The Member States should have made publicly available relevant information on the spatial protection measures within the programmes of measures by 2013 at the latest (Art. 13(6) MSFD).

Spatial Protection Measures

The programmes of measures are regulated in detail by Art. 13 et seq. of the MSFD. According to Art. 13(1) of the Directive, Member States shall identify measures which need to be taken in order to achieve or maintain GES. Therefore, any measure included in the programmes has to be chosen with regard to the initial assessment made pursuant to Art. 8(1) MSFD and the characteristics for GES on the basis of the qualitative descriptors in Annex I of the MSFD. Furthermore, the measures must be devised with reference to the environmental targets, established under Art. 10(1) and Annex VI of the MSFD.[5] Also worth mentioning here are the requirements laid down in Art. 13(3) MSFD: 'Member States shall give due consideration to sustainable development and, in particular, to the social and economic impacts of the measures envisaged.' Moreover, 'Member States shall ensure that measures are cost-effective and technically feasible, and shall carry out impact assessments, including cost–benefit analyses, prior to the introduction of any new measure'.

Spatial protection measures represent a special category within the aforementioned programmes (Art. 13(4) MSFD). They contribute 'to coherent and representative networks of marine protected areas, adequately covering the diversity of the constituent ecosystems'. Although, as described above, the programmes of measures have to be developed by 2015, Art. 13(6) MSFD obliged the Member States to make the relevant information on these areas publicly available by 2013. This had to take place in respect of the marine regions or subregions defined by Art. 4 MSFD.

Background: Global International Law

From the perspective of global international law, it appears that two conventions are of particular importance with regard to the establishment of MPAs. Firstly, in 1993 the Convention on Biological Diversity (CBD) entered into force. All Member States and the EU itself have become members of this treaty during recent years.[6] According to Art. 8 lit. (a) CBD, parties should '[e]stablish a system of protected areas or areas where special measures need to be taken to conserve biological diversity'. The term

protected area is defined in Art. 2 CBD as 'a geographically defined area which is designated or regulated and managed to achieve specific conservation objectives', and biological diversity means 'the variability among living organisms from all sources including, inter alia, terrestrial, marine and other aquatic ecosystems and the ecological complexes of which they are part', comprising diversity within species, between species and of ecosystems. It appears that it has become widely recognized that the obligation in Art. 8 lit. (a) CBD not only extends to territorial seas, but also to marine areas where coastal states only exercise sovereign rights and jurisdiction. This follows from the relationship of the convention with UNCLOS as laid down in Art. 22 CBD and Art. 237 UNCLOS. It also follows from these provisions that UNCLOS takes priority in its application in the marine area.[7]

Secondly, the general obligation of Art. 192 in Part XII of the UNCLOS requires parties to protect and preserve the marine environment. This rule represents a codification of customary international law and is therefore binding for all states. An outstanding provision with regard to MPAs in the UNCLOS is Art. 194(5) which clarifies that 'measures taken in accordance with this Part shall include those necessary to protect and preserve rare or fragile ecosystems as well as the habitat of depleted, threatened or endangered species and other forms of marine life' (Scovazzi, 2011). It is still under debate whether the contracting parties have to protect the features mentioned in Art. 194(5) UNCLOS not only from pollution (see the definition in Art. 1(4) UNCLOS) but also from negative impacts that result from other sources (Czybulka, 2016a). It appears that the European Commission tends towards the latter opinion.[8] Moreover, it can be anticipated that the decision of the Arbitral Tribunal in the 'Chagos MPA' Case[9] in 2015 will strengthen this position in future discussions.[10] Irrespective of this debate, there is no doubt that MPAs can serve as a very effective tool to protect and preserve eco-systems, habitats and species not only against pollution but also against other anthropogenic impairments.

While the prerogative of the coastal states to establish MPAs within their territorial seas has never been doubted since the UNCLOS entered into force, it is still being debated whether, and, if so, under which legal conditions MPAs may be established on the continental shelf and within EEZs. Although the current state of this discussion shall not be presented here in all its facets, it should be emphasized that any restriction of activities by a coastal state on the continen-tal shelf and within an EEZ may only be car-ried out with due regard to the rights of other states. These include, in particular, the 'rights of communication' (Art. 79 UNCLOS for the continental shelf and Art. 58(1), 87 UNCLOS for the EEZ).[11]

It should be noted that the rights and duties of the states with regard to the use and protection of the marine environment including the marine resources are specified in regional seas conventions (RSCs) for dif-ferent marine regions. The RSCs with a scope of application extending to marine regions falling under Art. 4 MSFD will be discussed below.

Networks of MPAs

On the one hand, requirements with regard to networks of MPAs resulting from the general provisions concerning the achieve-ment and maintenance of GES influence the design of MPA networks. On the other hand, specific requirements follow from Art. 13(4) MSFD.

General Requirements of the MSFD

On the one hand, the relevance of networks of MPAs as an element of the programmes of measures under the MSFD results from their contribution to GES. On the other hand, the general provisions concerning the

programmes of measures are important in addition to the detailed regulations regard-ing the description of GES.

A spatial protection measure contributes to GES if it has an effect on environmental status that fosters its development towards being considered as 'good'. The MSFD defines the environmental status in Art. 3(4) as 'overall state of the environment in marine waters, taking into account the structure, function and processes of the constituent marine ecosystems together with natural physiographic, geographic, biological, geo-logical and climatic factors, as well as physi-cal, acoustic and chemical conditions, including those resulting from human activ-ities inside or outside the area concerned'. According to Art. 3(5) MSFD, GES 'means the environmental status of marine waters where these provide ecologically diverse and dynamic oceans and seas which are clean, healthy and productive within their intrinsic conditions, and the use of the marine envi-ronment is at a level that is sustainable, thus safeguarding the potential for uses and activities by current and future generations'. The basis for the determination of GES in the marine regions or subregions is the list of qualitative descriptors in Annex I. As spa-tial protection measures have an effect that is limited to specially selected areas within marine regions or subregions, they are most likely to improve the status of certain local features in line with the objectives of some of the descriptors, such as:

(1) Biological diversity is maintained. The quality and occurrence of habitats and the distribution and abundance of species are in line with prevailing physiographic, geographic and climatic conditions.

(6) Sea-floor integrity is at a level that ensures that the structure and functions of the ecosystems are safeguarded and benthic ecosystems, in particular, are not adversely affected.

Spatial protection measures are also well suited to promote the defined goals of other descriptors, including:

(3) Populations of all commercially exploited fish and shellfish are within safe biological limits, exhibiting a population age and size distribution that is indicative of a healthy stock.

(4) All elements of the marine food webs, to the extent that they are known, occur at normal abundance and diversity and levels capable of ensuring the long-term abundance of the species and the retention of their full reproductive capacity.

For the remaining descriptors, it appears that spatial protection measures are more or less likely to contribute to their goals where specific areas are concerned. It should be noted that the criteria and methodological standards for the descriptors have to be laid down according to Art. 9(3) MSFD. Consequently, the Commission adopted Decision 2010/477/EU with detailed specifications for each of the descriptors, for which a revision process has already been announced (European Commission, 2014).

When determining GES, Member States shall take into account the indicative list of elements set out in Table 1 of Annex III to the MSFD as well as, in particular, 'physical and chemical features, habitat types, biological features and hydromorphology' (Art. 9(1) MSFD). The elements listed in Table 1 are relevant for spatial protection measures in so far as the description of the characteristics of habitat types and biological features, according to the detailed criteria, serves as a basis for the selection of these measures. The requirements with regard to the representativeness of MPA networks (see below) greatly depend on this data.

Article 9(1) MSFD requires Member States to take into account the pressures or impacts of human activities in each marine region or subregion, having regard to the indicative lists set out in Table 2 of Annex III, when

determining GES. The kinds of (negative) pressures and impacts on the marine environment which are best addressed by spatial protection measures can thus be inferred; indeed, area-based protective approaches are suitable for reducing the majority of the listed pressures and impacts.

Attributes of MPA Networks

According to Art. 13(4) MSFD, networks of MPAs should be coherent and representative. Moreover, the diversity of the constituent marine ecosystems must be covered by these networks.

The attribute 'coherent' is not explicitly defined for the purposes of the MSFD, but it is variously used in other contexts within the Directive. For example, a coherent legislative framework is required to achieve the envisaged objectives (recital (9) of the preamble) and the Directive should further enhance coherent contribution of the EU and the Member States with regard to international agreements (recital (16) of the preamble). Moreover, Art. 1(4) MSFD refers to the coherence of policies, agreements and legislative measures. It appears that in these cases coherence is to be understood as a call for common and coordinated political actions. It is therefore possible to derive some idea about the design requirements of a coherent network from the usage of the term elsewhere in the Directive.

Nevertheless, the meaning of coherence and representativity as attributes of networks of MPAs still remains rather unclear. The strategic document including a work programme for 2014 and beyond, within the Common Implementation Strategy (CIS) for the MSFD, by the Member States and the European Commission lists a 'common understanding on coherence and representativeness of MPAs in support of GES' as an activity to be undertaken. One plan is to benefit from work undertaken within the framework of the Convention for the Protection of the Marine Environment of the

North-east Atlantic (OSPAR Convention) on the assessment of coherence of MPAs. The assessment criteria developed by OSPAR can provide indications for the interpretation of the network attributes in Art. 13(4) MSFD. Ardron (2008) acknowledges that the ecological coherence of the OSPAR network can be assessed under the general criteria of adequacy/viability, representativity, replication and connectivity. Based on this scheme, the explicit reference to representativity in the wording of Art. 13(4) MSFD appears only to be a special emphasis, as it is covered anyway by the overarching aim of a coherent network. The four subcriteria have been developed by scientists at a global level and their main characteristics are commonly accepted. Therefore, the OSPAR framework follows a recognized methodological concept. This is also evident from the fact that the Conference of the Parties (COP) of the CBD decided in 2008 to select MPAs on these criteria, although the terms used differ slightly. Thus, COP Decision IX/20 on 'Marine and coastal biodiversity', *inter alia*, lists and defines in Annex II:

Representativity is captured in a network when it consists of areas representing the different biogeographical subdivisions of the global oceans and regional seas that reasonably reflect the full range of ecosystems, including the biotic and habitat diversity of those marine ecosystems.

Connectivity in the design of a network allows for linkages whereby protected sites benefit from larval and/or species exchanges, and functional linkages from other network sites. In a connected network individual sites benefit one another.

Replicated ecological features means that more than one site shall contain examples of a given feature in the given biogeographic area. The term 'features' means 'species, habitats and ecological processes' that naturally occur in the given biogeographic area.

Adequate and viable sites indicate that all sites within a network should have size and protection sufficient to ensure the ecological viability and integrity of the feature(s) for which they were selected.

As the EU and its Member States are all parties to the CBD, these criteria (including their respective definitions) should be taken into account as subcriteria of the term 'coherence' mentioned in Art. 13(4) MSFD. Even so, it remains a peculiarity of Art. 13(4) MSFD that the attribute 'representativity' is given equal weight to coherence and is not subordinated. In practice, however, this does not make any difference with regard to the substantive requirements to be met by MPA networks.

Types of Spatial Protection Measures

The term 'spatial protection measures' is not defined either within Art. 13(4) or in Art. 3 MSFD. However, the former provision lists examples of types of MPAs that fall under the term ('such as'), covering MPAs designated under EU law or within the framework of international or regional agreements.

MPAs Designated under EU Law

Many years before the MSFD entered into force, the 1979 Birds Directive (codified in 2009/147/EC) and the Habitats Directive (92/43/EEC) together obliged Member States to establish special protection regimes for certain areas. These protected areas are designated under the national law of the Member States in accordance with the duties created by the above-mentioned EU Directives. After being reported to the EU, the areas designated under either Directive belong to the Natura 2000 network (Art. 3(1) Habitats Directive). Although not explicitly laid down in the Directives, it cannot be denied that the area of application of both Directives extends to marine areas under the jurisdiction of the Member States. This was most notably the result from the 'Gibraltar Decision' taken by the European Court of Justice (ECJ) in 2005.[12] Therefore, the Natura 2000 network can be extended into the marine area and contribute to the objectives of the MSFD. Following from Art. 13(4) MSFD, the existing marine Natura 2000 sites are incorporated into the programmes of measures.

The spatial protection established by Natura 2000 is limited to certain natural features, and aims to maintain these features in a favourable conservation status within their natural range. Special Protection Areas in accordance with Art. 4(1) of the Birds Directive may only be designated for birds listed in Annex I and for regularly occurring migratory birds not listed in this Annex. Special Areas of Conservation, designated under Art. 3(1) of the Habitats Directive, host natural or semi-natural habitat types listed in Annex I and significant populations of the plant and animal species listed in Annex II.

MPAs Established in the Framework of Global and Regional International Agreements

The second type of MPAs falling under Art. 13(4) MSFD concerns sites which are agreed by the EU or Member States in the framework of 'international or regional agreements'.[13]

The CBD obliges its contracting parties on a global level to achieve targets for the designation of MPAs (Dudley and Hockings, this volume). However, the convention does not introduce a special type of protected area. Instead, the obligation is met by the designation of MPAs under regional international law or the national law of a contracting party.

The term regional (international) agreements in the sense of Art. 13(4) MSFD includes the 'regional sea conventions'

defined in Art. 3(10) MSFD but is more far reaching. Therefore, MPAs in the framework of the RSCs are of particular relevance with respect to spatial protection measures. Article 3(10) MSFD mentions some of the RSCs applicable in the different marine regions of the EU waters (see Art. 4 MSFD). This includes the Convention on the Protection of the Marine Environment of the Baltic Sea (Helsinki Convention), the Convention for the Protection of the Marine Environment of the North-east Atlantic (OSPAR Convention) and the Convention for the Marine Environment and the Coastal Region of the Mediterranean Sea (Barcelona Convention). The Convention on the Protection of the Black Sea Against Pollution (Bucharest Convention) is not included, although the Black Sea is one of the marine regions listed in Art. 4(1) MSFD.

Since 2014, MPAs within the framework of the Helsinki Convention have been established according to HELCOM Recommendation 35/1 'On a System of Coastal and Marine Baltic Sea Protected Areas'.[14] By taking these measures, the contracting parties contribute to their obligation to 'conserve natural habitats and biological diversity and to protect ecological processes' stated in Art. 15 of the Helsinki Convention. Within the system of the OSPAR Convention there exists a similar Recommendation 2003/3 'on a Network of Marine Protected Areas'. In contrast, the Parties of the Barcelona Convention decided to sign a separate agreement which is dedicated to MPAs in the Mediterranean Sea area. The 1982 Protocol Concerning Mediterranean Specially Protected Areas was replaced in 1995 by the Protocol Concerning Specially Protected Areas and Biological Diversity in the Mediterranean (SPA Protocol). The revised version allows the designation of two different types of MPAs: Specially Protected Areas (SPAs) under Art. 5 and Specially Protected Areas of Mediterranean Importance (SPAMIs) under Art. 8. It should be noted that MPAs of the second type may be established in the high seas or may extend over two or more marine areas belonging to different coastal states. The contracting states to the Bucharest Convention adopted the Biodiversity and Landscape Conservation Protocol to the Convention on the Protection of the Black Sea Against Pollution in 2002 which entered into force in 2007.[15] Due to Art. 4(1) lit. (a) of this protocol, each contracting party shall take all necessary measures to 'protect, preserve, improve and manage in a sustainable and environmentally sound way areas of particular biological or landscape value, notably by the establishment of protected areas [...]'. Moreover, the 'Strategic Action Plan for the Environmental Protection and Rehabilitation of the Black Sea' of 2009 states explicitly the need for MPAs.

Besides the RSCs, there exist conventions for the protection of certain species at the regional level. Two examples, concluded within the framework of the Bonn Convention on Migratory Species, are the Agreement on the Conservation of Cetaceans in the Black Sea, Mediterranean Sea and Contiguous Atlantic Area (ACCOBAMS) and the Agreement on the Conservation of Small Cetaceans in the Baltic, North East Atlantic, Irish and North Seas (ASCOBANS). ACCOBAMS constitutes an explicit obligation to conserve whales using the instrument of MPAs. This obligation may be fulfilled by the protection of certain areas under national law or within the framework of an RSC. One example for this is the Pelagos Sanctuary for Mediterranean Marine Mammals. This MPA was initially established by a trilateral agreement between France, Monaco and Italy[16] and was some years later listed as a SPAMI in accordance with the SPA Protocol mentioned above.[17]

Finally, the agreements establishing the regional fisheries management organizations (RFMOs) have to be considered as regional (international) agreements in the sense of Art. 13(4) MSFD. It is possible that fishing closures in certain areas on the basis of these agreements qualify as MPAs, but this depends on an assessment case by case (see below).

Further Spatial Protection Measures

Besides the types of MPAs explicitly mentioned in Art. 13(4) MSFD, there exist different kinds of spatial protection measures which do not necessarily offer all the characteristics of MPAs but nevertheless may contribute to networks of MPAs in accordance with this provision.

The first group is made up of area-based restrictions with regard to fishing activities. The exploitation of stocks of fish, crustaceans and molluscs not only affects the stocks as a component of marine biodiversity, but food webs and the integrity of the seafloor can also be substantially impaired (UNEP/MAP, 2012). Fisheries restricted areas can address these problems as an integral part of MPA networks. The restrictions may be focused on certain fishing methods or gears, either throughout the year or restricted to seasonal periods. The marine regions covered by the MSFD fall within the competence of different RFMOs on the basis of international conventions. Although the name may suggest otherwise, the General Fisheries Commission for the Mediterranean (GFCM) is also responsible for the Black Sea. Another RFMO which covers a larger geographical area, including the Mediterranean and Black Seas, is the International Commission for the Conservation of Atlantic Tunas (ICCAT). The scope of ICCAT is limited to tuna and tuna-like species. This is different in the case of the North East Atlantic Fisheries Commission (NEAFC). But although the 'Regulatory Area' of the NEAFC covers parts of the North Sea, it does not decide over binding management measures within EEZs or territorial seas.[18] Therefore, the area of application does not interfere with one of the MSFD's marine regions. An instructive example of a spatial protection measure regarding the regulation of fisheries is the fisheries restricted area (FRA) as provided for in Art. 8 (a) (iv) of the GFCM Agreement. Its purpose is 'the protection of vulnerable marine ecosystems, including but not limited to nursery and spawning areas'.[19]

Area-based fisheries measures are also an element of the EU's Common Fisheries Policy (CFP). Recital (39) of the MSFD addresses the importance of a 'full closure to fisheries of certain areas, to enable the integrity, structure and functioning of ecosystems to be maintained or restored and, where appropriate, in order to safeguard, inter alia, spawning, nursery and feeding grounds' within the framework of the CFP. Recital (11) of Regulation (EU) No 1380/2013 of the European Parliament and of the Council of 11 December 2013 on the Common Fisheries Policy (the 'Basic Fisheries Regulation', BFR) consequently highlights the contribution of the CFP to the aims of the MSFD, namely achieving GES. The BFR institutes two types of area-based fisheries measures. Firstly, the EU shall, after identification of suitable areas by the Member States, establish fish stock recovery areas under Art. 8 BFR. These areas primarily concern the protection of 'heavy concentrations' of fish below minimum conservation reference size and spawning grounds. Areas suitable to form part of a coherent network of protected areas have to be taken into account. Secondly, Art. 11 BFR empowers the Member States to adopt conservation measures with regard to the regulation of fisheries within waters under their sovereignty or jurisdiction to comply with their obligations resulting from Art. 13(4) MSFD. Appropriate measures within the territorial seas may be taken in accordance with Art. 20 BFR. Moreover, the EU's fisheries legislation provides the EU and the Member States with area-based instruments at the regional level. Council Regulation (EC) No 1967/2006 of 21 December 2006 concerning management measures for the sustainable exploitation of

fishery resources in the Mediterranean Sea (MFR), for example, allows the establishment of 'Community fishing protected areas' (Art. 6 MFR) and 'national fishing protected areas' (Art. 7 MFR).

Another human activity that requires limitation in certain areas is shipping, which is controlled by the International Maritime Organization (IMO) established in 1948.[20] This organization has the competence to establish different kinds of protecting measures, including Particularly Sensitive Sea Areas (PSSAs). These areas are based on IMO Resolution 982 and do not constitute restrictions on maritime traffic. This follows from the definition of a PSSA in IMO Resolution 398 as 'an area that needs special protection through action by IMO because of its significance for recognized ecological, socio-economic, or scientific attributes where such attributes may be vulnerable to damage by international shipping activities'. With respect to PSSAs, other regulations may serve as 'associated protective measures'. These include areas to be avoided (ATBAs) and no anchoring areas (NAAs). ATBAs are expressly provided for in Regulation 8 (a) of Chapter V of the International Convention for the Safety of Life at Sea (SOLAS). The definition for NAAs is laid down in IMO Resolution 572, which defines them as '[a] routeing measure comprising an area within defined limits where anchoring is hazardous or could result in unacceptable damage to the marine environment'. These IMO measures must not be confused with areas regulated in Art. 211(6) UNCLOS which covers special cases where coastal states believe that a clearly defined area within their EEZ needs special protection against pollution from ships, because international rules and standards are inadequate. The IMO is involved in the decision concerning the establishment of such areas.

Furthermore, maritime spatial planning instruments may play an important role as spatial protection measures (Schachtner, this volume). The EU recently introduced a

Maritime Spatial Planning Directive (MSPD, 2014/89/EU). According to Art. 8(1) of this Directive, Member States are obliged to 'set up maritime spatial plans which identify the spatial and temporal distribution of relevant existing and future activities and uses in their marine waters'. In particular, nature and species conservation sites and protected areas represent interests mentioned in Art. 8(2) MSPD that qualify as spatial protection measures within the framework of maritime spatial planning. This is further supported by the fact that the programmes of measures required by Art. 13(1) MSFD, which include those falling under Art. 13(4) MSFD, shall *inter alia* contain spatial and temporal distribution controls as listed in Annex VI No (3) to the MSFD.

Last but not least spatial protection measures are not limited to binding restrictions. These include, for instance, (voluntary) codes of conduct regarding the exercise of certain uses in specific marine areas. Other examples include the application of economic instruments to encourage the use of marine areas in an environmentally sound manner (MSCG, 2014; Ojea *et al.*, this volume). Moreover, Member States or authorized persons may actively eliminate factors which have a negative effect on the environmental status within a marine area, for example by rehabilitating contaminated sites.

Initiation of Spatial Protection Measures

In the foregoing discussion, reference was made to measures that may not be taken by the Member States due to a lack of competence. For these cases, a procedure is laid down in Art. 13(5) MSFD. This provision makes clear that Member States must not remain inactive. If the management of a human activity at European or international level is likely to have a significant impact on the marine environment, the competent

authority or international organization shall be addressed. This obligation refers to spatial protection measures in particular. As far as regional international organizations are concerned, the Member States operate within the framework of their general obligation contained in Art. 6(1) MSFD to make use of existing regional institutional cooperation structures, including those under RSCs. Especially when it comes to measures of regional organizations, Member States have significant influence on decisions over protective measures in their position as contracting parties to the underlying conventions.

Closing Remarks

The huge achievement of the MSFD, including with respect to networks of MPAs, is its integrative approach. Article 13(4) MSFD places new demands for establishing coherent networks of protected areas within European waters that did not exist in EU law before. Thus, MPA networks must contribute to GES taking account of the ecosystem approach, and the spatial protection measures taken shall give due consideration to sustainable development including social and economic impacts. These requirements for the network necessitate the integration of measures which go beyond Natura 2000 in several respects. To begin with, MPAs agreed within the framework of global and regional international organizations have to be added. Moreover, a variety of possible spatial protection measures have to be integrated if necessary, although these may not meet all characteristics of MPAs (however defined). In conjunction with the reformed CFP, the MSFD fosters the interaction of EU fisheries and environmental law, especially if it comes to the protection of certain areas within MPA networks. A number of sectoral regulatory measures are not covered by the EU's (exclusive) competence, namely those with regard to shipping, yet even these have to be applied within the MPA networks through consultation with the responsible body. Although the MSFD is only applicable within EU waters, the GES of marine regions will depend crucially on appropriate measures taken by third parties. Article 6(2) MSFD requires coordination between Member States and third countries that share the same marine region. This includes measures with respect to transboundary MPA networks. The tight timetable leading to the envisaged achievement of GES should thus accelerate their establishment, and not only limited to EU waters.

However, the MSFD does have some intrinsic problems which complicate the implementation of its ambitious goals by 2020. In particular, the application of Art. 13(4) MSFD is hampered because the meanings of essential terms, or their relation to each other, have not been defined in the Directive. This appears astonishing in view of the fact that Art. 3 MSFD contains a long list of detailed definitions for the purpose of the Directive. Sometimes it can be advantageous to refrain from the definition of certain terms since it allows their flexible interpretation with due regard to developments in science and politics during the implementation process. But when it comes to Art. 13(4) MSFD, Member States would benefit from more detailed interpretative guidance *within* the Directive. This issue is especially pertinent for questions such as: To what degree is the term 'spatial protection measure' wider in its scope than the term 'marine protected area'? What are the demands placed on the (natural) coherence of a network of MPAs and how do they relate to the attribute of representativity, and how does this differ from the meaning of the term 'coherence' in the general provisions of the MSFD? The EU must make every effort to foster a common understanding among the Member States concerning the answers to these questions.

Notes

1 See also Recital (4) of the MSFD.
2 See also Recital (3) of the MSFD.
3 See also Recital (15) of the MSFD.
4 First stated by T.T.B. Koh, President of UNCLOS III, available under http://www.un.org/Depts/los/convention_agreements/texts/koh_english.pdf; see also Proelß (2012).
5 The Member States have to pay special attention to their obligation resulting from Art. 13(7) MSFD.
6 A regularly updated list of parties is provided at https://www.cbd.int/information/parties.shtml
7 It is not possible to present all aspects of the discussion here, but reference should be made to the examination by CBD/SBSTTA (2003).
8 Recital (2) of Council Regulation (EC) No 734/2008 of 15 July 2008 on the protection of vulnerable marine ecosystems in the high seas from the adverse impacts of bottom fishing gears reads: 'The absence of a regional fisheries management organisation or arrangement does not exempt States from their obligation under the law of the Sea to adopt with respect to their nationals such measures as may be necessary for the conservation of the living resources of the high seas, including the protection of vulnerable marine ecosystems against the harmful effects of fishing activities.'
9 Decision of the Arbitral Tribunal Constituted under Annex VII of the United Nations Convention on the Law of the Sea of 18 March 2015 in the Matter of the Chagos Marine Protected Area – Republic of Mauritius./. United Kingdom of Great Britain and Northern Ireland –.

10 See the discussion of this decision by Czybulka (2016b).

11 This conflict is discussed in detail by Proelß (2012).

12 ECJ, Judgement of the Court of 20 October 2005 (Commission of the European Communities./. United Kingdom of Great Britain and Northern Ireland), C-6/04.

13 To be precise, the correct formulation would be 'global and regional international agreements', since agreements on a regional level are also 'international'.

14 This Recommendation superseded Recommendation 15/5.

15 The protocol is available at http://www.blacksea-commission.org/_convention-protocols-biodiversity.asp

16 Agreement concerning the Creation of a Marine Mammal Sanctuary in the Mediterranean of 25 November 1999, in force since 21 February 2002 (2176 UNTS 249).

17 A regularly updated list is provided under http://www.racspa.org/sites/default/files/doc_spamis/spamis_2015.pdf

18 See the map provided under http://archive.neafc.org/about/ra.htm

19 A regularly updated list is provided under http://www.fao.org/gfcm/data/map-fisheries-restricted-areas/en/

20 Convention on the International Maritime Organization of 6 March 1948, in force since 17 March 1958 (289 UNTS 3).

References

Ardron, J.A. (2008) The challenge of assessing whether the OSPAR network of marine protected areas is ecologically coherent. *Hydrobiologica*, **606**, 45–53.

Cacaud, P. (2005) Fisheries laws and regulations in the Mediterranean Sea: a comparative study. *GFCM Studies and Reviews*, **75**. 58 pp.

Convention on Biological Diversity (CBD)/SBSTTA (2003) *Marine and Coastal Biodiversity: Review, Further Elaboration and Refinement of the Programme of Work.* Study of the relationship between the Convention on Biological Diversity and the United Nations Convention on the Law of the Sea with regard to the conservation and sustainable use of genetic resources on the deep seabed (decision II/10 of the Conference of the Parties to the Convention on Biological Diversity). SBSTTA meeting, Montreal, 10–14 March 2003. 37 pp.

Czybulka, D. (2016a) Article 194, in *United Nations Convention on the Law of the Sea (UNCLOS)* (ed. A. Proelß). C.H. Beck, München.

Czybulka, D. (2016b) Paradise lost? Die Entscheidung eines Internationalen Schiedsgerichts vom 18.3.2015 in Sachen Chagos Marine Protected Area (Republik Mauritius/Vereinigtes Königreich). *EurUP Zeitschrift für Europäisches Umwelt- und Planungsrecht*, **14** (1), 1–16.

European Commission (2014) *The first phase of implementation of the Marine Strategy Framework Directive (2008/56/EC): The European Commission's assessment and guidance.* Commission Report to the Council and the European Parliament. Brussels. 194 pp.

Marine Strategy Coordination Group (MSCG) (2014) *Marine Strategy Framework Directive (MSFD), Common Implementation Strategy.* Thirteenth Meeting of the Marine Strategy Coordination Group (MSCG), Brussels. 38 pp.

Proelß, A. (2012) The law on the Exclusive Economic Zone in perspective: legal status and resolution of user conflicts revisited, in *Ocean Yearbook 26*. Martinus Nijhoff Publishers, Leiden/Boston. pp. 87–112.

Scovazzi, T. (2011) *Note on the establishment of Marine Protected Areas beyond national jurisdiction or in areas where the limits of national sovereignty or jurisdiction have not yet been defined in the Mediterranean Sea.* UNEP-MAP-RAC/SPA, Tunis. 47 pp.

United Nations Environment Programme Mediterranean Action Plan (UNEP/MAP) (2012) *State of the Mediterranean Marine and Coastal Environment.* UNEP/MAP – Barcelona Convention, Athens.

6

Socioeconomic Impacts of Networks of Marine Protected Areas

Elena Ojea[1,2], Marta Pascual[2,3], David March[4], Isabella Bitetto[5], Paco Melià[6], Margaretha Breil[7], Joachim Claudet[8] and Anil Markandya[2,3]

[1] University of Vigo, Spain
[2] Basque Centre for Climate Change (BC3), Bilbao, Spain
[3] Ikerbasque Foundation, Bilbao, Spain
[4] SOCIB – Balearic Islands Coastal Observing and Forecasting System, Palma, Spain
[5] COISPA Tecnologia & Ricerca, Bari, Italy
[6] Dipartimento di Elettronica, Informazione e Bioingegneria, Politecnico di Milano, Milano, Italy and Consorzio Nazionale Interuniversitario per le Scienze del Mare, Roma, Italy
[7] Fondazione Eni Enrico Mattei (FEEM) and Euro-Mediterranean Center for Climate Change, Venezia, Italy
[8] National Center for Scientific Research, CRIOBE, Perpignan, France

Introduction

Marine ecosystems have been recognized as one of the most important natural resources (Costanza, 1999; Beaumont *et al.*, 2007) as they offer a wide range of ecosystem services (Beaumont *et al.*, 2007; Atkins *et al.*, 2011; Burkhard *et al.*, 2011). This makes their conservation and management highly valuable for human well-being.

Marine Protected Areas (MPAs) can enhance fish size and abundance inside their borders across a variety of species, ecosystems and geographic regions (Roberts *et al.*, 2001; Lester *et al.*, 2009; Gaines *et al.*, 2010; Halpern *et al.*, 2010; Abbot and Haynie, 2012), as well as economic profit (White *et al.*, 2008), with potential positive spillover effects for adjacent fisheries (Russ *et al.*, 2004). Marine Protected Areas can also act as a safeguard against uncertain future environmental conditions. Given that the local and regional magnitude of climate change

impacts is difficult to project, marine reserves can provide an 'insurance factor' that buffers against some of these unknowns. For example, multiple MPAs in networks can spread the risk of impacts (such as catastrophic storms) that are spatially large relative to individual reserves but small relative to the scale of the network (Allison *et al.*, 2003; Game *et al.*, 2008; McLeod *et al.*, 2009; Gaines *et al.*, 2010; Gleason *et al.*, 2010). Protecting portions of stocks inside MPAs can buffer losses from management failure (Gell and Roberts, 2003) as well as provide reference areas for assessing climate impacts (Bohnsack, 1998), thus providing especially valuable insight in data-poor settings on stock fluctuations driven by factors other than fishing (Wilson *et al.*, 2010).

Given the broad range of ecological and socioeconomic impacts of MPAs, and the need to design and manage MPAs in their socioeconomic context, research is imperative in this area. During the EU-funded

Management of Marine Protected Areas: A Network Perspective, First Edition. Edited by Paul D. Goriup.
© 2017 John Wiley & Sons Ltd. Published 2017 by John Wiley & Sons Ltd.

research project on 'Towards coast to coast networks of marine protected areas (from the shore to the high and deep sea), coupled with sea-based wind energy potential' (CoCoNet), a series of virtual and regular workshops were held between researchers and practitioners to discuss the socioeconomic aspects of marine conservation in the Mediterranean and the Black Seas. The goal of these workshops was to provide a platform for experts from different origins and disciplines to debate specific questions concerning the state and the future of marine conservation in Southern European seas.

The first workshop was held online in December 2012, involving a total of 90 participants. The following areas were discussed: (i) the socioeconomic impacts of MPAs; (ii) methodologies for socioeconomic assessment; (iii) drivers of change; and (iv) future MPA networks and policies. Following a period of follow-up research, a second workshop on MPA network management was conducted in Mallorca (Spain) in October 2014. External experts, stakeholders and policy advocates from the project area (Mediterranean and Black Seas) and additional regional seas were also invited. The workshop focused on (i) establishing MPAs and MPA networks; (ii) managing MPAs and MPA networks; and (iii) monitoring MPAs and MPA networks. Finally, a third online workshop was organized in December 2014, with the objective of combining the existing experience from the socioeconomic analysis of MPAs conducted within the CoCoNet project.

This chapter presents the main findings and lessons learned from these series of scientific exchanges, and provides recommendations for the management of MPAs in the Mediterranean and Black Seas. Special emphasis is given to the socioeconomic aspects of MPAs, as it is now widely recognized that MPAs must be designed to address social and economic considerations as well as conservation goals. We present the state of the art concerning the study of the socioeconomic impacts of MPAs, and present tools for MPA socioeconomic assessments; we then introduce case studies gathered in the workshops from the Mediterranean and the Black Seas; summarize the literature and expert discussions by presenting lessons learned; and conclude with some final remarks.

State of the Art

Socioeconomic Impacts of MPAs

The impacts of establishing MPAs have been widely studied from a biological and conservation point of view, but less evidence exists about socioeconomic impacts. Literature has chiefly focused on the impacts of protected areas on activities such as fisheries and tourism, while other potential positive and/or negative impacts have received less attention.

A vast literature exists concerning the effects of MPAs on fisheries. Research shows that fisheries benefit from MPAs as protected eggs, larvae and adult fish spill over into adjacent fishing grounds, benefiting fishermen and their catches (Claudet and Guidetti, 2010). Thus, fishermen see MPAs as positive initiatives and might become involved in their management. For example, fishermen participating in managing a reserve in Torre Guaceto (Brindisi, Italy), where fishing was allowed in part of the reserve, obtained yields which were consistently about double those obtained from fishing grounds outside the reserve.

Apart from benefits to fisheries, however, MPAs also provide many other benefits. In 2009, a study by Lester *et al.* (2009) reviewed reports from 1224 no-take marine reserves in 29 countries and found documented increases in biomass, species richness and population size within the boundaries of the reserves.

Social sciences have contributed a relatively small but steadily growing body of literature that examines the social and economic implications of MPAs (Sanchirico and Wilen, 2002). In Southern European seas, a seminal paper by Badalamenti *et al.* (2000) remains the main source on the socioeconomic impacts of MPAs on the Mediterranean Sea. More recently, Rossetto *et al.* (2013) presented a synthetic review of the empirical evidence of benefits and costs of MPAs, in order to inform the planning of future protected areas. Pascual *et al.* (2016) updated the Badalamenti *et al.* (2000) analysis and expanded it to the Black Sea as well.

Based on a literature review on socioeconomic impacts of MPAs, together with the input from the workshop participants, Table 6.1 provides a comprehensive list of the different potential positive and negative impacts expected from a protected area, from a user perspective, for each socioeconomic activity. The table captures information from the Black and Mediterranean Seas to elucidate the main positive and negative impacts of activities on users.

Instead of looking at the impacts of economic uses on MPAs, the focus is the other way round: we are trying to understand the implications of MPAs for the society and economy of an area. For this purpose, we recommend ranking impacts according to: (i) the importance of the activity in the region; (ii) the importance of the stakeholder groups in the region; (iii) the socioeconomic context; and (iv) the magnitude of the impact. All these factors are of course related. For example, it emerged from the discussions that mineral extraction may be an activity considerably affected by MPA creation in the Black Sea, more so than commercial fisheries which have severely declined in the last few decades. The importance of recreational impacts in Mediterranean MPAs was highlighted, and regional differences are very relevant for understanding the dynamics.

Tools for MPA Socioeconomic Assessments

We now review some of the main methodologies and conceptual frameworks used for MPA research and management which we find useful for assessing the socioeconomic impacts of MPAs.

Social-Ecological Systems (SES) Approach

Novel conceptual frameworks address marine management from a social-ecological perspective. In her pioneering work, Ostrom (e.g. Ostrom, 2009) identifies a set of variables that affect the likelihood of self-organization in efforts to achieve a sustainable SES, such as cooperation in sustainable fisheries management. From this more holistic social-ecological perspective, marine resources are understood as an intertwined system where ecological and socioeconomic factors interact. Recent work has adopted this framework to investigate fisheries where resource system, resource users, resource units, governance, interactions and outcomes from the systems are analysed in order to understand the system's complexity and address management in a more sustainable way. Leslie *et al.* (2015), for example, apply a SES approach to artisanal fisheries in Baja California. For MPAs, Jones *et al.* (2013) rely on this framework to study in detail governance factors for 20 MPAs worldwide.

However, despite the recent growing body of case studies and recommendations on the benefits of adopting a social-ecological framework for resource management, as well as the potential for MPA design and management, such a framework remains very difficult to apply, whether in fisheries management (Kittinger *et al.*, 2013) or more generally. Further research and additional illustrative case studies are needed to explore the benefits of adopting a SES approach in MPA management.

Table 6.1 Comprehensive list of potential socioeconomic impacts of MPAs.

Type of activity	Sub-types of activity	Potential positive impacts on users	Potential negative impacts on users
Fisheries	Artisanal fisheries (small scale)	Improved catch mix Income and employment increase, for professional and pleasure fisheries and for diving Exclusive access (less competence)	Closure of areas to fisheries If retention rates inside the MPA are high (dispersal ability is low compared to MPA size) there might be no benefit for nearby fisheries
	Commercial fisheries (large scale)	Improved catch mix Increased catch ('spillover effect' and also by 'recruitment effect') Income and employment increase, for professional and pleasure fisheries and for diving Increased biomass (reserve effect) Increased fish size (reserve effect)	Closure of areas to fisheries If retention rates inside the MPA are high (dispersal ability is low compared to MPA size) there might be no benefit for nearby fisheries
	Recreational fisheries	Income and employment increase, for professional and pleasure fisheries and for diving	Closure of areas If retention rates inside the MPA are high (dispersal ability is low compared to MPA size) there might be no benefit for nearby fisheries
Aquaculture	Offshore aquaculture (longlines)	Economic benefits of employment and income	Impacts on local ecosystems?
	Offshore fish-farms	Economic benefits of employment and income	Impacts on local ecosystems?
Navigation and communications	Commercial shipping	Not available	Effect on shipping lanes Increased transport time due to reduced speed limits
	Ports and harbour service area	Not available	Negative effects of anchoring on seabed (e.g. seagrass)
	Communication cables	Not available	Limitation of allocation

Mineral, water and energy resources	Offshore oil/gas platforms, resources extraction, pipelines and cables	Not available	Limitation of extraction and allocation
	Offshore wind-farms	Not available	Limitation of allocation
	Sailing	Not available	Damage to ecosystem from tourist congestion (e.g. anchoring)
	Marine cruising	Increase in marine cruises related to cetacean or seabird sightseeing	Negative effects of anchoring on seabed (e.g. seagrass)
	Diving, snorkelling, nautical activities	Increase in diver visits; Income and employment increase, for professional and pleasure fisheries and for diving	Damage to ecosystem from tourist congestion; Negative non-consumptive diver impacts on the natural environment; Closure of areas
	Cetacean sighting cruising, seabird watching	Increase in demand	Negative effects on cetaceans
Management	MPA management	Economic benefits to scientists and biologists (budget for research, projects, etc.)	Economic cost for public finances: of administration, supervision, monitoring, scientific information policies, prohibitions with financial compensation

Source: First table draft from Pascual (2013), adapted to incorporate comments from participants in the CoCoNet workshops.

Ecosystem Services and Economic Valuation

The Millennium Ecosystem Assessment (MEA) uses a conceptual framework for documenting, analysing and understanding the effects of environmental change on ecosystems and human well-being. It views ecosystems through the lens of the services they provide to society, how these services in turn benefit humanity, and how human actions alter ecosystems and the services they provide (Carpenter *et al.*, 2009).

Assessing ecological processes and resources, in terms of the goods and services they provide, translates the complexity of the environment into a series of functions which can be more readily understood, for example by policy-makers and non-scientists (Beaumont *et al.*, 2007). As a consequence, the focus on ecosystem services has been widely adopted among the scientific and policy communities (Carpenter *et al.*, 2009), including those concerned with marine management and MPAs (Roncin *et al.*, 2008; Fletcher *et al.*, 2011).

Building on the ecosystem services framework, The Economics of Ecosystem Services and Biodiversity (TEEB) has recently applied a more mainstream economic approach to ecosystem services (Costanza *et al.*, 2014). TEEB adopts the MEA framework, but adapts it by including ecosystem functions. Ecosystem functions are defined as a subset of the interactions between structure and processes that underpin the capacity of an ecosystem to provide goods and services. The building blocks of ecosystem functions are the interactions between structure and processes, which may be physical (e.g. water infiltration, sediment movement), chemical (e.g. reduction, oxidation) or biological (e.g. photosynthesis, denitrification), and biodiversity is involved in all of them to varying degrees. Ecosystem services are defined in TEEB as the direct and indirect contributions of ecosystems to human well-being. Identifying and separating ecosystem processes and services avoids the risk of double counting benefits. Figure 6.1 shows the links between functions, services and well-being adopted by TEEB (2012).

Based on TEEB and the MEA frameworks, together with recent work on ecosystem services in the Mediterranean (Sardá, 2013), Table 6.2 summarizes methodologies from the economic literature that can be applied

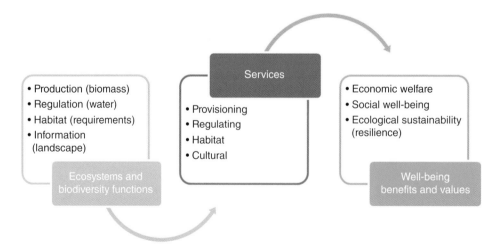

Figure 6.1 Links between functions, services and well-being adopted by The Economics of Ecosystem Services and Biodiversity (TEEB).

Table 6.2 Valuation techniques available for economic valuation of ecosystem services in MPAs.

Value	Function	Ecosystem good or service	Common valuation technique
Use value	*Direct use value:*		
	Provisioning or production services	Production of valuable food and fibre for harvest	NFI, PF, MP
		Pharmaceuticals	NFI, MP
		Raw materials	NFI, MP
	Cultural services	Recreational opportunities	NFI, TC, CV, CE
		Education and scientific knowledge	CV, CE
	Indirect use value:		
	Regulating services	Water quality control	NFI, RC, CV, HP, CE
		Waste treatment	NFI, RC, HP
		Flood control and storm buffering	NFI, RC, AD
		Biological regulation	CE, CV, PF
		Human disease control	NFI
	Supporting services	Climate regulation	RC
		Nutrient cycling	RC
	Option value:		
	Option value	Future benefit for direct and indirect uses	CV, CE
Non-use value	Existence value	Intrinsic value of species, habitat, biodiversity	CV, CE

NFI, Net Factor Income; PF, Production Function; MP, Market Price; TC, Travel Cost; CV, Contingent Valuation; CE, Choice Experiments; RC, Replacement Cost; HP, Hedonic Pricing; AD, Avoided Damage.

under an ecosystem services framework to study the socioeconomic impacts of MPAs and MPA networks. Also included is the type of values the methods can measure. The methods are outlined below:

Net Factor Income (NFI): this method estimates the value of ecosystem services as an input in the production of a marketed good. That is, NFI estimates the value of an ecosystem input as the total surplus between revenues and the costs of other inputs in production. For example, the value of a coral reef in supporting reef-based diving recreation should be calculated as the revenue received from selling diving trips to the reef, minus the labour, equipment and other costs of providing the service (Van Beukering *et al.*, 2007).

Production Function (PF): this method estimates the value of a non-market ecosystem product or service by assessing its contribution as an input into the production process or a commercially marketed good. It is different from the NFI method in that it estimates a functional relationship between inputs and outputs. A PF describes the relationship between inputs and outputs in production. This method could be useful when considering aquaculture, for example.

Market Prices (MP): these methodologies use market prices to estimate marginal economic values. This is feasible for those ecosystem goods and services that have a price in existing markets, such as seafood, fish or commercial algae, or revenues from outdoor recreational demand. A major disadvantage of the method is that many environmental goods and services are not traded directly in well-functioning markets and readily observable prices are not available. Additionally, if markets exist but are highly distorted, the available price information will not reflect true social and economic values.

Travel Cost (TC): this method is based on actual consumer or producer behaviour and preferences and values are 'revealed' in complementary or surrogate markets. It employs existing market data to derive the indirect value of nature. An example would be assessing the expenses incurred in visiting an MPA, as an estimate of how much the experience is worth.

Contingent Valuation (CV) and Choice Experiments (CE): these methodologies are used for those services that are not traded in a regular market and therefore have no market price. For such goods and services, usually the individual willingness to pay for a change in the level of provision of the service is estimated. These are the only methods capable of deriving economic values for highly valued species or cultural ecosystem values not related to direct use. This can be done through conducting surveys to collect data about individual preferences in relation to an environmental good. While the CV method asks for willingness to pay for specific changes in environmental quality, the CE method asks respondents to rank attributes of the ecosystem service or to choose among alternative scenarios.

Replacement Cost (RC): this method estimates the value of ecosystem services as the cost of replacing them with alternative goods and services. For example, the value of a wetland that acts as a natural reservoir can be estimated as the cost of constructing and operating an artificial reservoir of a similar capacity.

Hedonic Pricing (HP): this method employs existing market data to derive the indirect value of nature, for example by using property values, on the assumption that the price of a property will indirectly reflect any environmental benefits the property enjoys from an ecosystem service.

Avoided Damage (AD): this method uses the cost of actions taken to avoid damage to the system as a measure of the benefits provided by the ecosystem. For example, if a coastal wetland provides protection from inland flooding, the value of the protection afforded may be estimated by the damage to their properties avoided by local residents and government.

Ecosystem-Based Management

The rapid increase in the size and number of MPAs has been accompanied by a similar increase in implementation of marine ecosystem-based management (EBM) measures. In fisheries, for example, EBM focuses on controlling bycatch, protecting critical habitats, and recognizing predator–prey and other ecological relations, within the framework of traditional population-specific fisheries management (McCay and Jones, 2011). However, although fisheries managers may close some areas to fishing either permanently or temporarily, MPAs are still poorly integrated into ecosystem-based fisheries management (Halpern *et al.*, 2010).

One reason for resistance to MPAs as a central component of an ecosystem-based fishery may be that they are a relatively new approach, whereas species-specific fisheries management has a long, if not always successful, history. Moreover, decisions about size, site selection, and disturbance levels within MPAs are technically difficult, particularly given the relatively high degree of variability and complexity in marine ecosystems (McCay and Jones, 2011). For a detailed review and discussion on EBM see Sardá *et al.* (this volume).

Marine Spatial Planning

Marine Spatial Planning (MSP) allows the creation and establishment of a more rational organization of the use of marine space and the interactions between its uses, to balance demands for development with the need to protect the environment, and to achieve social and economic objectives in an open and transparent way (Douvere, 2008; Schachtner, this volume). Marine Spatial Planning operates at multiple spatial and temporal scales, by considering the three-dimensional nature of the sea, and addressing static and dynamic maritime activities from local to regional scales (Gilbert *et al.*, 2015). Designation of MPAs is an integral part of MSP and the achievement of ecosystem-based management (Crowder and Norse, 2008). Therefore, when establishing MPAs it is important to know how the spatial regulation of human activities within MPAs will affect marine users (Cárcamo *et al.*, 2014). In MPAs, marine uses may be subject to stringent conditions or even totally excluded depending on the location and type of the MPA. The specific location of the MPA determines how marine uses are positively or negatively impacted.

Marine Spatial Planning allows comprehensive analyses of MPAs and MPA networks, which are spatially explicit. Through MSP, it is possible to identify and quantify human activities surrounding an MPA network, to assess the compatibilities among activities and their environmental impacts. In fact, MPA design and consideration in a marine system is inherent to MSP. For MPA and MPA networks, MSP constitutes a framework that can be applied at multiple scales. For example, marine spatial plans can be conducted at local or regional level (i.e. trans-national), and MSP is expected to have much potential after the implementation of the EU Maritime Spatial Planning Directive (2014/89/EU).

DPSIR and DPSWR Environmental Indicator Frameworks

The DPSIR (Drivers–Pressures–State–Impact–Response) environmental indicator framework is a systems-based approach which captures key relationships between society and the environment (Lewison *et al.*, 2016), and is regarded as a philosophy for structuring and communicating policy-relevant research about the environment, for example by the European Environment Agency. Recent work on the DPSIR model has improved the framework to incorporate the welfare component of environmental factors, developed under the KNOWSEAS FP7 project (http://www.msfd.eu/). This improvement involves replacing 'Impact' by 'Welfare' in what is known as the DPSWR framework. In this new framework 'Welfare' is measures of changes (the 'costs') to human welfare as a result of State changes, and it thus provides a conceptual model that is a useful starting point for analysing coupled social and ecological systems (Cooper, 2013).

MPAs in the Mediterranean and Black Seas

We now present the evidence on the socioeconomic impacts of MPA networks gathered during the expert workshops for the Mediterranean and Black Seas. The majority of the works cited are ongoing research documents that participants to the workshop were engaged in, and therefore some are not yet published. Some of the working documents were originally in different languages from the Mediterranean basin and the Black

Sea area, and the workshops allowed beneficial exchange of knowledge about work going on in these areas.

Mediterranean Sea

A literature review was conducted during the workshops to collect case studies relating to MPAs in the Mediterranean Sea, and a total of 15 case studies were shared among participants. Most assessed the effects of MPAs on artisanal and commercial fisheries, with a few looking at recreational impacts. There was a clear dearth of studies of impacts on other ecosystem services that affect society, such as regulation services (e.g. climate, storm protection, salinization, carbon sinks). The studies are briefly described below.

---

1 Economic valuation of five marine and coastal protected areas in the Mediterranean (Plan Bleu, 2012)

Keywords: cost–benefit analysis, Mediterranean, ecosystem services, tourism, net present value

Summary: The study focuses on the valuation of costs and benefits for Mediterranean MPAs linked to ecosystem services, including professional and non-professional fishing, tourism, boating, diving, and carbon capture. The costs comprise the management body budget and the economic activities within, or related to, the MPAs. The case study areas are: Cap de Creus National Park (Spain), Kuriat Islands (Tunisia), Kaş Kekova (Turkey), Zakynthos National Park (Greece) and Mount Chenoua (Algeria). The analysis employs three scenarios of the potential evolution of MPA management: (i) more protection; (ii) less protection; and (iii) no change in management. As a general result they find that tourism accounts for 90% of the benefits of the MPAs. The balance between tourism and fishing seems to be the key to MPA acceptance. The net present value is highest for scenario (i) (increasing protection). However, the lack of information in some of the areas limited a wider analysis in the Mediterranean Sea, including additional locations. Also, the study produces estimates of costs and benefits from existing MPAs, but it was not possible to isolate the benefits of establishing additional MPAs.

---

2 Effects of habitat on spillover from marine protected areas into artisanal fisheries (Forcada *et al.*, 2009)

Keywords: MPA, artisanal fisheries, habitat connectivity, spillover, Mediterranean Sea

Summary: This is a case study on the effects of MPAs in artisanal fisheries in three marine reserves in the Mediterranean: Tabarca Marine Reserve (Spain), Carry-le-Rouet Marine Reserve (France) and Cerbère-Banyuls Marine Reserve (France). It finds that the spillover effect is localized to specific sectors and that MPAs provide benefits to artisanal fisheries in this case. The authors conclude that spillover effects are not a universal consequence of siting MPAs in temperate waters and depend on the distribution of habitats inside and around the protected spaces.

3 A review of marine protected areas in the north-western Mediterranean region: Siting, usage, zonation and management (Francour *et al.*, 2001)

Keywords: fishing, spear-fishing, MPA impacts, management, enforcement

Summary: This paper reviews MPAs in the north-western Mediterranean. It finds that semi-protected areas where professional fishing is still allowed clearly demonstrate the negative impact of spear-fishing, and the limited impacts from regulated professional activities in fish assemblages. The authors also conclude that the most important factor underlying whether or not an MPA is successful and beneficial is the presence of dedicated staff.

4 Marine protected areas in the Mediterranean Sea: Objectives, effectiveness and monitoring (Fraschetti *et al.*, 2002)

Keywords: effectiveness, research, reserve effect, environmental impacts

Summary: The authors of this study argue that in the Mediterranean Sea the lack of appropriate sampling designs and a proper set of experimental procedures prevent any scientific demonstration of MPA effectiveness. This lack of suitable data may be a result of several factors: field investigations of sub-tidal marine reserves are generally confounded by intrinsic ecological differences between the sites investigated, both inside and outside reserves; site and reserve replication is absent; or no information about the biota was collected before the reserve was established. As a result, the authors recommend the use of experimental procedures widely used for detecting environmental impacts.

5 Designing a network of marine reserves in the Mediterranean Sea with limited socio-economic data (Giakoumi *et al.*, 2011)

Keywords: MSP, Mediterranean, economic costs, MPAs, Natura 2000

Summary: This study identified priority areas for MPAs using spatial prioritization software in the eastern Mediterranean Sea, using different types of available data from visual census surveys (fish species abundance, presence of various habitat types, and per cent coverage of seagrasses and canopy algae). This approach can also be applied even if spatially explicit information is limited, through socioeconomic cost indices taking into account fisheries (including information on the location of ports and areas often inaccessible to fishermen due to high wind exposure) and tourism (on the basis of availability of beds for tourists). The paper examined how the spatial priorities for marine reserves varied using different combinations of these socioeconomic cost metrics, and compared the model outcomes with two non-systematic methods, the Natura 2000 proposed marine reserves and sites that local fishermen proposed for protection. In fact, only a few sites identified in the paper coincided with those recommended as part of Natura 2000 or the fishermen's proposals. This suggests that much more work is needed to harmonize the proposals in the paper with the principles of efficient systematic conservation planning.

6 Spillover from six western Mediterranean marine protected areas: Evidence from artisanal fisheries (Goñi et al., 2008)

Keywords: MPAs, spillover effect, artisanal fisheries, catch analysis

Summary: This study investigated the spillover (or biomass export) around six MPAs in the western Mediterranean based on catch and effort data from artisanal fisheries. The selected MPAs were Cerbère-Banyuls and Carry-le-Rouet in France, and Medes, Cabrera, Tabarca and Cabo de Palos in Spain. The authors found evidence of effort concentration and high fish production near closed areas for all fishing gear analysed. The authors concluded that coastal MPAs can be an effective management tool for artisanal fisheries in the region and that this could be extended to the rest of the western Mediterranean, as the fishing gear studied in this region were typical of the entire basin.

7 Potential of marine reserves to cause community-wide changes beyond their boundaries (Guidetti, 2007)

Keywords: spillover effects, fisheries ecology, economic impacts, Torre Guaceto

Summary: This study looked at the impact of marine reserves on fish ecology and their socioeconomic implications. The case study concerned the Torre Guaceto Marine Reserve (Italy). Results suggested that no-take marine reserves can promote community-wide changes beyond their boundaries. The effects on fishing communities may impact the earnings from fishing as there were shifts of target species and sizes, as well as other factors.

8 Mediterranean marine protected areas: Some prominent traits and promising trends (Harmelin, 2000)

Keywords: artisanal fisheries, Mediterranean, gear regulations

Summary: Small-scale artisanal fishing by trammel nets could persist at moderate level without affecting the spectacular replenishment of fish populations in shallow rocky areas when other fishing methods such as trawling and spear-fishing were controlled or banned. This result has a particular social and cultural interest in the Mediterranean context, considering the slow decline of this traditional fishery. The paper argues for a more active integration of professional fishermen in the preparation of new MPA projects.

9 Gradients of abundance and biomass across reserve boundaries in six Mediterranean marine protected areas: Evidence of fish spillover? (Harmelin-Vivien et al., 2008)

Keywords: spillover effect, fish ecology, MPA impacts, ecological impacts

Summary: Six Mediterranean MPAs were analysed in terms of their impact on fish biomass and abundance. The authors found fish spillover from reserves which was beneficial to local fisheries. This spillover effect occurred mostly at a small spatial scale (hundreds of metres). The existence of regular patterns of negative fish biomass gradients from within MPAs to fished areas was consistent with the hypothesis of processes of adult fish biomass spillover from marine reserves, and could be considered as a general pattern in this Mediterranean region.

10 Biological and socioeconomic implications of recreational boat fishing for the management of fishery resources in the marine reserve of Cap de Creus (NW Mediterranean) (Lloret *et al.*, 2008)

Keywords: tourism, MPAs, management, recreational value, recreational fisheries, angling

Summary: This study looked at recreational fisheries in the marine reserve of Cap de Creus (Spain). It found that recreational fisheries had a large effect on the local economy since the majority of fishermen were visitors on holiday in one of the villages belonging to the park, where most of the expenditure related to angling activities was made.

11 The impact of human recreational activities in marine protected areas: What lessons should be learnt in the Mediterranean Sea? (Milazzo *et al.*, 2002)

Keywords: recreation, MPAs, monitoring, tourism

Summary: The paper reviewed the worldwide impacts of recreational activities on marine communities in MPAs and highlighted the gaps in the relevant available literature. These gaps should be filled in order to facilitate research, monitoring and management of MPAs in the Mediterranean Sea. The study analysed the different recreational activities in MPAs that, when intensive, could modify marine communities at a local scale. More effort should be put into understanding the impact of 'marine-based' activities by assessing the habitats that most attract tourists, quantifying the cause–effect relationship between the biological impact and the amount of recreational activity in the MPAs, and, whenever possible, predicting the future impact of recreational activities on spatial and temporal scales to assist the MPA management process.

12 Integrating conservation and development at the National Marine Park of Alonissos, Northern Sporades, Greece (Oikonomou and Dikou, 2008)

Keywords: Greek MPA, preferences, costs and benefits, stakeholder analysis

Summary: The paper analysed the degree of acceptance of the MPA by local stakeholders through time, after its establishment 13 years earlier. The authors used questionnaires to collect stakeholders' views. They found that different groups had different perceptions of the MPA: for example, fishermen perceived costs due to restrictions while recreational companies reported benefits. The study illustrated the need for stakeholder analysis in order to understand perceptions and heterogeneity in the actors involved with and/or affected by an MPA.

13 Uses of ecosystem services provided by MPAs: How much do they impact the local economy? A southern Europe perspective (Roncin *et al.*, 2008)

Keywords: ecosystem services, socioeconomic impacts, recreation, stakeholder analysis

Summary: The paper reviewed 12 case studies in the Mediterranean looking at the main socioeconomic impacts of MPAs. An assessment was carried out, including stakeholder interviews. A variety of situations were identified in the different MPAs, from MPAs where commercial fishing was the major economic stake, to MPAs where recreational activities had a dominant economic role. The second situation was more typical. However, due to the lack of baseline data, the question of distinguishing the 'reserve effect' from the 'site effect' could only be addressed with the help of survey results concerning perceptions and attitudes of users.

---

14 Long-term and spillover effects of a marine protected area on an exploited fish community (Stobart *et al.*, 2009)

Keywords: spillover effect, MPAs, fishing, benefits, economic impact

Summary: The study analysed the spillover effect for artisanal fisheries in the Columbretes

Islands Marine Reserve (Spain). It concluded that the reserve establishment had had a positive effect on the exploitable fish community and that there was evidence of biomass export to the surrounding fishery.

---

15 Perspectives of economic effects of fisheries exclusion zones: A Sicilian case study (Whitmarsh *et al.*, 2002)

Keywords: marine reserves, fishery reserve, trawl, artisanal fisheries, spillover effects, impacts

Summary: The paper reported the results of a European project investigating the effects of a trawl ban introduced in the Gulf of Castellammare, north-west Sicily, in 1990. The results indicated that the prohibition on

trawling led to stock recovery and improved financial returns for the artisanal fishermen who had been permitted to operate within the restricted area. However, there was evidence that the displacement of trawlers to the outer periphery of the exclusion zone had impacted adversely on artisanal operators located immediately outside the trawl ban area.

Black Sea

A total of 15 case study reports and documents were obtained concerning the Black Sea. While for the Mediterranean Sea, studies on MPA impacts are numerous and cover many different areas of research, for the Black Sea scientific publications are scarce and more information can be found in the grey literature. Four of the documents gathered consisted of general background about the current state of fisheries, biodiversity, environment and transboundary diagnostic analysis of pollution in the Black Sea. The remaining 11 contained various levels of information on MPAs and their socioeconomic impacts. From these, three reports were representative of case study areas and are briefly described here.

---

1 Danube Delta, Romania and Ukraine

The case study of the Danube Delta, an area located at the boundary of Romania and Ukraine, was analysed in some detail. It was a good example of the geopolitical context problems that arise in some MPAs. The paper on boundaries and margins in the Danube Delta (Van Assche *et al.*, 2008) and the decision of the International Court of Justice on the delimitation of the maritime boundary between the two countries in the Black Sea (Zmeiny Island, ICJ Order 2009) detailed these transboundary problems, while the paper on transformations of knowledge/power and governance of the Danube Delta (Van Assche *et al.*, 2011) considered the potential for citizen participation in environmental governance as a possible means for solving these issues in transboundary areas.

2 Vama Veche, Romania

Some participants of the first workshop provided information on the protection and management of MPAs in Romania. They stated that expanding the European ecological network (Natura 2000) in Romania could lead to conflicts between the marine sites and fishery interests, especially at the Vama Veche – 2 Mai Reserve. In order to solve this conflict, they considered that measures should be taken including: (i) the legal control of demersal fisheries in the Romanian coast; (ii) protection of high economic value fish species by taking strong measures to stop illegal fishing and prohibit fishing at certain times of year; (iii) special protection of spawning grounds; (iv) development of fishing regulations; and (v) education/training of fishermen in the proper recording, handling and release techniques for dolphins accidentally caught in fishing gear. In fact, all these measures exacerbated the situation with local fishermen and the situation was resolved not by consensus but by application of law enforcement.

3 Karkinitsky Bay, Ukraine

A case study of Karkinitsky Bay off north-west Crimea (the largest bay in the Black Sea) concerned the socioeconomic impacts of protecting an area for the recovery of the red alga *Phyllophora crispa*. This alga was once harvested for agar and was an important nursery area for fish, both resources having declined since the 1970s. However, new protection measures to restore these resources have potential impacts for navigation as well as gas and mineral extraction which now take place in the bay. Accordingly, the boundaries of the MPA declared in November 2011 had to be drawn to avoid conflicts with these economic activities instead of following the ideal scientific extent (as would be required under EU legislation).

Lessons Learned for MPAs

Participants in the workshops shared their experiences on the implementation of MPAs and MPA networks in the Mediterranean and Black Seas. A main concern shared by all participants – and one that is also evident in the literature – is the level of effectiveness of the MPAs in these regions. This perception revolved around five main issues: (i) the mismatch between regulations and actual implementation and management performance; (ii) the protection level set; (iii) the simplicity of naturalistic approaches as opposed to socioeconomic-ecosystem (network) approaches; (iv) the importance of stakeholder involvement in governance and management from the early design of MPAs; and (v) the lack of resources (including political will) needed to reduce human pressures.

Mismatch between Regulations and Actual Implementation and Management Performance

The role of an MPA is universally recognized and therefore non-negotiable in its essence. However, participants stated that the implementation of the mechanisms and operations of MPAs is difficult; that there are not enough data; and that our knowledge is limited. Participants agreed with the ideas developed by Colloca *et al.* (2015) on no-take zones for nurseries, and the different effects MPAs can have depending on the way they are designed and managed.

A study by Mabile (2007) was proposed to help understand the implications of designing an MPA system in the context of decentralization, with examples from Italy and Spain. It showed that the legislative intervention for the creation of MPAs is a weak procedure which does not facilitate the necessary responsiveness or permit the rapid creation of new sites. This study also highlighted a second aspect: that MPAs are usually limited essentially to a naturalistic approach, which does not favour the acknowledgement of MPAs as a tool for local people, who also usually have no right of participation.

In most Black Sea countries today there exist many conflicts between national legislation, international commitments affecting MPAs, and decisions made about resources that could be exploited in the protected areas. In Ukraine, for example, following the state's nature protection legislation (Law on the Nature Protected Fund 1992), different levels of various activities, including the extraction and use of mineral and biological resources, were allowed in the Zernov's Phyllophora Field (in the central part of the north-western shelf of the Black Sea) and the Small Phyllophora Field (Karkinitsky Bay). However, a 'real' defence of MPAs in Ukraine only began after the introduction of the National Natural Park designation and the establishment of the Institute for Protection of MPAs. After this, in order to promote the formation of a transboundary networks of MPAs, it became necessary to strengthen the protected status of sites across the whole of Ukraine.

Regarding the law, participants believed that analysis cannot be limited to the legal norms alone, as legal standards are worthless if the administrative machinery for their implementation is not put in place. The effectiveness of laws and regulations should be measured in a 'public policies evaluation'.

Especially in developing countries, many laws have only been adopted following international pressure (and EU pressure – for example in technical assistance programmes before the integration of eastern countries, and now under the European Neighbourhood and Partnership Instrument); and even after being adopted they have so far had little application. The evaluation of governance effectiveness is an essential aspect of neo-institutional and social science research, but unfortunately we have very few data on governance effectiveness in the case of strengthening environmental laws at national and international level; on different management plans and best practice; and on the development of optimal action plans.

Protection Level

There are still many questions about the different levels and types of protection. It is not clear that the highest category of protection (IUCN Category I, strict nature reserve; Dudley and Hockings, this volume) could guarantee the conservation of biota and habitat diversity in MPAs. Furthermore, national 'Red Data Books' usually comprise just a list of endangered species and their basic biology and status; they seldom provide recommendations for conservation, or for recovery of species and their habitats.

For both artisanal and recreational fisheries there is literature regarding the potential of 'partial MPAs'. These can have some positive aspects, both economic (e.g. reduction of surveillance costs) and social (e.g. fishermen are allowed to fish on some days).

One of the difficulties noted for designing offshore marine reserves with higher protection levels is the cost of surveillance. Widespread use of electronic monitoring, such as the Automatic Identification System or Global Fishing Watch (http://globalfish ingwatch.org/), can contribute to reducing the costs of surveillance.

Naturalistic Approaches versus Socioeconomic-Ecosystem (Network) Approaches

The creation of MPAs in the Mediterranean and Black Seas invariably focuses on narrow biological aspects (e.g. presence of legally protected species, Red List species, attractive underwater seascapes or important resource species). However, for networking MPAs, the focus should be on higher biological community levels: this way MPA networks can protect the functions of ecosystems and not just single species (Boero, this volume).

Furthermore, participants generally agreed that to be effective, there should be legal, socioeconomic and functional MPA typologies, rather than typologies based only on biological criteria (Beal *et al.*, this volume). On the other hand, the EU Marine Strategy Framework Directive (2008/56/EC) encourages reaching Good Environmental Status through maintaining biodiversity and does not directly address livelihoods (Braun, this volume).

The inclusion of both natural and anthropogenic aspects is believed to be the most cost-effective way of addressing the socioeconomic impacts that MPAs and MPA networks might create. Ways of achieving this goal include stakeholder participation and methodologies such as multi-criteria analysis (Melià, this volume).

Stakeholder Involvement in Governance and Management from the Early Design of MPAs

A decentralized management model for MPAs is an important aspect of MPA effectiveness. However, cost comparisons should be based on MPAs with similar functions (e.g. no-take sanctuaries, regulating fisheries, recreational MPAs, MPAs with a large pelagic area of scientific importance). For instance, let us compare two examples from the French Mediterranean, namely the marine reserve of Banyuls (close to the Spanish border) and the Côte Bleue fisheries reserve (west of Marseille):

Banyuls-sur-Mer is a public institution area of 600 ha, of which 60 ha are no-take (full reserve). Management costs are estimated at €600 000 per year. It attracts a large amount of tourist activity related to diving and an underwater trail. The bulk of the expenses are monitoring, and it provides the data for a public biological laboratory (the costs of which are not included in the management costs given above).

Côte Bleue is a fishery reserve managed by a small fishermen's organization based on a traditional decentralized model: the *Prud'homies de pêcheurs*. It extends over 10 000 ha with 30 ha of no-take. This reserve was first established to protect the area against fishing trawlers coming from Marseille. The annual monitoring costs are estimated at €150 000, with the monitoring performed by professional fishermen (although they have difficulty with tracking navigation and recording recreational fishing).

These two cases are interesting because: (i) the functions are different – recreation and scientific purposes on the one hand, and responding to fisheries management and protection against larger scale fishing on the other; (ii) the legal framework for management is different: Banyuls has a bureaucratic, scientific and 'fonctionnarisée' administration by the district, while the Côte Bleue is a decentralized, empirical community; and (iii) the cost/area ratio is very high in Banyuls and low in the Côte Bleue. To be effective, therefore, we should have legal, socioeconomic and functional MPA typologies in addition to biological criteria.

A participatory process is needed for the establishment or extension of some MPAs, because without involving interest groups or specific users and local decision-making, it is likely these small economic structures will disappear.

Researchers with experience as custodians of marine reserves were aware that is very important to strengthen the legal framework. However, they considered that it was just as necessary to involve the stakeholders in the process of the management – to have participatory management. It is essential to take a structured approach that fully involves and engages the key (or primary) stakeholders (i.e. those whose livelihoods directly depend on the area, have ownership of it, or who have a statutory role in managing it). The preparation of a management plan is a good way of doing this, bearing in mind that the process of preparation is as important as the final result. How the management plan finally resolves the conflicts and is implemented depends on the legislation, political will, finance, and scientific and management expertise available.

Lack of Resources (Including Political Will) Needed to Reduce Human Pressures

Marine and coastal biodiversity is under increasing stress from intense human pressures, including rapid coastal population growth and development, over-exploitation of commercial and recreational resources, loss of habitat, and land-based sources of pollution. Marine Protected Areas are probably not the best instrument to address the impact of pollution and perturbations; other policies and institutions – such as urban policies, integrated coastal zone management, industrial policies, and investment in environmental protection measures – are better suited to protect the sea from these. However, the management level at which these policies and instruments are decided might not be sufficient for tackling these problems.

Apart from anthropogenic pressures, MPAs are also subject to the influence of natural environmental factors, making it difficult to separate the influence of environmental and anthropogenic factors when determining the source of effects on an MPA. A good example is the shallow-water Black Sea shelf in Ukraine, where two MPAs exist (the Zernov Phyllophora and the Small Phyllophora fields). These areas are under huge anthropogenic pressures (including freshwater inflow from coastal rice-irrigation schemes, sand and gas extraction, shipping, tourism, fisheries and military activities), whilst also being subject to natural geomorphological processes (such as huge sediment inputs from the Danube, Dniester and Dnepr rivers) that significantly influence benthic and pelagic communities, as well as building new areas of habitat.

Thus, improved research and monitoring techniques, as well as ex-ante analysis, are needed to gain a better understanding of the true scale of human impacts and damage to MPA ecosystems in order to argue for the resources needed to address them.

It has been stated that problems related to MPAs can be solved through targeted legislative instruments that must be strictly applied in protected areas. However, in Romania, for example, there is considerable nature protection legislation but it can easily be ignored, especially due to lack of involvement of local authorities. Furthermore, while a management plan is essential for an MPA, financial resources are also very important to put the conservation measures into practice. We should stimulate the decision processes and decision-makers in order to find those resources.

Concluding Remarks

We have provided an overview of recent trends in socioeconomic research on impacts of MPAs in the Mediterranean and Black Seas. We have collated and presented information provided by expert participants to a series of workshops in the EU

CoCoNet project, together with a review of published literature and unpublished documents provided by the participants. From the discussions in these forums and careful analysis of the materials exchanged, we have distilled some key messages and lessons learned for future MPA management. The main message is to consider the socioeconomic dimension of MPA creation and management in the areas concerned. These impacts will vary in magnitude and effect depending on the area and socioeconomic activity involved, as well as on the MPA purpose(s) and design. We have illustrated how different conceptual frameworks, such as ecosystem services or the social-ecological systems framework, can help to elucidate the complex relationships between the ecological and the social systems. We have also provided a review of the state of the art of current approaches to MPA management, including Marine Spatial Planning, stakeholder analysis, ecosystem-based management, and the DPSIR environmental indicator framework. We have summarized evidence arising from case studies of MPAs in the Mediterranean and Black Seas that resulted from the exchange of materials during the workshops, as a way of illustrating success stories. Finally, we provided a discussion on the main requisites for successful MPA management in these regions.

Acknowledgements

We acknowledge all workshop reviewers and participants of the two virtual workshops on the socioeconomic dimensions of MPAs, including: Lauretta Burke (WRI); Salman Hussain (SRUC); Pino Lembo (COISPA); Valerian Melikidze (TBILISI); Paul Goriup, Stephen Beal and Melanie Gammon (NatureBureau); Marisa Rossetto and Giulio A. De Leo (CONISMA); François Feral and Laurence Marill (CNRS); Kakha Bilashvili (TSU); Zaharia Magda (INCDN); George Gogoberidze and Julia Levnova (RHU); Areti Kontogianni and Aleksandar Shivarov (AEGEAN); Tatiana Pankeeva, Ekaterina Karishina, Olga Shakhmatova and Nataliya Milchakova (IBSS); Maud-Anaïs Claudot and Julien Le-Telier (PLAN BLEAU); Lilija Bondareva (NASU); Galina Minicheva and Evgeny Solokov (OBIBSS); Tatiana Begun (GEOECOMAR); Tania Zaharia, Victor Nita, Magda Menciu and Mariana Golumbeanu (INCDM); Levent Bat (SNU-FF); Enrique McPherson and Rafael Sardá (CSIC); Ferdinando Boero (UniSalento); Sylvaine Giakoumi and Mairi Maniopoulou (HCMR); Eva Schachtner (Rostock); Violeta Zuna (UNDP); Salit Kark (CEED); Nicola Beaumont (PML); Victor Karamushka, Mihail Son and Maialen Garmenda (BC3); and others who contributed with comments, revision and organization. Marta Pascual's Postdoctoral research grant was supported by the Basque Government and Ikerbasque Foundation. David March's Postdoctoral research grant was supported by the 'Juan de la Cierva' Programme from the Spanish Ministry of Economy and Competitiveness.

References

Abbot, J.K. and Haynie, A.C. (2012) What are we protecting? Fisher behavior and the unintended consequences of spatial closures as a fishery management tool. *Ecological Applications*, **22** (3), 762–777.

Allison, G.W., Gaines, S.D., Lubchenco, J. and Possingham, H. (2003) Ensuring persistence of marine reserves: catastrophes require adopting an insurance factor. *Ecological Applications*, **13** (1), 8–24.

Atkins, J.P., Burdon, D., Elliott, M. and Gregory, A.J. (2011) Management of the marine environment: integrating ecosystem services and societal benefits with the DPSIR framework in a systems approach. *Marine Pollution Bulletin*, **62**, 215–226.

Badalamenti, F., Sánchez Lizaso, J., Mas, J. *et al.* (2000) Cultural and socioeconomic effects of marine reserves in the Mediterranean. *Environmental Conservation*, **27**, 110–125.

Beaumont, N.J., Austen, M.C., Atkins, J.C. *et al.* (2007) Identification, definition and quantification of goods and services provided by marine biodiversity: implications for the ecosystem approach. *Marine Pollution Bulletin*, **54**, 253–265.

Bohnsack, J.A. (1998) Application of marine reserves to reef fisheries management. *Australian Journal of Ecology*, **23**, 298–304.

Burkhard, B., Kroll, F., Nedkov, S. and Müller, F. (2011) Mapping ecosystem service supply, demand and budgets. *Ecological Indicators*, **21**, 17–29.

Cárcamo, P.F., Garay-Flühmann, R., Squeo, F.A. and Gaymer, C.F. (2014) Using stakeholders' perspective of ecosystem services and biodiversity features to plan a marine protected area. *Environmental Science & Policy*, **40**, 116–131.

Carpenter, S.R., Mooney, H.A., Agard, J. *et al.* (2009) Science for managing ecosystem services: beyond the Millennium Ecosystem Assessment. *Proceedings of the National Academy of Sciences*, **106** (5), 1305–1312.

Claudet, J. and Guidetti, P. (2010) Fishermen contribute to protection of marine reserves. *Nature Letters*, **464**, 673.

Colloca, F., Garofalo, G., Bitetto, I. *et al.* (2015) The seascape of demersal fish nursery areas in the North Mediterranean Sea, a first step towards the implementation of spatial planning for trawl fisheries. *PLoS ONE*, **10** (3), 10.1371/journal.pone.0119590

Cooper, P. (2013) Socio-ecological accounting: DPSWR, a modified DPSIR framework, and its application to marine ecosystems. *Ecological Economics*, **94**, 106–115.

Costanza, R. (1999) The ecological, economic, and social importance of the oceans. *Ecological Economics*, **31**, 199–213.

Costanza, R., de Groot, R., Sutton, P. *et al.* (2014) Changes in the global value of ecosystem services. *Global Environmental Change*, **26**, 152–158.

Crowder, L. and Norse, E. (2008) Essential ecological insights for marine ecosystem-based management and marine spatial planning. *Marine Policy*, **32** (5), 772–778.

Douvere, F. (2008) The importance of marine spatial planning in advancing ecosystem-based sea use management. *Marine Policy*, **32**, 762–771.

Fletcher, S., Saunders, J. and Herbert, R. (2011) A review of the ecosystem services provided by broadscale habitats in England's Marine Protected Area network. *Journal of Coastal Research*, **64**, 378–383.

Forcada, A., Valle, C., Bonhomme, P. *et al.* (2009) Effects of habitat on spillover from marine protected areas to artisanal fisheries. *Marine Ecology Progress Series*, **379**, 197–211.

Francour, P., Harmelin, J.G., Pollard, D. and Sartoretto, S. (2001) A review of marine protected areas in the northwestern Mediterranean region: siting, usage, zonation and management. *Aquatic Conservation in Marine and Freshwater Ecosystems*, **11**, 155–188.

Fraschetti, S., Terlizzi, A., Micheli, F. *et al.* (2002) Marine protected areas in the Mediterranean Sea: objectives, effectiveness and monitoring. *Marine Ecology*, **23**, 190–200.

Gaines, S.D., White, C., Carr, M.H. and Palumbi, S.R. (2010) Designing marine reserve networks for both conservation and fisheries management. *Proceedings of the National Academy of Sciences*, **107**, 18286–18293.

Game, E.T., Watts, M.E., Wooldridge, S. and Possingham, H.P. (2008) Planning for persistence in marine reserves: a question of catastrophic importance. *Ecological Applications*, **18** (3), 670–680.

Gell, F.R. and Roberts, C.M. (2003) Benefits beyond boundaries: the fishery effects of marine reserves. *Trends in Ecology & Evolution*, **18** (9), 448–455.

Giakoumi, S., Grantham, H.S., Kokkoris, G.D. and Possingham, H.P. (2011) Designing a network of marine reserves in the

Mediterranean Sea with limited socio-economic data. *Biological Conservation*, **144**, 753–763.

Gilbert, A.J., Alexander, K., Sardá, R. *et al.* (2015) Marine spatial planning and Good Environmental Status: a perspective on spatial and temporal dimensions. *Ecology and Society*, **20** (1), 64.

Gleason, M., McCreary, S., Miller-Henson, M. *et al.* (2010) Science-based and stake-holder driven Marine Protected Area network planning: a successful case study from north central California. *Ocean Coastal Management*, **53**, 52–68.

Goñi, R., Adlerstein, S., Alvarez-Berastegui, D. *et al.* (2008) Spillover from six western Mediterranean marine protected areas: evidence from artisanal fisheries. *Marine Ecology Progress Series*, **366**, 159–174.

Guidetti, P. (2007) Potential of marine reserves to cause community-wide changes beyond their boundaries. *Conservation Biology*, **21** (2), 540–545.

Halpern, B.S., Lester, S.E. and McLeod, K.L. (2010) Placing marine protected areas onto the ecosystem-based management seascape. *Proceedings of the National Academy of Sciences*, **107** (43), 18312–18317.

Harmelin, J-G. (2000) Mediterranean marine protected areas: some prominent traits and promising trends. *Environmental Conservation*, **27** (2), 104–105.

Harmelin-Vivien, M., Le Direach, L., Bayle-Sempere, J. *et al.* (2008) Gradients of abundance across reserve boundaries in six Mediterranean marine protected areas: evidence of spillover? *Biological Conservation*, **141** (7), 1829–1839.

Jones, P.J.S., Qiu, W. and De Santo, E.M. (2013) Governing marine protected areas: social-ecological resilience through institutional diversity. *Marine Policy*, **41**, 5–13.

Kittinger, J.N., Finkbeiner, E.M., Ban, N.C. *et al.* (2013) Emerging frontiers in social-ecological systems research for sustainability of small-scale fisheries. *Current Opinion in Environmental Sustainability*, **5**, 352–357.

Leslie, H.M., Basurtoc, X., Nenadovic, M. *et al.* (2015) Operationalizing the social-ecological systems framework to assess sustainability. *Proceedings of the National Academy of Sciences*, **112** (19), 5979–5984.

Lester, S.E., Halpern, B.S., Grorud-Colvert, K. *et al.* (2009) Biological effects within no-take marine reserves: a global synthesis. *Marine Ecology Progress Series*, **384** (2), 33–46.

Lewison, R.L., Murray, A.R., Al-Hayek, W. *et al.* (2016) How the DPSIR framework can be used for structuring problems and facilitating empirical research in coastal systems. *Environmental Science and Policy*, **56**, 110–119.

Lloret, J., Zaragoza, N., Caballero, D. and Riera, V. (2008) Biological and socioeconomic implications of recreational boat fishing for the management of fishery resources in the marine reserve of Cap de Creus (NW Mediterranean). *Fisheries Research*, **91** (2–3), 252–259.

Mabile, S. (2007) *The Development of a System of MPAs in the Context of Decentralization: Italian and Spanish Examples*. IUCN, Gland.

McCay, B.J. and Jones, P.J.S. (2011) Marine protected areas and the governance of marine ecosystems and fisheries. *Conservation Biology*, **25** (6), 1130–1133.

McLeod, E., Salm, R., Green, A. and Almany, J. (2009) Designing marine protected area networks to address the impacts of climate change. *Frontiers in Ecology and the Environment*, **7** (7), 362–370.

Milazzo, M., Chemello, R., Badalamenti, F. *et al.* (2002) The impact of human recreational activities in marine protected areas: what lessons should be learnt in the Mediterranean Sea? *Marine Ecology*, **23**, 280–290.

Oikonomou, Z.-S. and Dikou, A. (2008) Integrating conservation and development at the National Marine Park of Alonissos, Northern Sporades, Greece. *Perception and Practice of Environmental Management*, **42**, 847–866.

Ostrom, E. (2009) A general framework for analyzing sustainability of social-ecological systems. *Science*, **325**, 419–422.

Pascual, M. (2013) *Ecosystem-based Marine Spatial Management in the Basque Country: linking human activities, biodiversity valuation and ecosystem services in supporting European Directives implementation*. PhD thesis, Department of Zoology and Animal Cell Biology, UPV-EHU.

Pascual, M., Rossetto, M., Ojea, E. *et al.* (2016) Socioeconomic impacts of marine protected areas in the Mediterranean and Black Seas. *Ocean & Coastal Management*, **133**, 1–10.

Pew Charitable Trusts (2015) *Virtual Watch Room*. http://www.pewtrusts.org/en/research-and-analysis/fact-sheets/2015/01/virtual-watch-room. Accessed 11 June 2015.

Plan Bleu (2012) *Water and Climate Change: Which Adaptation Strategy for the Mediterranean*. UNEP-MAP, Blue Plan Notes, No. 23.

Roberts, C.M., Halpern, S.B., Palumbi, S.R. and Warner, R.R. (2001) Designing networks of marine reserves: why small, isolated protected areas are not enough. *Conservation Biology in Practice*, **2** (3), 10–17.

Roncin, N., Alban, F., Charbonnel, E. *et al.* (2008) Uses of ecosystem services provided by MPAs: how much do they impact the local economy? A southern Europe perspective. *Journal for Nature Conservation*, **16**, 256–270.

Rossetto, M.F., Micheli, G.A., De Leo, P. *et al.* (2013) Socioeconomics of marine protected areas: a review of empirical evidences. *Rapport Commission Internationale pour l'Exploration Scientifique de la Méditerranée*, **40**, 626.

Russ, G.R., Alcala, A.C., Maypa, A.P. *et al.* (2004) Marine reserve benefits local fisheries. *Ecological Applications*, **14**, 597–606.

Sanchirico, J.N. and Wilen, J.E. (2002) The impacts of marine reserves on limited-entry fisheries. *Natural Resource Modeling*, **15** (3), 380–400.

Sardá, R. (2013) Ecosystem services in the Mediterranean Sea: the need for an economic and business oriented approach, in *Mediterranean Sea: Ecosystems, Economic Importance and Environmental Threats* (ed. Terrence B. Hughes). Nova Publishers, New York. pp. 1–33.

Stobart, B., Warwick, R., Gonzalez, C. *et al.* (2009) Long-term and spillover effects of a marine protected area on an exploited fish community. *Marine Ecology Progress Series*, **384**, 47–60.

The Economics of Ecosystems and Biodiversity (TEEB) (2012) *The Economics of Ecosystems and Biodiversity in Business and Enterprise* (ed. Joshua Bishop). Earthscan, London and New York.

Van Assche, K., Teampău, P., Devlieger, P. and Suciu, C. (2008) Liquid boundaries in marginal marshes: reconstructions of identity in the Romanian Danube Delta. *Studia Sociologia*, **53** (1), 115–133.

Van Assche, K., Duineveld, M., Beunen, R. and Teampău, P. (2011) Delineating locals: transformations of knowledge/power and the governance of the Danube Delta. *Journal of Environmental Policy & Planning*, **13** (1), 1–21.

Van Beukering, P., Brander, L., Tompkins, E. and McKenzie, E. (2007) *Valuing the Environment in Small Islands: An Environmental Economics Toolkit*. Joint Nature Conservation Committee, Peterborough.

White, C., Kendall, B.E., Gaines, S. *et al.* (2008) Marine reserve effects on fishery profit. *Ecology Letters*, **11** (4), 370–379.

Whitmarsh, D., James, C., Glenn, H. and D'Anna, G. (2002) Perspectives of economic effects of fisheries exclusion zones: a Sicilian case study. *Marine Resource Economics*, **17**, 239–250.

Wilson, J.R., Prince, J.D. and Lenihan, H.S. (2010) A management strategy for sedentary nearshore species that uses marine protected areas as a reference. *Marine and Coastal Fisheries*, **2** (1), 14–27.

7

Multi-criteria Decision-Making for Marine Protected Area Design and Management

Paco Melià

Dipartimento di Elettronica, Informazione e Bioingegneria, Politecnico di Milano, Milano, Italy and Consorzio Nazionale Interuniversitario per le Scienze del Mare, Roma, Italy

Introduction

Marine Protected Areas (MPAs) are a cornerstone of most marine conservation strategies, as they are expected to consistently provide ecological, economic and social benefits (Klein *et al.*, 2013). Effective planning and management of MPAs and MPA networks raises multifaceted problems: protected areas designed to deliver effective conservation benefits may, in fact, act at the expense of socioeconomic activities such as fisheries, whereas protected areas accommodating social needs may result in outcomes that do not adequately address the conservation of marine ecosystems (Klein *et al.*, 2008). The wide and complex range of relationships linking human activities and the surrounding environment make it crucial to integrate conservation science and the analysis of societal beliefs, customs, attitudes and practices into a unique framework (Voyer *et al.*, 2012). Designing MPA networks able to trade off biodiversity and socioeconomic goals is thus a major challenge for systematic conservation planning (Stewart and Possingham, 2005).

The goals that planning of future MPAs, as well as management of existing ones, should pursue are manifold, and ultimately

depend on the specific ecological, cultural or socioeconomic problems they are meant to improve; therefore, it is crucial that the local context is well understood and taken into consideration in the identification of goals and objectives of an MPA (Heck *et al.*, 2011). Clear statements need to be developed from the beginning of the planning process, to indicate the expected achievements for a new MPA and the ways its effectiveness can be monitored over time (Day, 2008). Identifying desired MPA performance, and adopting practices to demonstrate their effective management, increases the likelihood that established MPAs will not just remain so-called 'paper parks' (Thompson *et al.*, 2008). In addition, the definition of desired performances in the early planning stages allows the collection of data that can be used to characterize the current state of the area and measure the effectiveness of protection as the difference between the initial state and the outcomes achieved through the establishment of an MPA (Day, 2008).

In this sense, addressing social and economic aspects is key to effective MPA implementation, as ineffective social assessment can alienate local communities and undermine the success of existing and future

MPAs (Voyer *et al.*, 2012). In his comprehensive review on objectives, selection, design and management of MPAs, Jones (2002) listed 10 general objectives for inshore MPAs, primarily focused on environmental and cultural conservation:

- Protect rare and vulnerable habitats and species
- Conserve a representative set of habitat types
- Maintain and restore ecological functions
- Promote research and education
- Provide harvest refugia (no-take zones)
- Control tourism and recreation
- Promote integrated coastal management
- Maintain aesthetic values
- Maintain traditional uses
- Promote the cultural symbolic value of set-aside areas.

Jones observed that, while those objectives may appear justifiable and achievable (through the designation and management of MPAs) to marine conservationists and scientists, a large body of literature from around the world indicates that MPA proposals can generate different types of conflicts. He identified two major types of conflict: internal and basic. Internal conflicts are caused by the clash between different user interests and emerge when a social group feels discriminated against in favour of others. Basic conflicts, on the other hand, arise from deeper differences in ethical views, such as those emerging from the debate between 'nature protectionists' (i.e. those considering that conservation should focus primarily on the protection of biodiversity) and 'social conservationists' (i.e. those arguing that conservation should be focused on human welfare) (Miller *et al.*, 2011; Voyer *et al.*, 2012).

Therefore, the success of existing and future MPAs critically depends on the possibility that a consensus on MPA objectives is reached among the stakeholders involved about the environmental and socioeconomic impacts of an MPA. Failure to do this is argued to undermine the case for the establishment of MPAs and exacerbate conflicts among stakeholders during the formulation and implementation of management policies (Jones, 2002). On the contrary, including stakeholders in the decision process adds to the perceived legitimacy of the selection of preferred alternatives for proposed actions (Wadsworth *et al.*, 2014).

The concept of success in MPA management is inherently a social construct (Himes, 2007): expectations often diverge among user groups, depending on their background, values and affiliation (Heck *et al.*, 2011). Information needs for the evaluation of MPA performances cover a broad range of issues, including context, statutory requirements, planning, resources, processes, outputs and outcomes; consequently, they differ between sites, reflecting the unique management context of each case (Dahl-Tacconi, 2005). The active involvement of stakeholders from the early stages of the planning process helps to identify major issues concerning the establishment of an MPA, encourages the exchange of ideas and the reciprocal understanding about the issues at hand, and promotes the generation of new options and solutions (Pomeroy and Douvere, 2008), eventually increasing the likelihood that policy decisions are based upon accurate understanding of the local social and environmental conditions. Pomeroy and Douvere (2008) have synthesized the main reasons to involve stakeholders into the following five points:

1) Better understanding of the complexity of the ecosystem
2) Understanding of the human influence on the ecosystem and its management
3) Examining the compatibility and/or (potential) conflicts of multiple use objectives
4) Identifying, predicting and resolving areas of conflict
5) Discovering existing patterns of interaction.

To reduce both real and perceived conflicts, planning and management of MPAs should hence be driven by both ecological and socioeconomic factors (Klein *et al.*, 2013). During the last few decades, the notion of ecosystem services (Ehrlich and Ehrlich, 1981) has emerged as a key concept to understand the links between ecological and socioeconomic systems and a way to incorporate socioeconomic factors into decision-making. The Millennium Ecosystem Assessment (MEA, 2005) promoted the application of this concept, by providing a conceptual framework to analyse how ecological processes contribute to human well-being; the concept has also been used to quantify the socioeconomic benefits delivered by marine conservation and to provide guidelines for MPA management (e.g. Fletcher *et al.*, 2011). More recently, another international initiative, The Economics of Ecosystems and Biodiversity (TEEB, 2010) has further fostered this approach, bringing ecosystem services to a broader audience (Costanza *et al.*, 2014). The focus of research on ecosystem services has gradually shifted from investigating the links between biodiversity, ecosystem structure, functions and services (Ehrlich and Mooney, 1983) to developing and tuning methods to estimate the economic value of those services (Gómez-Baggethun *et al.*, 2010; Braat and de Groot, 2012). The monetary valuation of ecosystem services, however, raises a range of philosophical, ethical and methodological issues which have triggered a passionate debate (e.g. Gatto and De Leo, 2000; Costanza, 2006; McCauley, 2006). As economists themselves acknowledge that economic assessments are not the only way to express the importance of ecosystems and their links with human activities and human well-being (Costanza *et al.*, 2014), decision-making in the environmental field requires methodologies that can also account for non-market values

and are able to integrate them into the decision process.

This chapter aims to provide an overview of the main techniques that have been developed to support decision-making from a multi-criteria perspective, and how they have been applied to MPA design and management. The pros and cons of different approaches are examined, with particular regard to their ability to facilitate the interaction with stakeholders, and promote the elicitation of their objectives and preferences. A special focus is also given to their capacity to explicitly incorporate uncertainty. Finally, the extension of multi-criteria methods to the spatial dimension is critically discussed as an aspect of prominent relevance for marine conservation planning.

Multi-criteria Decision Methods

While techniques such as cost–benefit analysis can be effective when the objective can be expressed in monetary terms and reduced to maximizing economic efficiency alone, multi-criteria analysis can be more appropriate when the social implications and the environmental impacts of decisions are also important to decision-makers (Gregory and Slovic, 1997). Excluding non-monetary goods and services from the valuation, as well as failing to involve key stakeholders in the process, reduces the legitimacy of the decisions and can lead to poor implementation (Brown *et al.*, 2001).

In contrast, multi-criteria analysis (see Zionts, 1979 for a brief, non-technical review of the basic concepts) provides a conceptual framework to support decision processes by allowing decision-makers to address a number of objectives that cannot be reduced to a single dimension, such as the monetary one, and to highlight possible trade-offs among conflicting viewpoints.

Considering several objectives at once provides a comprehensive framework for the decision process, promotes the engagement of stakeholders as well as a more appropriate role in the process for the analyst, and usually generates a wider range of alternatives than those produced by single-objective analyses (Bevacqua *et al.*, 2009). Furthermore, multi-criteria analysis allows criteria that can hardly be expressed in quantitative terms to be explicitly incorporated into the analysis (Van Huylenbroeck and Coppens, 1995). The application of multi-criteria techniques has experienced a constant growth in the last half century, broadening its range from the original field of operations research to neighbouring disciplines such as engineering and information and communications technology, and on to social, economic and natural sciences (Bragge *et al.*, 2010).

The key concept of multi-criteria analysis is Pareto efficiency. A decision (for instance, a specific management plan for a protected area) is called Pareto-efficient when it is not possible to modify any decision variable to improve a performance indicator (e.g. the protection of biodiversity) without necessarily worsening at least another one (e.g. the income of fishers exploiting a fish stock). Decisions for which there exists at least an alternative choice that guarantees both higher biodiversity and higher income to fishers are called Pareto-dominated. The set of all non-dominated decisions is called the Pareto boundary (or Pareto set) and represents the range of alternative choices from which the decision-maker can reasonably select. An example of Pareto analysis is reported in Bevacqua *et al.* (2007), who analysed different options for the management of the eel fishery in the Camargue National Reserve (France). They compared alternative management rules with respect to two objectives (maximizing the number of adult spawners escaping from the lagoons towards the ocean, and maximizing the yield of the

fishery), and found that there was a wide range of options dominating the current management of the fishery (Figure 7.1).

The Pareto boundary and associated trade-offs provide important information and a useful reference for decision-makers. Moreover, it allows decision-makers to choose from a set of efficient policies, rather than just a single optimal policy derived through a one-dimensional optimization (such as cost–benefit analysis), offering a wider range of opportunities to manage potential conflicts among diverse stakeholders. Although only a specific decision must eventually be taken, it is desirable that the final choice is not the result of a formal maximization problem, but rather of a subjective appraisal of the decision-makers, reflecting the relative importance they give to the different decision objectives. The multi-criteria approach should therefore concentrate on providing clear and unbiased information to decision-makers regarding the range of effective choices and their expected consequences, rather than suggesting a single optimal solution (Gatto and De Leo, 2000).

Two major categories of problems involving multiple evaluation criteria can be distinguished (Wallenius *et al.*, 2008): optimization problems, which involve an infinite, or at least a very large, number of alternative choices (usually defined by a system of equations and inequalities that identify a feasible region for the decision variables), and discrete alternative problems, in which the set of alternative choices is reasonably small. In many applications, the number of alternatives that can be generated and evaluated in depth (in terms of their expected environmental and socioeconomic impacts) are relatively few. For this reason, in the following sections attention will be focused mainly on discrete alternative problems, although most methodological points that will be discussed can easily be extended to optimization problems. Another good reason to focus on discrete problems is that optimization problems, due to their complexity, usually require more

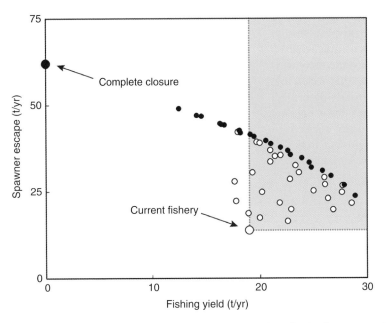

Figure 7.1 Pareto analysis of different management options for the Camargue eel fishery with respect to two objectives (maximizing adult spawner escape and maximizing fishing yield). Each symbol represents a management policy; filled symbols indicate Pareto-efficient policies (Pareto boundary). Policies within the shaded area dominate the current fishery management with respect to both objectives. *Source*: Adapted from Bevacqua *et al.* (2007).

computational resources than discrete problems, making it more difficult to explicitly incorporate uncertainty into the analysis. As discussed later, uncertainty has a critical influence on the robustness of the analysis, due to our limited knowledge and ability to predict the behaviour of the complex system under study, as well as to the intrinsic subjectivity of the evaluation process.

The vast range of problems that have been addressed through multi-criteria approaches has led to the evolution of different families of methods, each one resting on different philosophical foundations (see Köksalan *et al.*, 2013 for an historical perspective). Different methods rely on different techniques to measure and compare the performances of the options under scrutiny with respect to the considered criteria: some methods produce a ranking of the options, which can be complete or not (to allow for

incomplete comparability), some identify a single optimal alternative, while others differentiate only between acceptable and unacceptable alternatives (Levner *et al.*, 2005). Multi-attribute utility theory (MAUT) and the analytic hierarchy process (AHP) are based on the idea of quantifying the performances of alternative options with respect to each decision criterion and aggregating them into a single overall score. As low performances on one criterion can be compensated for by high performances on other criteria, they are classified as 'compensatory' methods (Linkov *et al.*, 2006). Unlike compensatory methods, outranking methods (which include the family of ELECTRE methods, as well as the PROMETEE method) are known as 'partially compensatory', since they do not allow for a full compensation in performances across criteria and do not presuppose that a single best alternative can be identified. Outranking techniques are best suited to problems in which performance metrics cannot be easily aggregated because units are incommensurate or incomparable. In the following, two of these methods are briefly described: MAUT and the AHP. They have become very popular, and have been applied to a number of problems involving fisheries management and MPA management and design.

Multi-attribute Utility Theory

Multi-attribute utility theory has been developed and formalized by Keeney and Raiffa (1993) and relies on the idea that decision-makers attempt to maximize their expected utility, that is, their overall satisfaction with respect to the expected consequences of their choice on a number of independent attributes, each one representing an objective of the decision. Utility is a cardinal function allowing the decision-maker to rank the possible impacts of a choice in order of preference. It can be viewed as the level of desirability, or satisfaction, associated to a given value of a specific indicator (Keeney, 1977): it maps the value

of each indicator into a range between 0 (minimum satisfaction with respect to that indicator) and 1 (maximum satisfaction). The utility function of each indicator can be built by (i) identifying its range of variation; (ii) determining its functional form (monotonically increasing, decreasing, or non-monotone); and (iii) defining the indicator values to be associated with minimum, maximum and/or intermediate levels of utility on the basis of specific reference points. Figure 7.2 shows some examples of utility functions: a piecewise linear function (Figure 7.2a), a non-linear, monotonically increasing function (Figure 7.2b), a non-monotonic function (Figure 7.2c), and a non-linearly (sigmoid) decreasing function (Figure 7.2d). While linear functions are often preferred for their simplicity, non-linear functions allow for a more realistic description of changes in the natural human propensity to risk across the range of variation of an attribute.

When there are several attributes, the overall utility U (representing the overall satisfaction of the decision-maker with respect to the whole set of management objectives) can be calculated as the weighted sum (assuming a fixed substitution rate

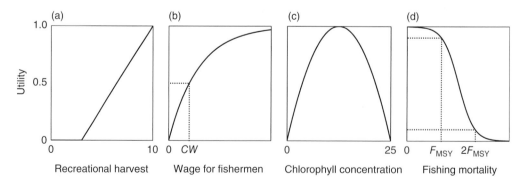

Figure 7.2 Examples of utility functions: (a) economic value (million dollars) of recreational harvest in Lake Erie; (b) average wage of fishermen employed in the demersal fishery of the southern Adriatic Sea (*CW*, current wage); (c) chlorophyll-a concentration in Lake Erie (μg/l); (d) fishing mortality of demersal species in the southern Adriatic Sea (*F*$_{MSY}$, fishing mortality at maximum sustainable yield). *Source*: (a) and (c) adapted from Anderson *et al.* (2001); (b) and (d) adapted from Rossetto *et al.* (2015).

between any two attributes) of the partial utilities associated to the different attributes:

$$U(x) = \sum_{i=1}^{n} w_i u_i (x_i)$$

where $u_i(x_i)$ (with $0 \leq u_i \leq 1 \ \forall \ i = 1, 2, ..., n$) is the utility associated to the value x_i taken by the i-th attribute, and w_i is the weight associated to the utility of the i-th attribute, subject to the constraint

$$\sum_{i=1}^{n} w_i = 1.$$

When the hypothesis of a fixed substitution rate among attributes is not likely to apply, a multiplicative utility function can be used to aggregate single-attribute utilities (Keeney and Raiffa, 1993). This aspect, which is generally intrinsic to the problem, should be investigated in the preliminary steps of the analysis (Mardle and Pascoe, 1999).

While specific applications of the multi-attribute approach to MPA design and management have been scarce up to now, Mardle and Pascoe (1999) reviewed seven applications to fisheries management. All the reviewed studies noted the potential usefulness of this technique and the wide range of management options that can be considered. In particular, a reported key feature of the MAUT approach is that it allows focusing on the points of agreement and disagreement between different interest groups (Boutillier *et al.*, 1988).

More recently, Anderson *et al.* (2001) used a multi-attribute approach to analyse the problem of nutrient management in Lake Erie. They identified a set of six attributes representing the performances of a decision along three main dimensions: social (recreation, aesthetics), ecological (balance of the fish community, eutrophication), and economic (fishing yield, cost of phosphorus removal). They then derived a utility function for each attribute to quantify the relative desirability of various attribute levels (see Figure 7.2). Finally, they ranked

alternative management policies with respect to their overall desirability (expressed as a weighted sum of partial utilities). Utility functions were built by interviewing two fisheries biologists.

Bryan *et al.* (2011) mapped and compared a range of social and ecological values for natural areas in the South Australian Murray-Darling Basin. Social values were mapped by interviewing community members involved in natural resource management in the study area. Interviewed people were asked to identify places they valued for the existence of natural capital assets and the delivery of ecosystem services, to locate and map their spatial extent, and to indicate the relative value associated to each site. Ecological value was characterized by a suite of 12 indicators commonly used in setting spatial conservation priorities and grouped into five major categories: climate change, patch metrics, protection status, species richness, and support to native vegetation. Multi-attribute utility theory was then used to combine the spatial distribution of the selected indicators into two spatial layers representing the social and ecological value, respectively.

Read *et al.* (2011) compiled a list of planning criteria for optimizing compliance in MPAs, and compared the perceptions of recreational fishers and officers to manageability and voluntary compliance in the Port Stephens–Great Lakes Marine Park (SE Australia) through a simplified multi-attribute method, the so-called Simple Multi Attribute Rating Technique (SMART; see Mardle and Pascoe, 1999). Recreational fishers indicated zone identification (i.e. designing protection zones with a simple shape, so that they can easily be understood and enforced), compliance education and capacity building (e.g. training skippers in the use of GPS equipment), impacts of protection on important fishing grounds, and legitimacy (appropriate justification of the zoning) as the most important criteria for

MPA design. Compliance officers placed similar importance on identification, education and legitimacy, but were much less interested in minimizing impacts on marine uses. Read *et al.* (2011) used the weighting set produced by the analysis to associate a manageability score to existing no-take zones with respect to the identified criteria, and observed a significant negative correlation between the score assigned to a zone and the number of enforcement actions recorded for that zone. Therefore, they pointed out that designing MPAs taking appropriate account of their manageability is crucial to ensure proper enforcement.

Rossetto *et al.* (2015) used MAUT to assess the performances of alternative fishing management policies in Mediterranean demersal fisheries. They identified eight attributes ascribed to four major categories of fisheries objectives: economic efficiency (gross value added, ratio of revenues to break-even revenues), social well-being (employment and wage), biological conservation (fishing mortality rate and spawning stock biomass) and biological productivity (fishing yield and discard rate). They then defined a set of utility functions to express the level of satisfaction associated with different values of the attributes on a standardized scale (see Figure 7.2 for an example). To overcome a commonly recognized critical point in the application of MAUT, that is the determination of the weighting set (Andalecio, 2010; Innes and Pascoe, 2010), they combined MAUT with the analytic hierarchy process (see next section).

Analytic Hierarchy Process

The analytic hierarchy process is a method developed by Saaty (1977, 1980) to facilitate the elicitation of individual preferences towards the different attributes, and their conversion into a set of weights. The method can be summarized into four major steps: (i) decompose the problem into a hierarchy encompassing the decision goal, the alternatives considered to reach it, and the criteria used to evaluate the alternatives; (ii) establish priorities among the elements of the hierarchy by making a series of judgements based on a pair-wise comparison of the elements; (iii) synthesize these judgements into a set of weights expressing the relative priority of each element in the hierarchy; (iv) aggregate the relative weights of the decision element to derive a rating for each alternative.

Thanks to its approach based on pair-wise comparisons, the AHP can be applied not only to quantitative but also to qualitative attributes (Heck *et al.*, 2011). So far, the AHP has been used to investigate different aspects of fisheries management, social acceptance of aquaculture, and stakeholder preferences for conservation versus development of wetlands (see Heck *et al.*, 2011 and references therein). In the last two decades, the AHP has been applied also to MPA management with the aim of prioritizing management objectives (Fernandes *et al.*, 1999; Himes, 2007), as well as to identify and assess planning alternatives and management options (Fernandes *et al.*, 1999; Villa *et al.*, 2002).

More recently, Heck *et al.* (2011) used the AHP to investigate the opinion of different stakeholder groups about the performances of a proposed National Marine Conservation Area on the west coast of Canada prior to its establishment. In order to elicit the importance given to performance criteria, a questionnaire-based survey was undertaken. Seven groups participated in the study, including two commercial user groups (marine tourism operators and commercial fishers), two recreational user groups (boaters and recreational fishers), the main governing agency, NGO members, and local governments. The questionnaire contained a closed question asking about the importance of a set of performance criteria on a nine-point AHP scale. Pair-wise comparisons were made between the four main criteria categories (environmental, social, economic and management) and between subcriteria under each 'parental' criterion (Figure 7.3).

Results revealed that, when only top-level criteria were considered, all stakeholders agreed that the proposed marine conservation area should mainly achieve environmental improvements in the area. However, when the analysis included subcriteria, local economic benefits were also given high relevance. In particular, greater importance was given to local income from tourism within the protected area than to income for local fishers, with comments by the respondents suggesting to 'promote tourism as an alternative economic driver to traditional fishing industries'. Also, while non-user groups focused almost entirely on environmental improvements, local user groups had more diverse expectations for the future of the area. The most important criterion for tourism operators, recreational fishers and commercial fishers was local economic benefits, but while preferences of recreational fishers were similar to those of tourism operators, commercial fishers put more importance on fishery income, enforcement of regulations, reduced pollution, and habitat restoration.

Yang *et al.* (2011) selected, through a literature review, 21 critical factors for sustainable use of MPA resources (organized along four dimensions: ecosystem, society, economy and policies) and interviewed four stakeholder groups in Green Island (Taiwan) to analyse their perceptions about the relative importance of those factors. All groups (managers, fishers, local business operators and tourists), which had relevant interests but different positions, indicated preservation of ecosystem integrity and the prevention of environmental pollution as the most important factors. On the other hand, a possible conflict among groups emerged with respect to policy implementation, with fishers disagreeing with the tough enforcement of restrictions on fishing.

Li *et al.* (2014) applied the AHP to assess the ecological status of the Haizhouwan Protected Area (Lianyungang, China). They used a set of 20 indicators of ecosystem health, organized into five major categories (environmental status, environmental disasters, environmental background, system structures and functions, system stability) to classify the area according to five levels of ecosystem quality (from 'very poor' to 'excellent') and identified two priority areas for ecological restoration.

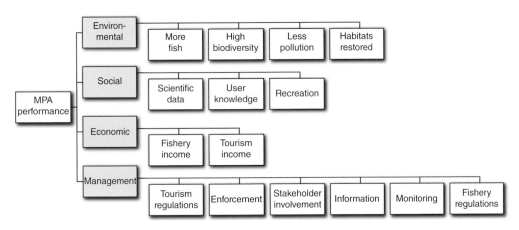

Figure 7.3 Hierarchy of performance criteria for the assessment of a candidate MPA in western Canada. *Source*: Adapted from Heck *et al.* (2011).

Wadsworth *et al.* (2014) used the AHP to identify marine research priorities in the Aleutian Islands. They involved an expert panel (including personnel of resource management agencies, academics, NGO representatives, and individuals engaged in the fishing and processing sectors) to prioritize 115 research needs organized into a hierarchy of five categories (catalogue organisms and identify habitats; identify indicators, monitor trends, and predict changes; determine the function and interrelationships of organisms in the ecosystem; understand factors that influence and control ecosystem dynamics; understand the significance of injurious agents, human activities and other perturbations on the ecosystem and mitigate impacts) and 16 subcategories. They also interviewed, via a web-based survey, a broader range of stakeholders from the same employment categories. In both groups, highest priority was given to increasing basic knowledge of the marine ecosystem. As noted by the authors, however, the rating of research priorities was probably not controversial enough to highlight conflicts among the stakeholder groups: if the survey had been on a more contentious topic, such as regulating fisheries, then collecting stakeholders' beliefs would possibly have

provided useful information to natural resource managers to predict the reactions of the stakeholders to different policies.

Tuda *et al.* (2014) integrated the AHP into a Marine Spatial Planning procedure to manage conflicts in the Mombasa Marine Nature Park and Reserve, a multi-use coastal area in Kenya. They gathered geographical information on coastal marine habitats and competing human activities, which were categorized into five primary coastal uses (habitat protection, sea access and anchorage, water recreation, beach activities and artisanal fishing). Then, they used the AHP to assess how each coastal use contributes to existing spatial conflicts among stakeholders. They eventually mapped (see section *Multi-criteria Analysis in a Spatial Dimension*) the intensity and location of conflicts and allocated spaces to competing users in order to minimize the intensity of conflict.

Dealing with Uncertainty

Multi-criteria assessments are affected by a wide range of uncertainties (Brown, 2004; Figure 7.4): the intrinsic variability characterizing all environmental systems, the imperfect knowledge of the specific system

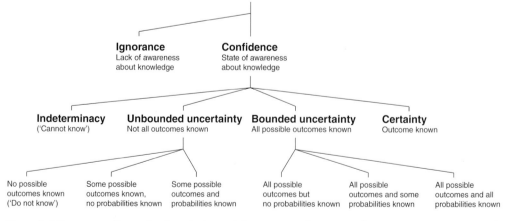

Figure 7.4 Taxonomy of imperfect knowledge and the corresponding range of uncertainty. *Source*: Adapted from Brown (2004).

under study, as well as the subjectivity of expert judgements (see e.g. Refsgaard *et al.*, 2007 for a review of sources of uncertainty and methods for their assessment) all have a critical influence on the reliability of the outcomes of the analysis. Therefore, it is very desirable that a sensitivity analysis is carried out to evaluate the robustness of the results with respect to the different sources of uncertainty.

Sensitivity analysis investigates 'how uncertainty in the output of a model (numerical or otherwise) can be apportioned to different sources of uncertainty in the model input' (Saltelli *et al.*, 2004). Depending on the complexity of the model outputs, methods for sensitivity analysis may range from simple to relatively complex (see Saltelli *et al.*, 2008 for a review). Sensitivity analysis provides valuable insights regarding the influence of input variation on model results, and helps to identify the most critical parameters affecting the reliability of the outcomes. A limitation of sensitivity analysis is that it usually takes the model structure and system boundaries for granted (Refsgaard *et al.*, 2007). Uncertainty about model structure can be addressed by multiple model simulation (Refsgaard *et al.*, 2007), which uses alternative models based on different process descriptions. This approach has the advantage that the effect of different model structures can be explicitly analysed, provided that the range of plausible interpretations of the process is reasonably well known.

When the major source of uncertainty is the future evolution of the system (for instance, because one wants to investigate the expected consequences of different protection measures), different alternative futures can be explored via scenario analysis. Different types of scenarios can be distinguished, for instance baseline vs. policy scenarios, exploratory vs. anticipatory scenarios, and qualitative vs. quantitative scenarios (Alcamo, 2001). Baseline scenarios

(often referred to as business-as-usual scenarios, reference scenarios, or benchmark scenarios) describe the expected trajectory of the system under current management (i.e. if no additional protection measures are implemented), while policy scenarios depict alternative futures under different protection policies. Exploratory (or descriptive) scenarios explore possible trends into the future starting from the present, while anticipatory (or prescriptive) scenarios go backwards from a prescribed vision of the future to devise the process through which that future could emerge. Qualitative scenarios (or storylines) describe possible futures in a narrative form, while quantitative scenarios provide numerical estimates of measurable attributes produced by predictive models.

Another important source of uncertainty is that affecting expert judgements: eliciting adequate quantitative information is, in fact, considered one of the major challenges within the field of decision analysis (Riabacke *et al.*, 2012). In particular, evaluating the sensitivity of a multi-criteria assessment to the weights assigned to the criteria is crucial to understand whether the ranking of the options at stake is stable to perturbations of the weighting set. This can be assessed, for instance, via Monte Carlo simulation, which is based on calculating the results (realizations) of a model for a large number of random draws from the (*a priori*) probability distributions of input data and/or model parameters, hence associating an empirical (*a posteriori*) probability distribution to the model output. The Monte Carlo approach can easily be applied to a vast range of different models and does not impose particular assumptions on probability distributions and correlations. However, it requires that the probability distributions of the inputs and/or parameters are known (or at least that reasonable hypotheses can be made on them).

Anderson *et al.* (2001) assessed, through a two-way sensitivity analysis, to what extent the choice of the optimal control policy for nutrient management in Lake Erie was affected by the relative importance of two management targets, as perceived by respondent stakeholders (Figure 7.5a). They concluded that, because the optimal choice was quite sensitive both to variation in the weights assigned to criteria and to specific assumptions on which the model used to estimate the outcomes of alternative policies was based, further analysis was essential before recommending a change from the current management. Rossetto *et al.* (2015) used a Monte Carlo approach to evaluate the robustness of the ranking of alternative fisheries management policies with respect to the uncertainty associated to the weights expressing the relative importance of management objectives used to rank the policies. They randomly perturbed the original weighting set to produce an empirical probability distribution of the overall utility associated to each management policy (Figure 7.5b). They found that the ranking of the alternative policies was robust with respect to a moderate level of uncertainty, but warned that the involvement of a broader range of stakeholders could substantially impact the outcomes of the analysis.

Choosing the Right Approach

Multi-criteria analysis can be carried out using a variety of methods; Cinelli *et al.* (2014) reviewed pros and cons of different multi-criteria approaches with respect to 10 comparison criteria that a method should satisfy to properly handle problems concerning sustainability. These criteria include, among others, the ability to deal with both quantitative and qualitative information; the level of compensation among objectives; the capability of coping with uncertainty; the robustness of the ranking to the addition or deletion of alternatives; and the ease of interaction with stakeholders and decision-makers. They concluded that MAUT and the AHP are fairly simple to understand, but they are cognitively demanding for the decision-makers. Mixed information and uncertainty can be managed by all the methods, while robust results can only be obtained with MAUT. On the other hand, non-compensatory approaches can adopt a strong sustainability perspective (in that they limit, partially or completely, the possibility of compensating performances across criteria) but suffer from rank reversal

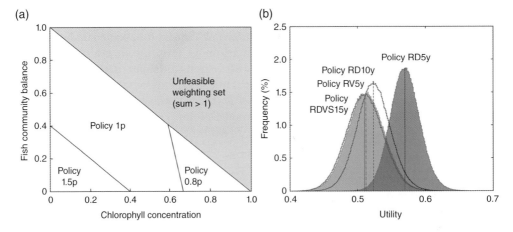

Figure 7.5 Examples of sensitivity analyses. (a) Optimal policy for nutrient management in Lake Erie as a function of the weights assigned to chlorophyll-a concentration and balance of the fish community. (b) Empirical frequency distribution of the overall utility of four management policies for the demersal fishery of the southern Adriatic Sea. *Source*: (a) adapted from Anderson *et al.* (2001); (b) adapted from Rossetto *et al.* (2015).

(i.e. the addition or removal of even a single alternative can change the ranking of the other ones; see e.g. Wang and Luo, 2009). Methods based on pair-wise comparison, such as the AHP, do not allow consideration of a very high number of criteria and/or alternatives, as this would require a very high number of comparisons, making the elicitation of stakeholder preferences very tiring (Heck *et al.*, 2011). In short, a single right method does not exist: the choice of the most appropriate one depends on a range of factors, including the specific characteristic of the problem at hand and the social and cultural context of the actors involved in the decision process. As different techniques can, in principle, provide different results, it is advisable, whenever possible, to test the robustness of the outcomes also with respect to the choice of the method.

With regard to MPA design and management in particular, a vast range of criteria specifically aimed to assess MPA performances from the different perspectives discussed in this chapter can be found in the scientific literature. Ecological criteria for preliminary evaluation of candidate sites for marine protection have been developed, for instance, by Roberts *et al.* (2003), and a list of environmental and cultural criteria for MPA design has been compiled by Jones (2002). Leslie (2005; see Table 1 therein) has reviewed primary conservation objectives in a number of marine conservation planning cases, while biophysical and socioeconomic objectives for the design of MPA networks have been listed by Klein *et al.* (2008, Table 1). The European Commission (EC, 2010) has produced a set of detailed criteria and indicators to help Member States monitoring the achievement of the Good Environmental Status envisaged by the Marine Strategy Framework Directive (2008/56/EC). As regards the socioeconomic consequences of MPAs, Ojea *et al.* (this volume) report a list of potential impacts of MPAs on human activities, and

Pascual *et al.* (2016) provide a comprehensive review of stakeholders' perceptions from the Mediterranean and Black Seas; criteria expressing key stakeholder objectives have been reviewed by Halpern and Warner (2003) and Himes (2007), while factors affecting fisheries profitability of MPAs have been discussed by Gaines *et al.* (2010). As for MPA management, a comprehensive list of criteria characterizing the manageability of MPAs has been compiled by Read *et al.* (2011, Table 1) for optimizing voluntary compliance.

Another important point is the selection of the indicators used to measure the level of achievement of the decision objectives. Keeney and Raiffa (1993) suggest five major properties that the set of indicators to be used in a multi-criteria analysis should have: completeness, operability, decomposability, non-redundancy, and minimal size. A set of indicators is complete if it covers all the aspects that are relevant to the decision and indicates the degree to which the overall objective is met. Indicators that are operable are meaningful to the decision-maker (in that they allow understanding the implications of the candidate alternatives) and facilitate explanation to the stakeholders. Thus indicators should be specific, concrete, and experienced instead of general, theoretical, and ambiguous (Anderson *et al.*, 2001). If a set of attributes is decomposable, the assessment process can be broken down into parts of smaller dimensionality, that is, involving only a subset of the indicators at a time. Non-redundant indicators neither overlap on dimensions of value nor are correlated with each other, so as to avoid double counting. Finally, it is desirable to keep the set of indicators reasonably small, so that the analysis remains manageable and the interaction with stakeholders relatively easy without losing the variety of perspectives that a multi-criteria approach can embrace. Other important features to be considered have been pointed out by Li *et al.* (2014), who based their selection of indicators to assess marine environmental status on four criteria: scientificity (i.e. the indicators are capable of objectively representing the components of an ecosystem and the interconnections among them), sensitivity (they reflect the response of the ecosystems to natural and anthropogenic disturbance), manoeuvrability (they can be measured on the basis of accessible data) and comparability (they can be compared at different times and spaces).

Multi-criteria Analysis in a Spatial Dimension

Particularly critical to MPA science is the spatial nature of the problems it must deal with. While conventional multi-criteria analysis has no explicit notion of geographical space, it can be easily extended to a spatially explicit context to make it suitable to environmental applications such as MPA planning (Villa *et al.*, 2002). To this end, it can be integrated into a geographical information system to identify and compare solutions on the basis of the combination of multiple factors that can be, at least partially, represented by maps (Malczewski, 2006). This approach (known as spatial multi-criteria analysis) takes advantage of both the capability of geographic information systems to manage and process spatial information, and that of multi-criteria analysis to structure decision problems, and assess and prioritize alternative decisions. Operationally, the steps of a spatial multi-criteria analysis are similar to those of a standard multi-criteria analysis, but with all or part of the criteria, as well as the final outputs, represented by maps.

In recent decades, spatial multi-criteria analysis has become a methodological basis for systematic conservation planning. Analyses in which the multi-criteria approach is applied only to the aggregation of assessment criteria, and in which the

number of protection scenarios considered is relatively small (e.g. Villa *et al.*, 2002; Bryan *et al.*, 2011; Li *et al.*, 2014), can be carried out using discrete multi-criteria methods. However, when the set of candidate solutions is open, such as in the case of determining the optimal spatial allocation of protection and/or competing marine uses, optimization algorithms are needed. Different software platforms have been developed to solve optimization problems for systematic conservation planning, including Marxan (Ball *et al.*, 2009), Marxan with zones (Watts *et al.*, 2009), ConsNet (Sarkar *et al.*, 2009), C-plan (Pressey *et al.*, 2009), and Zonation (Moilanen *et al.*, 2009). In the last few years, the number of case studies that have been investigated with these tools has rapidly increased (e.g. Klein *et al.*, 2013; Mazor *et al.*, 2014a, 2014b; Ruiz-Frau *et al.*, 2015). Because computing times rapidly increase with the dimension of the problem, software for spatial planning typically makes use of heuristic algorithms: these are more computationally efficient than optimal algorithms (such as integer linear programming) but cannot ensure a global optimal solution, nor inform the researcher about the level of suboptimality of the solution (Cheng *et al.*, 2015). In addition, spatial multi-criteria analyses based on optimization algorithms are not always able to promote effective stakeholder involvement during the comparison of alternative MPA design options, as the range of alternatives considered may be either too wide (before the optimization is carried out) or too narrow (after the optimization).

Tuda *et al.* (2014) provide an interesting example of a Marine Spatial Planning process in which stakeholders have been involved to resolve conflicts among different marine uses. They describe it as an iterative process (inspired by the guidelines of Ehler and Douvere, 2009) based on a tight interaction with stakeholders; the basic steps are summarized in Figure 7.6 and include

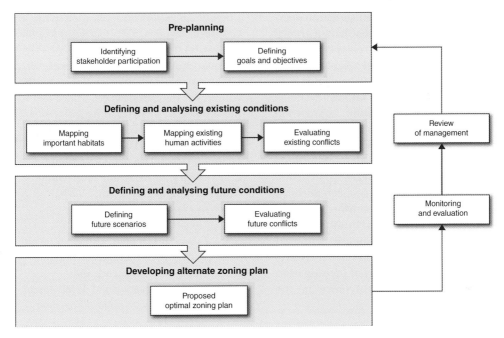

Figure 7.6 Basic steps of a Marine Spatial Planning process. *Source*: Adapted from Tuda *et al*. (2014).

(i) pre-planning, (ii) definition and analysis of the present conflicts, (iii) definition and analysis of future conditions, and (iv) development of alternative allocation plans.

The pre-planning phase aims to identify the key stakeholders to be involved in the process and to define the desired conflict resolution outcomes (goals and objectives). The second step consists of a spatial multi-criteria analysis: marine habitats and competing marine uses are mapped using a geographic information system, and a multi-criteria approach is then used to determine the relative contribution of the different uses to existing conflicts. In the third step, a new spatial multi-criteria analysis is carried out to answer questions on how management actions are expected to affect spatial conflicts in the future. The final phase allocates, using optimization techniques, spaces within the conflict areas to competing users so as to minimize the intensity of conflict. After the spatial management plan has been implemented and enforced, monitoring and

ex post evaluation will provide information for a critical review of the plan and feedback for adapting it through a new iteration of the process.

Conclusions

The quest for long-term sustainability of marine ecosystems and the human activities that depend on them requires an approach promoting the intersection and the integration of multiple perspectives. In this regard, multi-criteria analysis represents an improvement with respect to traditional, single-objective approaches to planning problems (such as cost–benefit analysis), because it provides a way to address several objectives that cannot be reduced to a single dimension (e.g. the monetary one). Multi-criteria methods are available in many different forms, and the choice of the best method depends on the characteristics of the specific problem at hand and the actors involved in the process. As marine conservation planning deals with problems that have a crucial spatial component, the classical multi-criteria approach should be extended, whenever available information allows it, to a spatially explicit context, taking advantage of the capability of geographic information systems to manage and process spatial information. In recent years, an increasing number of studies have tackled spatial multi-criteria problems with a range of different approaches. The active involvement of stakeholders in complex processes, such as Marine Spatial Planning in the presence of multiple objectives embracing both biological conservation and socioeconomic sustainability issues, remains a challenging field of investigation for marine conservation science in the near future.

References

Alcamo, J. (2001) *Scenarios as Tools for International Environmental Assessments.* Environmental Issue Report 24. European Environment Agency, Copenhagen.

Andalecio, M.N. (2010) Multi-criteria decision models for management of tropical coastal fisheries: a review. *Agronomy for Sustainable Development*, **30**, 557–580.

Anderson, R.M., Hobbs, B.F., Koonce, J.F. and Locci, A.B. (2001) Using decision analysis to choose phosphorus targets for Lake Erie. *Environmental Management*, **27**, 235–252.

Ball, I.R., Possingham, H.P. and Watts, M.E. (2009) Marxan and relatives: software for spatial conservation prioritization, in *Spatial Conservation Prioritization: Quantitative Methods and Computational Tools* (eds A. Moilanen, K.A. Wilson and H.P. Possingham). Oxford University Press, Oxford, UK. pp. 185–195.

Bevacqua, D., Melià, P., Crivelli, A.J. *et al.* (2007) Multi-objective assessment of conservation measures for the European eel (*Anguilla anguilla*): an application to the Camargue lagoons. *ICES Journal of Marine Science*, **64**, 1483–1490.

Bevacqua, D., Melià, P., Crivelli, A. *et al.* (2009) Assessing management plans for the recovery of the European eel: a need for multi-objective analyses. *American Fisheries Society Symposium*, **69**, 637–647.

Boutillier, J., Noakes, D., Heritage, D. and Fulton, J. (1988) Use of multiattribute utility theory for designing invertebrate fisheries sampling programs. *North American Journal of Fisheries Management*, **8**, 84–90.

Braat, L.C. and de Groot, R. (2012) The ecosystem services agenda: bridging the worlds of natural science and economics, conservation and development, and public and private policy. *Ecosystem Services*, **1**, 4–15.

Bragge, J., Korhonen, P., Wallenius, H. and Wallenius, J. (2010) Bibliometric analysis of multiple criteria decision making/ multiattribute utility theory, in *Multiple Criteria Decision Making for Sustainable Energy and Transportation Systems: Lecture Notes in Economics and Mathematical Systems* (eds M. Ehrgott, B. Naujoks, T.J. Stewart and J. Wallenius). Springer, Berlin and Heidelberg. pp. 259–268.

Brown, J.D. (2004) Knowledge, uncertainty and physical geography: towards the development of methodologies for questioning belief. *Transactions of the Institute of British Geographers*, **29**, 367–381.

Brown, K., Adger, W.N., Tompkins, E. *et al.* (2001) Trade-off analysis for marine protected area management. *Ecological Economics*, **37**, 417–434.

Bryan, B.A., Raymond, C.M., Crossman, N.D. and King, D. (2011) Comparing spatially explicit ecological and social values for natural areas to identify effective conservation strategies. *Conservation Biology*, **25**, 172–181.

Cheng, H.-C., Château, P.-A. and Chang, Y.-C. (2015) Spatial zoning design for marine protected areas through multi-objective decision-making. *Ocean and Coastal Management*, **108**, 158–165.

Cinelli, M., Coles, S.R. and Kirwan, K. (2014) Analysis of the potentials of multi criteria decision analysis methods to conduct sustainability assessment. *Ecological Indicators*, **46**, 138–148.

Costanza, R. (2006) Nature: ecosystems without commodifying them. *Nature*, **443**, 749; author reply 750.

Costanza, R., de Groot, R., Sutton, P. *et al.* (2014) Changes in the global value of ecosystem services. *Global Environmental Change*, **26**, 152–158.

Dahl-Tacconi, N. (2005) Investigating information requirements for evaluating effectiveness of marine protected areas: Indonesian case studies. *Coastal Management*, **33**, 225–246.

Day, J. (2008) The need and practice of monitoring, evaluating and adapting marine planning and management: lessons from the Great Barrier Reef. *Marine Policy*, **32**, 823–831.

Ehler, C. and Douvere, F. (2009) *Marine Spatial Planning: A Step-by-Step Approach Toward Ecosystem-Based Management.* Report by United Nations Educational, Scientific and Cultural Organization (UNESCO). 98 pp.

Ehrlich, P.R. and Ehrlich, A. (1981) *Extinction: The Causes and Consequences of the Disappearance of Species.* Random House, New York.

Ehrlich, P.R. and Mooney, H.A. (1983) Extinction, substitution, and ecosystem services. *BioScience*, **33**, 248–254.

European Commission (EC) (2010) *Commission Decision of 1 September 2010 on criteria and methodological standards on good environmental status of marine waters.* 2010/477/EU.

Fernandes, L., Ridgley, M.A. and van't Hof, T. (1999) Multiple criteria analysis integrates economic, ecological and social objectives for coral reef managers. *Coral Reefs*, **18**, 393–402.

Fletcher, S., Saunders, J. and Herbert, R. (2011) A review of the ecosystem services provided by broadscale habitats in England's Marine Protected Area network. *Journal of Coastal Research*, **64**, 378–383.

Gaines, S.D., White, C., Carr, M.H. and Palumbi, S.R. (2010) Designing marine reserve networks for both conservation and fisheries management. *Proceedings of the National Academy of Sciences*, **107**, 18286–18293.

Gatto, M. and De Leo, G.A. (2000) Pricing biodiversity and ecosystem services: the never-ending story. *BioScience*, **50**, 347–355.

Gómez-Baggethun, E., de Groot, R., Lomas, P.L. and Montes, C. (2010) The history of ecosystem services in economic theory and practice: from early notions to markets and payment schemes. *Ecological Economics*, **69**, 1209–1218.

Gregory, R. and Slovic, P. (1997) A constructive approach to environmental valuation. *Ecological Economics*, **21**, 175–181.

Halpern, B.S. and Warner, R.R. (2003) Matching marine reserve design to reserve objectives. *Proceedings of the Royal Society B*, **270**, 1871–1878.

Heck, N., Dearden, P. and McDonald, A. (2011) Stakeholders' expectations towards a proposed marine protected area: a multi-criteria analysis of MPA performance criteria. *Ocean and Coastal Management*, **54**, 687–695.

Himes, A.H. (2007) Performance indicator importance in MPA management using a multi-criteria approach. *Coastal Management*, **35**, 601–618.

Innes, J.P. and Pascoe, S. (2010) A multi-criteria assessment of fishing gear impacts in demersal fisheries. *Journal of Environmental Management*, **91**, 932–939.

Jones, P.J.S. (2002) Marine protected area strategies: issues, divergences and the search for middle ground. *Reviews in Fish Biology and Fisheries*, **11**, 197–216.

Keeney, R.L. (1977) A utility function for examining policy affecting salmon on the Skeena River. *Journal of the Fisheries Board of Canada*, **34**, 49–63.

Keeney, R.L. and Raiffa, H. (1993) *Decisions with Multiple Objectives: Preferences and Value Tradeoffs*. Cambridge University Press, Cambridge and New York.

Klein, C.J., Chan, A., Kircher, L. *et al.* (2008) Striking a balance between biodiversity conservation and socioeconomic viability in the design of marine protected areas. *Conservation Biology*, **22**, 691–700.

Klein, C.J., Tulloch, V.J., Halpern, B.S. *et al.* (2013) Tradeoffs in marine reserve design: habitat condition, representation, and socioeconomic costs. *Conservation Letters*, **6**, 324–332.

Köksalan, M., Wallenius, J. and Zionts, S. (2013) An early history of multiple criteria decision making. *Journal of Multi-Criteria Decision Analysis*, **20**, 87–94.

Leslie, H.M. (2005) A synthesis of marine conservation planning approaches. *Conservation Biology*, **19**, 1701–1713.

Levner, E., Linkov, I. and Proth, J.-M. (eds) (2005) *Strategic Management of Marine Ecosystems*. Springer-Verlag, Berlin and Heidelberg.

Li, F., Xu, M., Liu, Q. *et al.* (2014) Ecological restoration zoning for a marine protected area: a case study of Haizhouwan National Marine Park, China. *Ocean and Coastal Management*, **98**, 158–166.

Linkov, I., Satterstrom, F.K., Kiker, G. *et al.* (2006) Multicriteria decision analysis: a comprehensive decision approach for management of contaminated sediments. *Risk Analysis*, **26**, 61–78.

Malczewski, J. (2006) GIS-based multicriteria decision analysis: a survey of the literature. *International Journal of Geographical Information Science*, **20**, 703–726.

Mardle, S. and Pascoe, S. (1999) A review of applications of multiple-criteria decision-making techniques to fisheries. *Marine Resource Economics*, **14**, 41–63.

Mazor, T., Giakoumi, S., Kark, S. and Possingham, H.P. (2014a) Large-scale conservation planning in a multinational marine environment: cost matters. *Ecological Applications*, **24**, 1115–1130.

Mazor, T., Possingham, H.P., Edelist, D. *et al.* (2014b) The crowded sea: incorporating multiple marine activities in conservation plans can significantly alter spatial priorities. *PloS ONE*, **9**, e104489. doi:10.1371/journal.pone.0104489

McCauley, D.J. (2006) Selling out on nature. *Nature*, **443**, 27–28.

Millennium Ecosystem Assessment (MEA) (2005) *Ecosystems and Human Well-Being: Synthesis*. Island Press, Washington, DC.

Miller, T.R., Minteer, B.A. and Malan, L.-C. (2011) The new conservation debate: the view from practical ethics. *Biological Conservation*, **144**, 948–957.

Moilanen, A., Kujala, H. and Leathwick, J.R. (2009) The Zonation framework and software for conservation prioritization, in *Spatial Conservation Prioritization: Quantitative Methods and Computational Tools* (eds A. Moilanen, K.A. Wilson and H.P. Possingham). Oxford University Press, Oxford.

Pascual, M., Rossetto, M., Ojea, E. *et al.* (2016) Socioeconomic impacts of marine protected areas in the Mediterranean and Black Seas. *Ocean & Coastal Management*, **133**, 1–10.

Pomeroy, R. and Douvere, F. (2008) The engagement of stakeholders in the marine spatial planning process. *Marine Policy*, **32**, 816–822.

Pressey, R.L., Watts, M.E., Barrett, T.W. and Ridges, M.J. (2009) The C-plan conservation planning system: origins, applications and possible futures, in *Spatial Conservation Prioritization: Quantitative Methods and Computational Tools* (eds A. Moilanen, K.A. Wilson and H.P. Possingham). Oxford University Press, Oxford.

Read, A.D., West, R.J., Haste, M. and Jordan, A. (2011) Optimizing voluntary compliance in marine protected areas: a comparison of recreational fisher and enforcement officer perspectives using multi-criteria analysis. *Journal of Environmental Management*, **92**, 2558–2567.

Refsgaard, J.C., van der Sluijs, J.P., Højberg, A.L. and Vanrolleghem, P.A. (2007) Uncertainty in the environmental modelling process: a framework and guidance. *Environmental Modelling and Software*, **22**, 1543–1556.

Riabacke, M., Danielson, M. and Ekenberg, L. (2012) State-of-the-art prescriptive criteria weight elicitation. *Advances in Decision Sciences*, **2012**, 1–24.

Roberts, C., Branch, G., Bustamante, R. *et al.* (2003) Application of ecological criteria in selecting marine reserves and developing reserve networks. *Ecological Applications*, **13**, S215–S228.

Rossetto, M., Bitetto, I., Spedicato, M.T. *et al.* (2015) Multi-criteria decision-making for fisheries management: a case study of Mediterranean demersal fisheries. *Marine Policy*, **53**, 83–93.

Ruiz-Frau, A., Possingham, H.P., Edwards-Jones, G. *et al.* (2015) A multidisciplinary approach in the design of marine protected areas: integration of science and stakeholder based methods. *Ocean and Coastal Management*, **103**, 86–93.

Saaty, T.L. (1977) A scaling method for priorities in hierarchical structures. *Journal of Mathematical Psychology*, **15**, 234–281.

Saaty, T.L. (1980) *The Analytic Hierarchy Process: Planning, Priority Setting, Resource Allocation*. McGraw-Hill, New York.

Saltelli, A., Tarantola, S., Campolongo, F. and Ratto, M. (2004) *Sensitivity Analysis in Practice: A Guide to Assessing Scientific Models*. John Wiley & Sons.

Saltelli, A., Ratto, M., Andres, T. *et al.* (2008) *Global Sensitivity Analysis: The Primer*. John Wiley & Sons.

Sarkar, S., Fuller, T., Aggarwal, A. *et al.* (2009) The ConsNet software platform for systematic conservation planning, in *Spatial Conservation Prioritization: Quantitative Methods and Computational Tools* (eds A. Moilanen, K.A. Wilson and H.P. Possingham). Oxford University Press, Oxford.

Stewart, R.R. and Possingham, H.P. (2005) Efficiency, costs and trade-offs in marine reserve system design. *Environmental Modeling and Assessment*, **10**, 203–213.

The Economics of Ecosystems and Biodiversity (TEEB) (2010) *The Economics of Ecosystems and Biodiversity Ecological and Economic Foundations* (ed. P. Kumar). Earthscan, London and Washington.

Thompson, M.H., Dumont, C.P. and Gaymer, C.F. (2008) ISO 14001: towards international quality environmental management standards for marine protected areas. *Ocean and Coastal Management*, **51**, 727–739.

Tuda, A.O., Stevens, T.F. and Rodwell, L.D. (2014) Resolving coastal conflicts using marine spatial planning. *Journal of Environmental Management*, **133**, 59–68.

Van Huylenbroeck, G. and Coppens, A. (1995) Multicriteria analysis of the conflicts between rural development scenarios in the Gordon District, Scotland. *Journal of Environmental Planning and Management*, **38**, 393–408.

Villa, F., Tunesi, L. and Agardy, T. (2002) Zoning marine protected areas through spatial multiple-criteria analysis: the case of the Asinara Island national marine reserve of Italy. *Conservation Biology*, **16**, 515–526.

Voyer, M., Gladstone, W. and Goodall, H. (2012) Methods of social assessment in Marine Protected Area planning: is public participation enough? *Marine Policy*, **36**, 432–439.

Wadsworth, R.M., Criddle, K. and Kruse, G.H. (2014) Incorporating stakeholder input into marine research priorities for the Aleutian Islands. *Ocean and Coastal Management*, **98**, 11–19.

Wallenius, J., Dyer, J.S., Fishburn, P.C. *et al.* (2008) Multiple criteria decision making, multiattribute utility theory: recent accomplishments and what lies ahead. *Management Science*, **38**, 645–654.

Wang, Y.-M. and Luo, Y. (2009) On rank reversal in decision analysis. *Mathematical and Computer Modelling*, **49**, 1221–1229.

Watts, M.E., Ball, I.R., Stewart, R.S. *et al.* (2009) Marxan with zones: software for optimal conservation based land- and sea-use zoning. *Environmental Modelling and Software*, **24**, 1513–1521.

Yang, C.-M., Li, J.-J. and Chiang, H.-C. (2011) Stakeholders' perspective on the sustainable utilization of marine protected areas in Green Island, Taiwan. *Ocean and Coastal Management*, **54**, 771–780.

Zionts, S. (1979) MCDM: if not a Roman numeral, then what? *Interfaces*, **9**, 94–101.

8

Ecosystem-Based Management for Marine Protected Areas: A Systematic Approach

Rafael Sardá[1], Susana Requena[2], Carlos Dominguez-Carrió[2] and Josep Maria Gili[2]

[1] Centre d'Estudis Avançats de Blanes (CEAB-CSIC), Girona, Spain
[2] Institut de Ciències del Mar (ICM-CSIC), Barcelona, Spain

Introduction

The recent and ambitious Integrated Maritime Policy (IMP) of the European Union comprises two major pillars: the Marine Strategy Framework Directive (MSFD, 2008/56/EC) and the Maritime Spatial Planning Directive (MSP, 2014/89/EU). Proposed by two different European General Directorates, these IMP regulatory tools aim to coordinate and establish coherent decision-making in order to maximize the sustainable development, economic growth and social cohesion of EU Member States in the marine domain. In addition, regarding biodiversity and nature, the European Commission has adopted an EU Biodiversity Strategy to 2020 (European Commission, 2011) to halt the loss of biodiversity and ecosystem services in the EU by 2020.

The strategy addresses the Aichi Biodiversity Targets adopted by the Convention on Biological Diversity (CBD, 2010), and thus biodiversity protection has become a prerequisite in Europe for sustainable development. The first EU 2020 Biodiversity Strategy target is to 'fully implement the Birds and Habitats Directives' (which corresponds to Aichi targets 1, 11 and 12).

The extension of the Natura 2000 network to the offshore environment was particularly emphasized so as to assure the long-term survival of Europe's most valuable marine threatened species and habitats by conserving, 'through effectively and equitably managed, ecologically representative and well connected systems of protected areas and other effective area-based conservation measures', at least 10% of all marine European waters. All of these policy frameworks are based on the utilization of the Ecosystem Approach for their implementation.

The Ecosystem Approach (EA) strategic concept, which accepts that humans are part of the global ecosystem and not separate from it, has emerged as the dominant paradigm for managing coastal and marine ecosystems (Olsen *et al.*, 2009; Farmer *et al.*, 2012). At the heart of the EA is the assumption that coupled social and ecological systems can be studied and managed in a holistic manner. This approach offers new opportunities for sustainable use of the sea but requires better understanding of how marine social-ecological systems operate, how they generate goods and services, how well these benefits are captured, how human

Management of Marine Protected Areas: A Network Perspective, First Edition. Edited by Paul D. Goriup.

degradation of the systems affects human welfare and generates costs, and the complex social relations and value systems underpinning human governance of marine systems.

Despite the importance of the EA in a growing number of policy and guidance documents, the concept remains imprecise and this makes the EA appear nebulous, rendering it difficult to put into practice. Because of these difficulties, it has been noted that management applications of EA through Ecosystem-Based Management (EBM) frameworks are wholly dependent on the aspirational visions for the social-ecological systems that deserve to be managed, and that EA and/or EBM are not goals in themselves. Appropriate tools inside effective governance systems are required in order to guide EA implementation; for this to happen, the theory of ecosystem science must be reconciled with the practice of ecosystem management (deReynier *et al.*, 2010). In order for the EA to be more widely adopted in management, we have developed a standardized, stepwise process for management: the Ecosystem-Based Management System (EBMS) (Sardá *et al.*, 2014). The EBMS introduces a common set of tools and procedures and a common language that can facilitate knowledge transfer and capacity building for managers putting the EA into practice.

The conservation of ecosystem structure and functioning to maintain ecosystem services is a priority target of the EA. Genetic diversity is widely endangered and conservation measures need to be introduced rapidly to halt the loss of biodiversity. In the marine environment, we have launched some measures to prevent environmental degradation such as the MSFD that requires all EU marine waters to achieve Good Environmental Status by 2020 (Braun, this volume), and the construction of a large network of Marine Protected Areas (MPAs). An MPA is defined as 'any area or sub-tidal terrain, together with its overlying water and associated flora, fauna, historical and cultural features, which has been reserved by law or other effective means to protect part or all of the enclosed environment (IUCN, 1994). Despite recent large-scale efforts to protect marine waters, especially relatively unaltered pristine places (e.g. around the UK Dependent Territories of Pitcairn Island Marine Reserve – $834\,334\,km^2$; Chagos Marine Protected Area in the Indian Ocean – $640\,000\,km^2$; and Ascension Island Marine Reserve in the South Atlantic – $234\,291\,km^2$; as well as the Marine Reserve of Nazca-Desventuradas Islands in Chile – $297\,518\,km^2$; and the Palau National Marine Sanctuary – $500\,000\,km^2$), the total area of marine protected space is not very large.

Scientists have proposed that at least 20% of the entire ocean space should be protected ('Troubled Waters: A Call for Action' statement – https://marine-conservation.org/marine-reserve-statement/) but only 10% is reflected in official documents such as the Aichi targets. In the Mediterranean, one of the major global marine and coastal biodiversity hotspots (Coll *et al.*, 2010), the Contracting Parties to the Barcelona Convention, through the Protocol Concerning Specially Protected Areas and Biological Diversity in the Mediterranean (SPA/BD Protocol), have established a list of Specially Protected Areas of Mediterranean Importance (SPAMI) in order to promote cooperation in the management and conservation of natural areas, as well as in the protection of threatened species and their habitats (Webster, this volume). Despite this initiative, the entire area of Mediterranean MPAs is not large: they are mostly small and, apart from the Pelagos Marine Mammal Sanctuary ($87\,500\,km^2$; http://www.sanctuaire-pelagos.org/), they total around $30\,000\,km^2$, which is clearly not adequate (Gabrié *et al.*, 2012).

Effective marine biodiversity conservation is dependent on a clear scientific rationale for practical interventions (Hiscock, 2014) but also depends on appropriate management. Placing MPAs onto EBM frameworks is urgently needed; they are not two distinct strategies that can substitute each other as has sometimes been said (Halpern *et al.*, 2010), but rather EBM is a way to put MPA targets into practice. This chapter explains the advantages of using an EA strategy for the management of MPAs and describes the need to standardize planning methods and stakeholder engagement (especially if networks are to be built), as well as the importance of incorporating risk assessment for evaluating proposed management activities. Finally, we propose using the EBMS for the management of MPAs since using a standard management tool that allows nested applications improves the protection of the marine environment. We refer to some well-established MPA networks and particular MPA sites as examples of potential EBMS application.

Marine Protected Areas and Networks

For conservation purposes, a key strategy to address many issues affecting marine and coastal ecosystems and resources is the establishment of MPAs and linking them in global network systems. Marine Protected Areas typically support a single societal value – nature conservation – having clear targeted visions (Halpern *et al.*, 2010). Numerous publications have dealt with the design and implementation of MPA networks. In particular, Laffoley (2014) stated five biophysical and ecological principles to guide such efforts: (i) include the full range of biodiversity present in the biogeographic region; (ii) ensure that ecologically significant areas are incorporated; (iii) maintain long-term protection; (iv) ensure ecological

linkages; and (v) ensure maximum contributions of individual MPAs to the network. The construction of MPA networks should follow strategic decisions that set objectives for marine conservation as a whole (long-term objectives) and also the formulation of network policies and principles intended to govern those objectives, which may cover more than nature conservation alone (Beal *et al.*, this volume). To examine these issues in practice, two established MPA networks, and one of their component sites, are briefly discussed below.

Network of Marine Protected Area Managers in the Mediterranean (MedPAN)

The Mediterranean Sea is considered to be one of the world's priority ecoregions (UNEP-MAP RAC/SPA, 2010). The objective of MedPAN is to facilitate exchanges between Mediterranean MPAs in order to improve their management (Webster, this volume). Created in 1991, the MedPAN network acts to build the capacity of MPA managers around the Mediterranean basin through the exchange of best practice and the development of tools for the management of MPAs. MedPAN also contributes to the establishment of a representative and coherent ecological network of MPAs, which is a step beyond the more traditional approach of designing MPAs as single independent entities (http://www.aires-marines.com/International/Exchange-Networks/Medpan). MedPAN, in collaboration with the Regional Activity Centre for Specially Protected Areas (RAC/SPA) of the United Nations Environment Programme/Mediterranean Action Plan (UNEP-MAP), has recently reviewed the status of MPAs in the Mediterranean (Gabrié *et al.*, 2012) and made important recommendations for further work (Table 8.1). One of the points highlighted in the report was the low level of

Table 8.1 Main conclusions and recommendations concerning the status of MPAs in the Mediterranean Sea.

Main conclusions

1) Information on Mediterranean MPAs is more accurate than that for other areas. Details have been recorded in the MAPAMED database.

2) The target of 10% protection is far from being achieved.

3) There is still a disproportionate geographical distribution and MPAs are still mainly on the coast.

4) Representativity of ecological sub-regions, habitats and species is very variable.

5) The adequacy and viability of sites is very variable.

6) The ecological coherence is better in the western basin but still low on a Mediterranean scale.

7) MPA management is still insufficient.

Recommendations

1) Reinforce the development of the MPA network in order to achieve the target of 10% of Mediterranean surface area being protected.

2) Reinforce the effectiveness of protection management and evaluation measures in MPAs.

3) Reinforce the resources and tools to ensure evaluation of management effectiveness.

4) Promote the development of evaluation tools on a regional level.

5) Ensure a better management of threats to MPAs.

6) Enhance the international recognition of Mediterranean MPAs.

Source: Adapted from Gabrié *et al.* (2012).

management effectiveness and a lack of application of the EA in management, with a general recommendation that management tools should be better implemented.

Gabrié *et al.* (2012) identified 677 MPAs in the complex environment of the Mediterranean. From the answers of 80 respondents to a questionnaire, 42% had management structures and 84% had permanent staff. Application of a standard management tool could facilitate coordination and harmonization of conservation practices and clearly would facilitate dialogue. It could also work within a nested application, covering regional- and national-level networks as well as single MPAs. Below we consider the 'Cap de Creus' MPA, as an example.

The 'Cap de Creus' MPA

In the north-western part of the Mediterranean, the 'Cap de Creus' region exhibits environmental, social, economic and geographical characteristics that make it unique in the Mediterranean. It includes a large portion of the marine area located off Alt Empordà county (Girona, Spain) protected by two contiguous Sites of Community Importance (SCI) designated under the EU Habitats Directive. The 'Cap de Creus' SCI, which is also a maritime-terrestrial Natural Park (the first one established in Catalonia, in 1998), has an area of 13 844 ha of which 22% is marine. In 2014, the offshore marine waters around the platform and submarine canyons of the region were also proposed by the Spanish government as an SCI denoted 'Sistema de cañones submarinos occidentales del Golfo de León'. This SCI covers an area of 98 772 ha. Together, the two SCIs form one of the largest protected spaces in the Mediterranean Sea and will be referred to as the 'Cap de Creus' area from here on, although they are managed by different national and regional governance structures.

The 'Cap de Creus' area is located at the border between France and Spain where French authorities had already established different protected areas under their national regulations ('Parcs Naturels Marins') and Natura 2000 (Figure 8.1). Consequently it has unique characteristics as an area located in a transboundary region. Moreover, the 'Parc Naturel Marin du Golfe du Lion' goes beyond the median line which separates the French and Spanish territorial waters, which creates an added difficulty concerning the overlap of conservation schemes, planning and management. As a result, this example provides an extremely interesting case for further study and application of the EBM framework.

North-East Asian Marine Protected Areas Network (NEAMPAN)

The United Nations North-East Asian Sub-regional Programme for Environmental Cooperation (NEASPEC) launched its NEAMPAN project in 2012. It includes five countries: China, Democratic People's Republic of Korea, Republic of Korea, Japan and the Russian Federation. The project aims to establish an effective, functional and representative network of MPAs in the sub-region for better conservation of marine and coastal biodiversity and more efficient MPA management. The network focuses on (i) protection of key marine animals and their habitats, (ii) sustainable use of marine resources, and (iii) effective MPA management. NEAMPAN holds regular network meetings, and expects to conduct in-depth research, provide training courses for capacity building, and network with relevant regional and global mechanisms. Nevertheless, a recent assessment of its operations identified some severe limitations for the process, including use of different terminologies, inconsistency in MPA identification, deficiencies in national-level MPA networks, different institutional settings for management, and low level of international cooperation (http://www.neaspec.org/our-work/marine-protected-areas-mpa-north-east-asia). As observed earlier in the MedPAN network, management effectiveness is still far from being achieved, although some success in the region may pave the way for improvements; such is the case of the Suncheon Bay Wetland Protected Area.

Suncheon Bay Wetland Protected Area

The Suncheon Bay (3550 ha) and Muan Tidal Flats (3559 ha) protected areas in the Republic of Korea are recognized as wetlands of international importance under the Ramsar Convention, making them one of the most spectacular places in South Korea. Both sites support a range of threatened migratory birds and are also important for harvesting fish, seaweed and molluscs using traditional techniques. They have been incorporated into the NEAMPAN network and are subject to a large-scale master plan for Suncheon Bay Landscape Conservation. A set of policies have been implemented in the area, starting with setting up a Committee for Suncheon Bay Nature and Ecology that promotes networking activities between civil society groups, government bodies and specialists in the conservation of Suncheon Bay. The mid- to long-term master plan comprises: (i) roadmap of stages to enhance new values of the bay; (ii) analysis of ecological health and change of mudflats and reedbeds; (iii) development of community-based ecotourism and community well-being; (iv) adoption of nature protection priorities; (v) restoration of the mudflats ecosystem; and (vi) enlargement of business and civil society initiatives within the nature protection priorities. A clear governance system, coupled with the establishment of a Suncheon Bay Conservation Fund, makes this MPA a good place to implement an EBM framework approach.

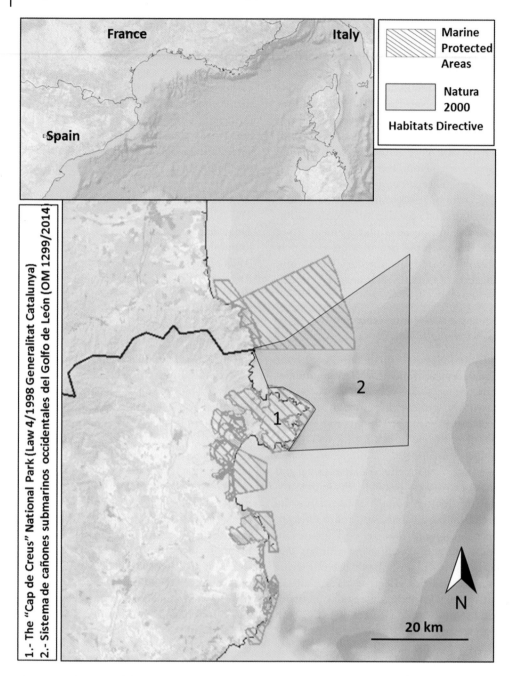

1.- The "Cap de Creus" National Park (Law 4/1998 Generalitat Catalunya)
2.- Sistema de cañones submarinos occidentales del Golfo de León (OM 1299/2014)

Figure 8.1 The MPAs of the 'Cap de Creus' region. (1) The 'Cap de Creus' SCI, a maritime-terrestrial Natural Park. (2) The 'Sistema de cañones submarinos occidentales del Golfo de León' SCI, an offshore Natura 2000 area.

The Use of Ecosystem-Based Management for MPAs and Networks

Once MPA sites and networks are planned, designed and implemented, and spatially bounded regions containing a particular ecosystem and social system interacting with each other are delimited, then appropriate management tools should follow. Management can be defined as the function that coordinates the efforts of people to accomplish goals and objectives by using the available resources efficiently and effectively. Applying EBM as a framework for managing MPAs and networks is desirable in order to fully incorporate MPAs and networks within a clear EA strategy (Sardá *et al.*, 2014). Essentially, EBM requires consideration of whole ecosystems at a scale that ensures that ecosystem integrity is maintained. It recognizes the complex interactions between species that make up marine ecosystems, and so is underpinned by principles of community biology and ecology. Ecosystem-based management also brings together the human, biological and physical parts of the system for which management action is needed. It adopts a new model of integrated management that addresses the Malawi principles of the EA (CBD, 1998).

In order to use an EBM framework under the EA strategy for MPAs, the Malawi principles need to be translated into management actions. Several aspects that relate EA principles with clear management actions can then be considered:

- Setting the scene of management (principle 6)
- Using a systems approach to management – enhancing participation, achieving a common view on societal choices (principle 1)
- Implementing adaptive management – targeted long-term visions with operational short ones (principle 8)

- Recognizing the importance of the ecosystem structure and function (principle 5)
- Working with decision-making procedures in a decentralized way (principles 2 and 4)
- Developing an environmental accounting framework (principles 3 and 11)
- Taking account of all scaling effects (principles 7 and 9)
- Considering humans as part of the global ecosystem – but having a clear site/network vision and involving all sectors of society (principles 10 and 12).

Setting the Scene of Management

The first task is to determine the area under management. In the case of MPAs this task is normally simple because the boundaries of the area, the social-ecological area to be managed, are precisely defined. After delimitation, management of the area should be based on measures derived from an initial assessment (departure stage) and a desired final vision (desired stage). The desired vision will establish the goals and timescales for environmental performance against which the effectiveness of the management system can be judged.

In formulating the desired vision, joint fact-finding is important in order to develop shared knowledge about the site and reach the best vision while avoiding conflicts. It is a way to guide the process of gathering information, analyse facts, and make informed decisions collectively. An absence of joint activities is very likely to lead to conflicts sooner or later.

Using a Systems Approach to Management

A systems approach to project management enables MPA managers to continuously evaluate the needs of the area; the end results to be achieved in line with the final vision for the MPA; and the needs in terms of resources, budget and time. In order to quantify the desired vision for the MPA, it is recommended to work with something similar to the 'Good Environmental Status' (GES) concept described in the MSFD. We strongly believe this strategic GES concept can be applied worldwide, although obviously, in the case of EU Member States, the descriptor and indicators used in the MPA application will be much stricter. Good Environmental Status should be established individually for every MPA, defined as 'the vision status of the MPA where these provide an ecologically diverse and dynamic environment which is clean, healthy and productive in accordance with its conservation status'. Then, depending on the MPA selected, possible uses made of its marine resources should take place at a sustainable level, ensuring their continuity for future generations, and an evaluation of pressures should be carried out. Figure 8.2 shows a schematic diagram of the GES descriptors set out in the MSFD; for these 11 descriptors, desired state indicators should be

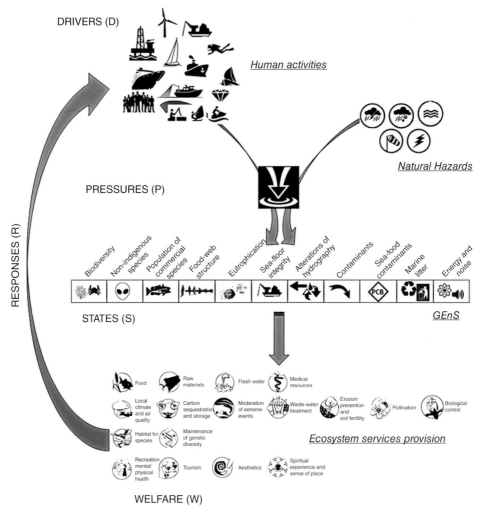

Figure 8.2 A DPSWR representation of the platform of indicators for the EBMS information pillar. Ecosystem service provision icons obtained from The Economics of Ecosystems and Biodiversity (TEEB).

selected in order to evaluate management effectiveness through time depending on the MPA concerned (governance structure, budget, pressures). Figure 8.2 also indicates how these state indicators can be linked with other indicators of the Drivers–Pressures–State–Welfare–Response (DPSWR) social-ecological accounting framework that can be used for aspects related to information (Cooper, 2013; Ojea *et al.*, this volume).

An initial assessment report should be drafted to develop a common understanding of the system under management. It should collate and synthesize all the relevant information (ecosystem overview, socio-economic pressure factors and related stakeholders to be considered in the management guidance) as well as an assessment of the ecosystem services provided by the area since this is an intrinsic part of an EA strategy. This report constitutes an ecosystem overview, the baseline 'status quo' of the MPA.

Implementing Adaptive Management

Having prepared and considered the above-mentioned policy elements (the definition of the present state of the MPA social-ecological system, its 'status quo' or ecosystem overview, and formulation of a desired vision in terms of GES with its provision of ecosystem services), the issue of using adaptive management as a tool both to change and to learn about a system comes to play a key role.

Adaptive management is a structured, iterative process of robust decision-making in the face of uncertainty, which aims to reduce uncertainty over time via system monitoring. Adaptive management offers a practical means of integrating knowledge over social and economic as well as ecological scales. It can accommodate unexpected events by encouraging approaches that build system resilience and is becoming accepted as a valuable tool for delivering the EA.

Adaptive management encourages managers to adopt policy cycles for a limited period, closely observing the outcomes of interventions through carefully focused monitoring. At the end of an initial learning period, the model can be further refined and new management objectives set.

Advocating the Use of the Ecosystem Services Vocabulary

One of the basic principles of the EA strategy is the conservation of ecosystem structure and functioning to maintain ecosystem services. Sustaining the long-term capacity of marine ecosystems, and in this case MPAs, to deliver a range of ecosystem services with a focus on both ecosystem health and human well-being is a key part of the management required. This necessitates the identification of how GES generates goods and services based on the MPA vision; how well these benefits are captured; how human activities and natural hazards may affect MPAs and generate costs; and the complex social relations and value systems underpinning human governance.

Although MPAs have the ultimate goal of biodiversity conservation, as Potts *et al.* (2014) have shown, MPAs can also provide direct or indirect benefits for society through the delivery of different ecosystem services. In this context, the mapping of ecosystem services is increasingly recognized as a valuable tool in the management of MPAs and building stakeholder appreciation of such services, and so taking them into account in collaborative approaches (Cárcamo *et al.*, 2014). The concept of ecosystem services is crucial when social and ecological issues need to be managed in a holistic way.

Working with Decision-making Procedures in a Decentralized Way

The inclusion of a risk management standard follows modern management best practice for environmental decision-making.

During recent years the inclusion of these risk management tools into decision-making applications when managing the marine environment has been advocated (MacDiarmid, 1997; Cormier *et al.*, 2013, 2015). In the case of MPAs, the aim is to assess the significant risks that could impede achievement and/or maintenance of GES. These risks basically fall into two main groups: (i) those derived from external pressures and/or events that can separate future and/or present environmental states from the desired ones; and (ii) those related to an evaluation of the capacity of the region and the human activities involved to provide the ecosystem services required by the MPA. Historically, decision-making at this level had been largely sectoral and more judgemental than analytical. Correct selection of the key social-ecological aspects and planning evaluation will increasingly favour programmes intended to reduce negative effects while moving towards GES.

A prioritization tool intended to help MPA managers determine which projects and/or activities should be carried out before others, based on a social trade-off analysis and the established MPA vision (GES), has been described in detail in Sardá *et al.* (2010). The tool works in three sequential stages: (i) the identification procedure, including the identification of the main components of the system and the risk identification process; (ii) the assessment phase, which is the initial prioritization procedure; and (iii) the final decision about priority objectives and targets for the implementation phase.

Developing an Environmental Accounting Framework

A prerequisite for correct environmental management is the comprehensive compilation and analysis of environmental information. This information must be combined with user-friendly tools to facilitate the decision-making process. Traditionally in MPAs, information about the area is linked to monitoring programmes in the context of marine reserves, and observational research that documents variability in natural systems by comparing them with manipulated systems over time. However, other types of information are also needed depending on the management policy cycle and indicators selected.

For example, the information system could employ indicators within the DPSWR social-ecological accounting framework (Figure 8.2). Since conservation of the ecosystem structure and functioning to maintain ecosystem services is the priority target of the EA (see Table 8.2), and the EBMS is based on an EA strategy, description of the ecosystem services desired in the visioning phase is a key point in the indicator analysis. Welfare indicators will be associated with the provision of these ecosystem services. Then, the provision of these ecosystem services will be related to state indicators raised using the GES framework. The vulnerability of services will be expressed as human activities and natural hazards (pressure indicators) that can potentially harm ecosystem state components and ultimately modify their provision of services. The relationship between these indicators is shown in Figure 8.2.

Considering Humans as Part of Global Ecosystems

The participation of society is an essential element of the EA. Normally, MPAs are established through consultation processes: national planning forums, expert panels and so on come up with a list of potential areas for protection that will require governmental approval. Scientific and preparatory work is needed to persuade governments to conserve and restore the richness of marine life and habitat. Once the MPA has been designated, an effective governance structure is

Table 8.2 Relationship between the Ecosystem Approach principles developed by the Convention on Biological Diversity[a)] and their application to MPA management frameworks.

CBD Ecosystem Approach principles	MPA management needs
1) The objectives of management of land, water and living resources are a matter of societal choices	• Use participatory planning: appropriate management schemes should ensure adequate and timely participation in a transparent decision-making process by local populations • Adopt a holistic methodology from a geographic perspective: MPAs cannot be isolated from one another and the regional network should be designed with societal approval
2) Management should be decentralized to the lowest appropriate level	• Develop an effective governance structure to guide MPA management implementation
3) Ecosystem managers should consider the effects (actual or potential) of their activities on adjacent and other ecosystems	• Elucidate the social-ecological dynamics and functioning of the MPA • Integrate all elements relating to the hydrological, geomorphological, climatic, ecological, socio-economic and cultural systems into the prevailing conservation view
4) Recognizing potential gains from management, there is usually a need to understand and manage the ecosystem in an economic context	• Accommodate and prioritize ecosystem services given by MPAs, but also consider the multiplicity of social-ecological activities/events that can be observed in these areas
5) Conservation of ecosystem structure and functioning, to maintain ecosystem services, should be a priority target of the Ecosystem Approach	• The concept of ecosystem services should be central in the management of MPAs
6) Ecosystems must be managed within the limits of their functioning	• MPA management should support natural processes and adopt a long-term perspective • Damage to the MPA should be strongly resisted and, where it occurs, appropriate restoration measure should be taken rapidly
7) The Ecosystem Approach should be undertaken at the appropriate spatial and temporal scales	• MPA management frameworks should be designed in a network context; the use of nested management structures is highly recommended to address this need
8) Recognizing the varying temporal scales and lag-effects that characterize ecosystem processes, objectives for ecosystem management should be set for the long term	• MPA management should be part of a vision-driven process • The ultimate aim is to align this management with obtaining a conservation goal for the MPA that promotes sustainable development in its surrounding area
9) Management must recognize that change is inevitable	• Adaptive management should be incorporated in the planning process to recognize change through a dedicated monitoring programme

(Continued)

Table 8.2 (Continued)

CBD Ecosystem Approach principles	MPA management needs
10) The Ecosystem Approach should seek the appropriate balance between, and integration of, conservation and use of biological diversity	• In the case of MPAs this is an intrinsic part of the conservation goal
11) The Ecosystem Approach should consider all forms of relevant information, including scientific and indigenous and local knowledge, innovations and practices	• An information system should be developed to harness results from monitoring and assist decision-making in the management process
12) The Ecosystem Approach should involve all relevant sectors of society and scientific disciplines	• Institutional coordination of the various administrative services and regional and local authorities competent in the coastal and/or marine zone should be required • An appropriate and effective governance structure is needed, from the site level through to established MPA networks

a) https://www.cbd.int/ecosystem/principles.shtml

needed in order to develop and implement management measures, together with identification of the main actors around the MPA that can pressure and/or work with it. In addition, the EA requires the adoption of a holistic attitude from a geographic perspective because the MPA (or MPA network) under management cannot be isolated from its surroundings. The governance structure must promote significant cooperation amongst government bodies, civil society and private interests in the pursuit of collective action. Clearly, a participatory policy can facilitate this work in order to overcome possible operational and financial barriers for management.

A Systematic Approach for Using the EBMS in MPA Management

The EBMS was developed by Sardá *et al.* (2014; http://www.msfd.eu) to integrate EA into management functions. It employs a standardized process for applying EA principles by ensuring the inclusion of essential components such as participation, planning and decision-making, and by promoting accountability and quality assurance. The EBMS seeks to achieve vision-based management objectives that follow sustainable development principles based on the provision of ecosystem services (CBD, 1998; Balvanera *et al.*, 2001; Cognetti and Maltagliati, 2010). The introduction of the EBMS into MPA management can (i) enhance management effectiveness as recommended by Gabrié *et al.* (2012), and (ii) address the need to include the principles of the EA into MPA management practice (see Table 8.2).

The EBMS has a three-pillar structure that facilitates the incorporation of the EA into the management of coastal and marine zones, regardless of the ecosystem or administrative scales (Figure 8.3). The general points to be emphasized are:

• The EBMS follows a vision-driven process: a societal vision needs to be developed prior to the use of the framework, in this case a clear conservation vision.

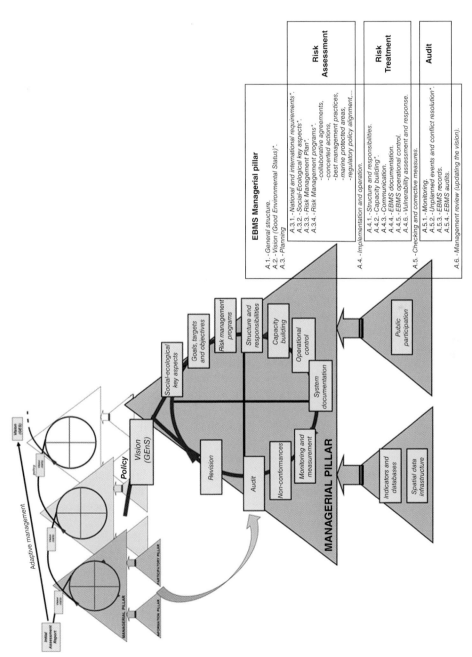

Figure 8.3 The managerial pillar of the EBMS showing the different stages used in the policy cycle, following the Deming cycle of management.

- The system identifies actions and/or activities to reach and/or to maintain this desired vision using risk management tools.
- The system prioritizes actions and/or activities that conserve ecosystem structure and functioning, to maintain ecosystem services. Evaluation measures are incorporated into each policy cycle.
- Information tools follow a DPSWR accounting framework.
- Participatory tools ensure the active involvement of stakeholders.

The *managerial pillar* is the foundation of the system: it is derived from classical environmental and risk management systems that include environmental considerations and objectives within a continuous improvement cycle of adaptive management. The managerial pillar has to be supported by governance structures that provide oversight and thereby ensure that planning and implementation activities adhere to modern environmental principles. It follows the main elements of a Deming cycle loop: policy baseline, planning preparedness, implementation and operation, checking and corrective actions (Deming, 1986). Formally, it works within the structure of ISO 14001 where most of its clauses had been replaced by those used in the previously developed ISO 31000:2009 for risk management (ISO, 2009a, 2009b, 2009c). The elements of the policy cycle have been adapted to work with the principles of the EA (see Table 8.2). A DEcision-MAking (DEMA) tool has been designed to intervene at this planning phase, as the conceptual procedure to bring the above clauses into practice. This iterative DEMA process follows the recent ISO framework for risk management (ISO 31000). The inclusion of a risk management standard follows modern management best practice for environmental decision-making (MacDiarmid, 1997; Cormier *et al.*, 2013, 2015).

Second, the *information pillar* ensures that data and scientific advice are grounded on best available knowledge. It employs the DPSWR social-ecological accounting framework to organize the information on aspects of social and ecological systems relevant to representing the interactions between them (Cooper, 2013). It is structured in line with the so-called Information Factory with two main support tools: a Spatial Data Infrastructure (SDI) that can be appended to a knowledge-based portal, and a platform of indicators linked to the desired provision of MPA ecosystem services (Cinnirella *et al.*, 2011). Both tools should be accessible in the system at any time.

Third, the *participation pillar* brings together institutional coordination, communication and consultation requirements. It is designed to accomplish three main tasks: (i) facilitation of stakeholder identification, (ii) allowing effective participation and conflict resolution, and (iii) enhancing capacity building. Tools are available for the identification of stakeholders (e.g. Sanó, 2009; Bainbridge *et al.*, 2011), and initiatives to generate informed networks of stakeholders and enhancing capacity are beginning to emerge. Without doubt, management faces its greatest difficulties when putting into practice this new paradigm of participative governance and conflict resolution due to the fact that different stakeholders around the MPA (including all national, regional and local authorities competent in the MPA as well as society in general) have different interests (Cormier-Salem, 2014). The use of the EBMS framework could introduce a common language and a common set of procedures facilitating dialogue, coordination, and capacity building between the different offices and public agents involved. At the same time, the use of the EBMS should allow clear statements of future visions for the MPA that could facilitate public engagement and participation.

The EBMS can be used at different scales, from individual MPA sites to different types of network (Beal *et al.*, this volume).

It can facilitate understanding and alleviate common problems related to the terminology used in management activities. Using the two networks considered above (MedPAN and NEAMPAN), good examples for pilot plan applications were identified.

In the 'Cap de Creus', scientific studies in the region (Gili *et al.*, 2011, 2012; Lo Iacono *et al.*, 2012; Sardá *et al.*, 2012) provide excellent baseline information to develop a GES vision for the area as well to set up its initial assessment. In addition, characterization of its conservation status could be made through an assessment of pressures, a stakeholder mapping structure and an ecosystem service review. The closest approximation to this task would be the ecosystem overview and socio-economic overview reports used by the Canadian Department of Fisheries and Oceans, which provide comprehensive descriptions of the current knowledge of ecological, cultural, social and economic considerations for a planning area (DFO, 2005). The 'Cap de Creus' could be a very good example for an EBMS implementation: information is available, management structures are in place, coordination is feasible and a general vision has been developed.

Effective MPA management is one of the main objectives for NEAMPAN. A recent presentation in the Suncheon Bay Wetland Protected Area explored how the EBMS could be applied to this coastal environment, as well as the Muan Tidal Flats. As participatory tools have already been implemented in the Suncheon Bay Wetland Protected Area and a Committee of Suncheon Bay Nature and Ecology drives the authorities to enhance conservation of the bay, this could favour an EBMS application. The use of the EBMS at governmental level to manage the coastal-marine network, and in the NEASPEC to enhance the intergovernmental cooperation mechanism for the region, is possible by applying a nested scaled application of the EBMS.

Discussion and Conclusions

The Ecosystem Approach strategic concept has emerged as the dominant paradigm for managing coastal and marine ecosystems with the main goal of maintaining and/or restoring marine biological integrity to ensure the adequate provision of ecosystem goods and services. Regarding conservation objectives, MPAs are planned and designed to meet long-term nature protection, a clear long-term objective under an EA strategy. Although the majority of MPAs combine protection and the sustainable development of activities, their ultimate vision is to conserve biodiversity, habitat structure and the functioning of the ecosystem. When designing the tactical and operational objectives for running MPAs to achieve visions, goals and targets in these areas, EBM frameworks should be considered. The area can be problematical due to the fact that sometimes the division between the EA and EBM is not clear (Halpern *et al.*, 2010). The Ecosystem Approach is a strategic concept whereas EBM is the tactical and operational means to implement the strategy. As MPAs have delimited boundaries around social-ecological systems where a clear vision has been defined, EBM frameworks ought to be deployed to manage these areas to achieve the declared vision. In this chapter we have proposed (i) adoption of the strategy of the MSFD by adopting a GES vision for every MPA under protection that can be linked to its ecosystem services provision, and (ii) use of an EBMS as the standard management tool to reach and/or to maintain this vision (Sardá *et al.*, 2014).

The EBMS is designed to be a standard adaptive management methodology to assist MPA managers by providing a common set of tools and procedures and a common language that can facilitate knowledge transfer and capacity building. In addition, the EBMS is easily scalable and can be hierarchically introduced at different spatial scales, which could facilitate the institutional coordination needed to solve the problem of policy fragmentation and differentiated responsibilities normally seen in reality. The EBMS is considered a quality assurance tool in itself, being used in a vision-driven process of continuous improvement (towards achieving GES), necessitating reaching of a prior consensus for the desired future conditions of the MPA environment under management – something that lies at the heart of these designated areas.

The use of the EBMS will allow authorities to manage, in an integrated way, the different functions of the MPA environment and the ecosystem services they provide. The EBMS adds new aspects not considered in a classical MPA management structure:

1) MPA management is part of a clear vision-driven process
2) It adopts a holistic approach from a geographical perspective
3) It requires pressure analysis and institutional coordination inside clear participatory planning
4) Planning is achieved through the use of risk management techniques
5) The concept of ecosystem services is central
6) It uses the DPSWR as its analytical accounting framework of indicators
7) A good final state is based on state indicators using the GES concept
8) It ensures timely participation by the local population.

The basic structure of the EBMS and related material is available as a web platform tool (http://www.msfd.eu) to facilitate training and capacity building.

A large number of MPAs worldwide have in place a management structure and associated permanent staff. Management of these areas, however, is normally carried out using informal systems and tools. It would be easy to conclude from this global pattern that every MPA constitutes a particular case, in which it takes time to understand how it is working and to accomplish the desired objectives. A correct management cycle for all of these areas should focus on measures (monitoring programmes) that allow managers to alleviate negative pressures for the correct functioning of the area, disclosing all the information following transparent sustainable principles. Effective governance structures and relevant tools are needed for this change, and the EBMS has been designed to facilitate this.

Acknowledgements

This work was carried out within the framework of the PLAYA+ project (CGL 2013-49061) of the National Research Plan of Spain in R+D+i, as well as the KnowSeas+project (201530E018) and the LIFE+INDEMARES project. We thank the United Nations North-East Asian Subregional Programme for Environmental Cooperation (NEASPEC) and the MedPAN network for the invitation to present this work in their facilities.

References

Bainbridge, J.M., Potts, T. and O'Higgins, T.G. (2011) Rapid policy network mapping: a new method for understanding governance structures for implementation of marine environmental policy. *PLoS ONE*, **6** (10), e26149. doi:10.1371/journal.pone.0026149

Balvanera, P., Daily, G.C., Ehrlich, P.R. *et al.* (2001) Conserving biodiversity and ecosystem services. *Science*, **291**, 2047.

Cárcamo, P.F., Garay-Flühmann, R., Squeo, F.A. and Gaymer C.F. (2014) Using stakeholders' perspective of ecosystem services and biodiversity features to plan a marine protected area. *Environmental Science & Policy*, **40**, 116–131.

Cinnirella, S., March, D., O'Higgins, T.G. *et al.* (2011) A multidisciplinary Spatial Data Infrastructure for the Mediterranean to support implementation of the Marine Strategy Framework Directive. *International Journal of Spatial Data Infrastructures Research*, **7**, 323–352.

Cognetti, G. and Maltagliati, F. (2010) Ecosystems service provision: an operational way for marine biodiversity conservation and management. *Marine Pollution Bulletin*, **60**, 1916–1923.

Coll, M., Piroddi, C., Steenbeek, J. *et al.* (2010) The biodiversity of the Mediterranean Sea: estimates, patterns, and threats. *PLoS ONE*, **5** (8), e11842. doi:10.1371/journal. pone.0011842

Convention on Biological Diversity (CBD) (1998) *Report of the Workshop on Ecosystem Approach, Lilongwe, Malawi, 26–28 January 1998*. UNEP/CBD/COP/4/Inf.9, p. 15.

Convention on Biological Diversity (CBD) (2010) *Strategic Plan for Biodiversity 2011–2020*. Decision X/2 of the 10th Conference of the Parties. https://www.cbd. int/decision/cop/?id=12268

Cooper, P. (2013) Socio-ecological accounting: DPSWR, a modified DPSIR framework, and its application to marine ecosystems. *Ecological Economics*, **94**, 106–115.

Cormier, R., Kannen, A., Elliott, M. *et al.* (2013) Marine and Coastal Ecosystem-Based Risk Management Handbook. ICES Cooperative Research Report 317. 59 pp.

Cormier, R., Kannen, A., Elliott, M. *et al.* (2015) *Marine Spatial Planning Quality Management System*. ICES Cooperative Research Report 327. 106 pp.

Cormier-Salem, M.C. (2014) Participatory governance of Marine Protected Areas: a political challenge, an ethical imperative, different trajectories. *S.A.P.I.EN.S*, **7** (2), 1–10.

Deming, E.W. (1986) *Out of the Crisis*. MIT Center for Advanced Engineering Study.

deReynier, Y.L., Levin, P.S. and Shoji, H.L. (2010) Bringing stakeholders, scientists and managers together through an integrated ecosystem assessment process. *Marine Policy*, **34**, 534–540.

DFO (2005) *Guidelines on Evaluating Ecosystem Overviews and Assessments: Necessary Documentation*. DFO Canadian Science Advisory Secretariat, Advisory Report, 2005/026.

European Commission (2011) *Our Life Insurance, Our Natural Capital: An EU Biodiversity Strategy to 2020*. Communication from the Commission. COM/2011/0244 final.

Farmer, A., Mee, L., Langmead, O. *et al.* (2012) *The Ecosystem Approach in Marine Management*. EU FP7 KNOWSEAS Project. ISBN 0-9529089-5-6.

Gabrié, C., Lagabrielle, E., Bissery, C. *et al.* (2012) *The Status of Marine Protected Areas in the Mediterranean Sea*. MedPAN, Marseilles and RAC/SPA, Tunis. MedPAN Collection. 256 pp.

Gili, J.M., Madurell, T., Requena, S. *et al.* (2011) *Caracterización física y ecológica del área marina del Cap de Creus Informe final área*. LIFE+ INDEMARES (LIFE07/ NAT/E/000732) Instituto de Ciencias del Mar/CSIC (Barcelona). Coordinación, Fundación Biodiversidad, Madrid. 272 pp.

Gili, J.M., Requena, S., Madurell, T. and Dominguez, C. (2012) *Memoria justificativa de la revisión de la zonificación y de la gestión propuesta por el CSIC para la zona indemares 'Cañón de Creus' plataforma y talud continental*. 15 pp.

Halpern, B.S., Lester, S.E. and McLeod, K.L. (2010) Placing marine protected areas onto the ecosystem-based management seascape. *Proceedings of the National Academy of Sciences*, **107** (43), 18312–18317.

Hiscock, K. (2014) *Marine Biodiversity Conservation: A Practical Approach*. Routledge, UK. 318 pp.

International Standard Organization (ISO) (2009a) *Risk Management: Principles and Guidelines.* International Standards Organization. ISO 31000:2009(E).

International Standard Organization (ISO) (2009b) *Risk Management: Vocabulary.* International Standards Organization. ISO GUIDE 73:2009(E/F).

International Standard Organization (ISO) (2009c) *Risk Management: Risk Assessment Techniques.* International Standards Organization. IEC/ISO 31010.

International Union for Conservation of Nature (IUCN) (1994) *Guidelines for Protected Area Management Categories.* IUCN, Cambridge, UK and Gland, Switzerland.

Laffoley, D. (ed.) (2014) *Guidelines to Improve the Implementation of the Mediterranean Specially Protected Areas Network and Connectivity between Specially Protected Areas.* RAC/SPA, Tunis. 32 pp.

Lo Iacono, C., Orejas, C., Gori, C. *et al.* (2012) Habitats of the Cap de Creus continental shelf and Cap de Creus canyon, Northwestern Mediterranean, in *Seafloor Geomorphology as Benthic Habitat.* Elsevier. pp. 457–469.

MacDiarmid, S.C. (1997) Risk analysis, international trade, and animal health, in *Fundamentals of Risk Analysis and Risk Management.* CRC Lewis Publications, Boca Raton. pp. 377–387.

Olsen, S.B., Page, G.G. and Ochoa, E. (2009) *The Analysis of Governance to Ecosystem Change: A Handbook for Assembling a Baseline.* LOICZ Reports and Studies 34. GKSS Research Center, Geesthacht. 87 pp.

Potts, T., Burdon, D., Jackson, E. *et al.* (2014) Do marine protected areas deliver flows of ecosystem services to support human welfare? *Marine Policy,* **44**, 139–148.

Sanó, M. (2009) *A systems approach to identify indicators for integrated coastal zone management.* PhD dissertation, Department of Water Science and Technology, Civil Engineering School, University of Cantabria, Santander, Cantabria, Spain.

Sardá, R., Diedrich, A., Tintoré, J. *et al.* (2010) *Development of Risk Assessment: Decision Making (DEMA) Tool and Demonstration.* Deliverable 6.2 EU FP7 KnowSeas project. http://www.msfd.eu. 54 pp.

Sardá, R., Rossi, S., Martí, X. and Gili, J.M. (2012) Marine benthic cartography of the 'Cap de Creus' (NE Catalan Coast, Mediterranean Sea). *Scientia Marina,* **76**, 159–171.

Sardá, R., O'Higgins, T., Cormier, R. *et al.* (2014) Proposal of a marine ecosystem-based management system: linking the theory of environmental policy with practice of environmental management. *Ecology and Society,* **19** (4), 51.

UNEP-MAP RAC/SPA (2010) *The Mediterranean Sea Biodiversity: State of the Ecosystems, Pressures, Impacts and Future Priorities* (eds H. Bazairi, S. Ben Haj, F. Boero *et al.*). RAC/SPA, Tunis. 100 pp.

9

Developing Collaboration among Marine Protected Area Managers to Strengthen Network Management

Chloë Webster

Mediterranean Protected Areas Network (MedPAN), Marseilles, France

Introduction

Effective collaboration among people is a prerequisite to the successful management of Marine Protected Areas (MPAs) and achievement of ecologically coherent MPA networks. A network of MPAs can be planned as an ecological one to conserve biological features and manage activities, but more often it is an opportunistic system of protected sites or an assortment of designated spaces that intends to manage pressures and protect biological diversity, cultural heritage and resources (see Beal *et al.*, this volume). Each site has objectives and all sites have different types of designations, or labels, which are chaperoned by different shades of legally binding provisions down to pure soft law.

In the Mediterranean, there are currently 1236 MPAs[1] recorded (out of 1458 entries registered in the database of MPAs and other sites of interest for conservation; MAPAMED, 2016). This chapter will explore how collaboration between MPA managers enhances management at the level of a single MPA and at the level of systems of MPAs. To do this, it will step back into the unnatural history of the Mediterranean basin, inspect human behaviour and the role of cultural identity, deconstruct social networks, touch on economics, look at how bottom-up initiatives meet top-down ones, and question what could come next. It will do this by referring to MedPAN (Mediterranean Protected Areas Network), a social network of MPA managers, but also to other analogous networks such as RAMPAO (*Réseau régional d'Aires Marines Protégées en Afrique de l'Ouest*) and CaMPAM (Caribbean Marine Protected Area Management Network and Forum).

Two questions will be answered:

1) If everywhere on our blue planet social networks of MPA managers sprout up, how come?
2) How can a social network of MPA managers successfully support the management of an ecological system of MPAs?

Management of Marine Protected Areas: A Network Perspective, First Edition. Edited by Paul D. Goriup.
© 2017 John Wiley & Sons Ltd. Published 2017 by John Wiley & Sons Ltd.

Once Upon a Time, the Mediterranean Basin Filled Up Again ... and Soon Emptied

In the late Miocene, the Mediterranean had become isolated and mostly dried up with only a few pockets of water here and there, an event referred to as the Messinian salinity crisis (Gautier *et al.*, 1994; Krijgsman *et al.*, 1996). With the Zanclean flood, the tap was turned on again (Clauzon *et al.*, 1996; Garcia-Castellanos *et al.*, 2009) and century after century the small semi-enclosed *Mare nostrum* became a principal hub for human civilizations, their trades and their enrichment. The marine environment got on with its own business, seizing all circumstantial conditions to diversify its habitats and species, thrive, and generously afford a 'gift economy'[2] to ambitious and often covetous human animals. All this happened along a virtual unconsulted parallel: human thinking and acting on one side, natural processes on the other. The 'tragedy of the commons'[3] was thus born long ago. And even with the predictable fall of the age of plenty lying behind us, hope continues to drive ocean prospection and research. There must still be untapped resources in that blue treasure chest!

While the non-expandable Mediterranean space has become the theatre of all trades, one can't help but marvel at the sight of free Mediterranean marine ecosystem services continuing to benefit growth, or the recycled concepts leading to Blue Growth or Blue Economy. In fact, many marvel at the sight of what life remains under the surface despite decades of abuse, many love the sight of marine and coastal natural landscapes, many have been and go on to be inspired by Mediterranean sights and wonders. For them, that is the true blue treasure chest.

But of course, paying back the sea for its services can't be done at the cashier desk! And fast emptying with a twist of trashing attitude leaves time neither to replenish the shelves nor to conduct some maintenance and cleaning-up.

How the 'Reasonable' Person Revelation Sparks Environmental Conservation Policies

These degradation facts led to a general sense of helplessness among different social groups and very much among the Mediterranean people...

So here steps in the good side of human nature. Although we have the immense capacity to be unreasonable, cognitive humans have repeatedly found niches for survival from an evolutionary perspective – a good enough motivation! As such, to break away from a corrosive sense of helplessness somewhat linked to altruism (Piliavin and Hong-Wen, 1990; Simpson and Willer, 2015) and to the 'economic man' theories,[4] the 'reasonable person' movement was conducive to defining solutions for bettering environmental behaviour in a participatory way (Kaplan, 2000). One could argue that this movement was characterized by various international and regional conventions,[5] national laws and more local rules and regulations (whether recognized juridically or agreed on within a social group), all to protect the natural marine environment. One could call these top-down processes in most cases, although some have adopted highly consultative approaches with field-based cornerstone representatives.

Still with reference to the 'reasonable person' movement, simply marvelling at the Mediterranean, loving it and understanding that quality of life relies on environmental quality are likely to be added motivations which condition the actions of key players and often the path they choose for a career. When these individuals connect, begin sharing experience and get together for conservation action, one could easily argue that they represent social networks of reasonable and responsible humans who pave the road towards healthier environmental behaviour. One could call these social groups part of a bottom-up process.

So, stepping back into the Mediterranean with its hectic weaving of countless overlapping human activities and good share of resulting clashes over the use of marine and maritime space and its resources: environmental responsibility is therefore expected to temper ecological abuse, serve conflict resolution over the natural 'wet' space, and continue to provide a measure of sustenance while ensuring sustainability.

Looking at how this responsibility evolved from a legal framework point of view, it is important to trace it back to the early part of the 20th century.

First looking at the top-down process, an initial attempt in 1930 by some nations to extend their ownership over the sea and fish stocks and thereby rein in some pressures over the marine environment was inconclusive (League of Nations, The Hague, 1930). During the next 25 years, a number of countries claimed their rights over the 200 nautical miles area from their coasts. It took that time-lapse to convene UNCLOS I (United Nations Conference on the Law of the Sea, 1956) and draw up four conventions that entered into force in the 1960s. While UNCLOS II was inconclusive, UNCLOS III (1973 to 1982) led to the Montego Bay Convention (1982, and effective in 1994) to which 167 nations, observers and the European Union are parties. This Convention, also known today as the United Nations Convention on the Law of the Sea (or Treaty on the Law of the Sea), together with the 1994 Agreement on the sea bed, established a more comprehensive set of provisions to organize 'ownership and use' and solve some conflicts. While the societal wish for free access to the commons persists

in many individual approaches, UNCLOS has contributed to reshaping some of the 1609 Grotius legal principles expressed in *Mare Liberum* (Grotius, 1609), from a top-down perspective. Yet, while this Convention rightly addresses pollution and living resources, it is more relevant to rights of access than to conservation.

It is only in the 1970s that unanimous international recognition of the need to protect the environment takes a sharp turn. It does so in the 1972 Stockholm Declaration, with the understanding that this cannot be done without integrating humankind back into its environment. The Stockholm Declaration came out of a conference suitably entitled 'the United Nations Conference on the Human Environment' and underlined precisely that 'no-growth' was not a viable solution. To spell things out bluntly, the traditional 'economic man' was required to bid farewell to the gift economy and start counting his cowries, looking at what they are and where they come from. Subsequent key international UN events that led to responsible decision making for the environment, and that are also relevant to the Mediterranean Sea, can be traced back through the following important steps:

1) The 1992 Rio Declaration from the UN Conference on Environment and Development (UNCED – its principles of Agenda 21).
2) The signature at the 1992 UN Earth Summit of the Convention on Biological Diversity (CBD) by 150 states to translate Agenda 21 into reality.
3) The 2000 UN Millennium Summit (with its commitment via the Millennium Declaration), and the Millennium Development Goals (MDG) that followed the summit.
4) The 2002 Rio + 10 UN World Summit on Sustainable Development (WSSD) – with the Johannesburg Declaration and United Nations Environment Programme (UNEP) Assessment.

5) The 2010 Convention on Biological Diversity (CBD) launch of a Strategic Plan for biodiversity 2011–2020 (with 20 ambitious yet achievable targets).

6) The UN General Assembly recognizing 2011–2020 to be the decade of biodiversity.

7) The 2012 Rio + 20 UN Conference on Sustainable Development with its Sustainable Development Goals (SDG) to reach beyond the agenda of the MDG.

Specific to the Mediterranean, and shortly after the Stockholm Ministerial Conference that set up UNEP, 16 countries and the European Union adopted the Mediterranean Action Plan (MAP) and the Barcelona Convention for the Protection of the Mediterranean Sea Against Pollution in 1976. Initially concerned chiefly with pollution, the parties ratified seven protocols over time, including one focusing on MPAs and marine biodiversity protection: the Specially Protected Area for Biological Diversity Protocol (SPA/BD) which was adopted in 1995 and ratified in 1999 (excepting four parties). This protocol aims to contribute to the CBD objectives, in particular the Aichi target 11 of 10% of the marine environment to be protected and well managed (CBD, 2008). At present, 22 parties strive to fulfil the latest phase of the MAP, including objectives relating to MPAs in order to contribute to the CBD targets.

Now looking at the same decades, but from a field-based bottom-up perspective, the above-mentioned 'responsible' people seized conservation opportunities arising from the favourable political context described above. Some of them also became organized as structured civil society groups (associations and NGOs). Their primary goal was to preserve the natural marine environment quickly and at all costs. Some of these people also pushed for more stringent and concrete conservation actions in the field, namely establishing MPAs, thus truly triggering the launch of area-based management measures.

The Escalating Establishment of MPAs in the Mediterranean since 1960

About half a century ago, in the early 1960s (a decade before the 1972 Stockholm conference), only a dozen MPAs had been created in the Mediterranean. The first one was born in Croatia (then part of former Yugoslavia) with the Mljet National Park in 1960, followed in 1963 by the establishment of the *Parc National de Port-Cros* off the southern shores of France. Alongside Croatia and France, Israel and Turkey contributed to establishing these 12 'early' MPAs. Looking just at nationally declared MPAs across the 20 countries that have created protected sites and which are parties to the Barcelona Convention,[6] there were establishment peaks in the early 1990s and then around 1997, 2002 and then 2010 (see Figure 9.1). From the late 1990s, the European Union (EU) network of Natura 2000 sites at sea comes in to boost numbers. Yet, peaks in numbers of sites do not mean that greater surface coverage was achieved at these dates (a peak can represent a large number of very small sites being created for example). From a surface coverage point of view, trends differ, with a take-off in the late 1980s and larger surface coverage contributions in 2000 and 2014 (MAPAMED, 2015). These coverage figures account for designation overlap and therefore refer to a total surface area by eliminating double counting.

At the beginning of 2016, MAPAMED (the database on MPAs and other sites of interest for conservation of the Mediterranean marine environment) reported 1236 MPA entries of varying levels of legal protection[7] (MAPAMED, 2016). However, many of

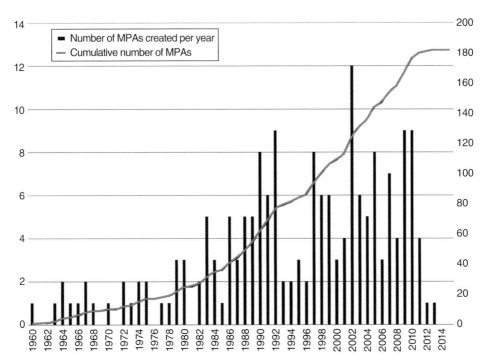

Figure 9.1 Number of national MPAs established since 1963 up to 2013, and cumulative number. *Source*: MAPAMED (2015). Reproduced by permission of MedPAN/MAPAMED.

these MPAs currently lack enough support to be effective, and in reality they are little more than 'paper parks'. Having said that, among those identified, there are the following: 176 nationally designated MPAs, 898 EU Natura 2000 sites at sea, 34 Specially Protected Areas of Mediterranean Importance (SPAMIs), four International Fisheries Restricted Areas (General Fisheries Commission for the Mediterranean (GFCM) FRAs), one Sanctuary for Marine Mammals, one Particularly Sensitive Sea Area (International Maritime Organization (IMO) PSSA), 97 Ramsar sites, 11 United Nations Educational, Scientific and Cultural Organization (UNESCO) Biosphere Reserves, and three UNESCO World Heritage sites.[8]

While MAPAMED has developed a definition for the generic term of MPA for inclusion of sites in the database (Claudet *et al.*, 2011) – hence including FRAs and PSSAs based on their conservation-driven

objectives – many other sites of interest for conservation of the Mediterranean marine environment which have currently been included under a different coding, along with future other sites (such as national fisheries reserves) could potentially be classified as Other Effective area-based Conservation Measures (OECMs). It could indeed be recognized that national fisheries reserves that are managed all year round, as well as the regional GFCM ban on bottom trawling below 1000 m depth (which could become a FRA) and the EU ban on towed gear within three nautical miles of the shore (or above the closest 50 m isobath to shore) bring ancillary conservation value to this system of MPAs. However, they are not considered in this analysis as the focus here rests on designations clearly targeting marine biological conservation as their prime motive, and not protection of food-related production resources.

From Marine Natural Features under a Bell Jar … to Participatory Measures for Managing Pressures

The early approach to conservation consisted of locking away parts of the natural marine environment to protect it, as if placed under a bell jar. The trend today has shifted to establishing more flexible multi-use MPAs which include zoning with targeted regulations attached to these sub-zones, at least in the Mediterranean (Webster *et al.*, in press). As such, the area-based conservation measures may, or may not, include no-take or even no-go zones and these may not necessarily cover the full 'surface' of the MPA. These multi-use MPAs increasingly involve collaborations (or even a sort of co-management) with the interested parties (stakeholders) in order to manage the space and especially threats to habitats and species, or the natural capital upon which an economic activity relies. These MPAs thus increasingly focus on curbing pressures and are consequently driven to integrate the territorial geographical and social dimension more markedly. In fact, they resemble Marine Spatial Planning (MSP) strategies that integrate the principles of sustainable development and where the needs of different users are taken into account. Today, this means many more multiple relationships at multidisciplinary levels. Furthermore, existing MPAs persistently express the need for adequate financial and human resources and capacity to be actually managed – let alone effectively managed – and for information and data exchange or flow. These very needs are central both to the development of social networks and to maintaining their vitality.

What is a Social Network?

To understand how collaborations among MPA managers can benefit the management of an individual MPA or of a system, or even true network, of MPAs from the ecological/biological stance, one needs to grasp (i) what social networks are, (ii) what may trigger their establishment, (iii) the role they can play, and (iv) how to sustain them.

Networks, generically, are sets of relationships that contain nodes and different types of relations and/or interactions between these nodes (static, directional, mutually directional or multi-directional). When there is more than one relationship, it is referred to as a multiplex relationship. To describe a network, one needs to look at these relationships and also at the distance between them. When all nodes are directly connected (whatever the nature of the relation), it is called a balanced network.

When this concept is transferred to people (through human nodes or organization nodes), a social dimension is introduced bringing along complex human reasons for the connections between the nodes. While social networks have existed since the dawn of time, thinkers and academics began to consider the concept and principles of social networks in the 19th century, gradually teasing out theories that have integrated several disciplines over time, and readjusting these in reaction to technological breakthroughs. Today, social network analysis allows us to better understand the individual and collective motivations behind an effective social network, what it can bring to an individual or groups of individuals that are part of the network, and what role the social network can play on a geographical scale and in a given thematic. It provides this understanding by mapping and measuring the ties (relationships) and edges (interactions) between people, groups of people or organizations, examining the nature of the flows between these nodes, and building metrics on the type of connections, distributions and segmentations.

The types of connections will include levels of similarities or 'assortativity', such as age, occupation and values, or level of friendship and frequency of interaction.

The types of distributions include how a particular node makes the link between other nodes, or how a node/group of nodes influences the rest of a network (centrality/popularity), or the proportion of direct interactions between nodes in relation to the total number possible. Researchers may also identify 'structural holes', where there are weak ties between several parts of a network, in order to determine networking needs. The types of segmentations comprise nodes that are likely to form a clique (where all nodes have a direct tie to one another) or structural cohesion, which can also lead to a break with the rest of the full network.

Following from the above, in the context of marine nature conservation, it is possible to identify entities that can serve as nodes of a social network, including: MPA managers; organizations that work with marine conservation (e.g. NGOs, consultancies, other private companies); international, regional, national and local institutions; and individuals that interact with any of the previously mentioned entities whether decision makers, entrepreneurs, fishermen, federations or business operators.

MedPAN: An Example of an MPA Social Network

MedPAN was set up officially under the name of Mediterranean Protected Areas Network during the Monaco Conference in 1990 (with the support of the World Bank and contributions from IUCN and the French government). The Secretariat was initially entrusted to *Parc National de Port-Cros* (France). The founders of this network, who were MPA managers, aimed for MedPAN to become a network open to all Mediterranean marine and coastal protected areas in order to facilitate exchange between managers to improve the efficiency of the management of these protected areas. To this end, a forum was organized. Yet in 1996, activities come to a halt for lack of funding. In 1999, the MedPAN members sought support from the UNEP-MAP and set up a non-profit association under French law. Following this, they requested technical support from WWF-France to take over operational responsibility for the network, and in 2004 activities resumed with secure funding from the European Regional Development Fund INTERREG IIC as of 2005. Funding dwindled yet again in 2007 for holding the 1st Mediterranean MPA Conference (Porquerolles, France). Consequently, in 2008, at the request of nine founding members (mainly MPA managers), a new legal entity with a self-governing Secretariat was created (under French law) for the MedPAN organization, independent from the leadership of a single existing entity, such as the Port-Cros MPA or WWF NGO.

Interestingly, the driving force for setting up the MedPAN network up to 2008 was a group of MPA managers and other people or organizations interested in Mediterranean marine conservation who got on well with one another and who had charisma along with fresh ideas: in social network analysis terms, these were nodes with strong ties and edges, centrality and good structural cohesion.

From 2008, the MedPAN organization had a Secretariat based in France which grew into a team comprising seven staff today, led by an Executive Secretary. Members of MedPAN are all MPA managers (while partners are organizations that support marine conservation and MPAs but are not the actual official managers). Members vote on a number of decisions such as the work plan, budget, and inclusion of new partners to the network at an annual General Assembly (GA). In 2015, the GA involved 63 members, who together managed close to 100 MPAs, and 36 partners, from 18 Mediterranean countries. A classical mapping of this network is shown in Figure 9.2.

Another key element to the governance of MedPAN is the elected Board of Directors

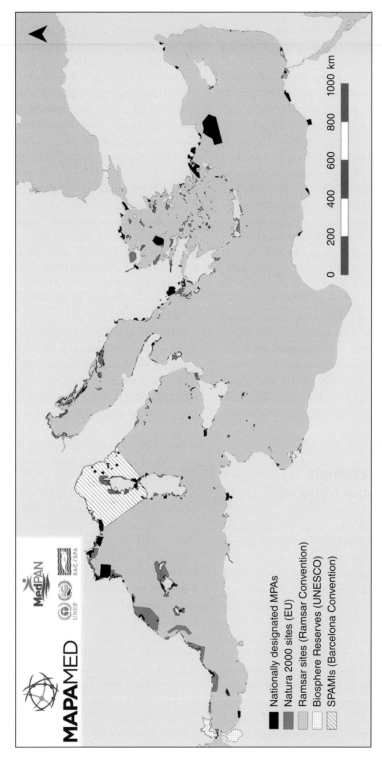

Figure 9.2 Map of all MPA entries from MAPAMED (top) and representation of MedPAN member MPAs (bottom). *Source:* MAPAMED (2015). Reproduced by permission of MedPAN/MAPAMED.

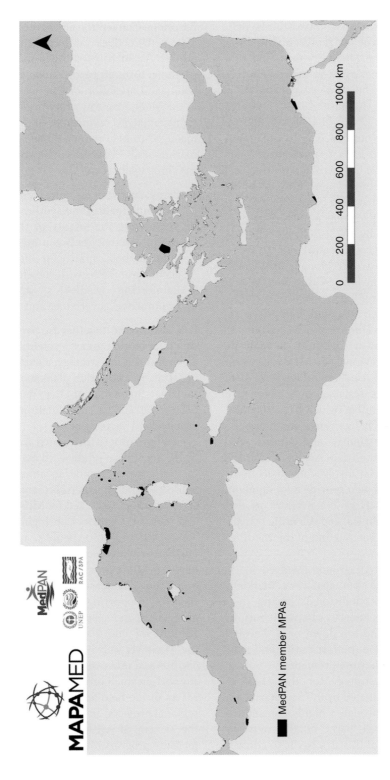

Figure 9.2 (Continued)

(BoD) which consists of two-thirds MPA management structure representatives from all shores of the Mediterranean, and one-third institutional partners' representatives. The latter are the strategic collaborations MedPAN has established with regional players such as UNEP-MAP RAC/SPA and WWF, to name but two. The 13 BoD members are elected for a term of three years, with one-third of the members retiring each year. They meet every month by telephone conference, and at least twice a year face to face, to approve (or not) a number of decisions proposed by the Secretariat. The BoD Chairperson plays a central role in representing the network regionally and worldwide in different fora. Within the BoD, an Executive Committee (or Bureau) is elected among the members and comprises the Chairperson, an Honorary Vice Chairperson, a Secretary and a Treasurer. Beyond the BoD, strategic regional partners, with whom Memoranda of Understanding have been signed in most cases, meet once a year in an Advisory Committee (which has a consultative role) to exchange views on all areas of the MedPAN strategy and annual work plan. Furthermore, MedPAN has a Scientific Committee which involves up to 15 nominated members who provide a fundamental consultative body of the network as they support the technical and scientific orientations of the Secretariat and of the MedPAN strategic actions.

All the different groups involved in the governance and, perhaps more markedly, some prominent individuals among these groups, serve as interfaces between bottom-up processes and top-down ones, between field-based action (both scientific and managerial) and policy requirements.

A five-year strategy for the MedPAN network was initiated by the Secretariat and BoD in 2012 through a highly consultative process with all parts of the network. The backbone of this strategy is to contribute to reaching targets set by the CBD, the Barcelona Convention and several European Union Directives and policies, among others. As such, MedPAN's mission is to promote, through partnerships, the sustainability and operation of a network of MPAs in the Mediterranean which are ecologically representative, connected and effectively managed to help reduce the current rate of marine biodiversity loss. The strategy itself has three main axes:

1) Being a network for knowledge, information, anticipation and synthesis
2) Reinforcing the vitality of the network and interactivity between members, and building their capacity for effective management of MPAs with stakeholders
3) Reinforcing MedPAN's sustainability, prominence, governance and resources.

It is also bound together by five cross-cutting intervention areas: a scientific strategy, a communication strategy, a capacity building strategy, and action plans for (i) the sustainable financing of the network and Secretariat, and (ii) consolidation of the governance and Secretariat.

Examples of activities which are organized by the Secretariat jointly with many partners in the network include an annual training workshop and an experience-sharing workshop on a topic selected by MPA managers and network partners, bringing together about 40 and 150 people, respectively. Further activities comprise exchange visits between MPAs for capacity building; developing tools, databases, guidelines and tutorials for supporting management actions; newsletters and a series of scientific special editions for information flow; maintaining the MedPAN website (as well as Facebook, Twitter and other informal e-mail groups)[9]; supporting small projects to directly benefit MPAs in the field and taking part in larger regional projects; an analysis of the status of the system of MPAs every four years to measure progress towards reaching regional mandatory targets (which also incorporates running MAPAMED, jointly with RAC/ SPA); paving the way with a roadmap for contracting parties to regional and international treaties to be followed in reaching conservation objectives; and prompting the establishment of a regional trust fund for Mediterranean MPAs.

Activities undertaken through the MedPAN strategy reach out to about 7000 individuals in the Mediterranean region and worldwide. The questions that then arise are how did this social network grow so quickly, how does it keep strong, and how does this influence or even strengthen the management of a single MPA, of sub-regional systems of MPAs and of regional systems of MPAs intending to become ecologically coherent?

Social Network Analysis: MedPAN Network under Scrutiny

At the beginning of the chapter it was noted how groups of 'reasonable' people sparked off environmental conservation policies. In the context of MedPAN, it is again the case that some 'reasonable' people sparked off a niche social network. A full analysis of the MedPAN social network, applying the appropriate modelling, statistical and algorithmic tools, is beyond the scope of this chapter. However, it is possible to identify some of the principal nodes through an array of metrics, and assign some qualitative characteristics to them.

In the early stages of MedPAN network development, the nodes that triggered the initiative had strong ties and edges, despite some geographical distance, as they met or communicated often enough to share their commonalities. In other words, they were highly assortative because they shared numerous salient characteristics, displayed multiplexity in the sense that they held a

professional relationship and also developed friendship, along with showing reciprocity of friendship and of professional interaction. These nodes also probably shared the same motivation to cherish the blue treasure chest and perhaps the desire to address a perceived lack of urgent action needed to curb environmental degradation by existing institutions. This could possibly be defined as the healthy need for cognitive closure, that is, to find answers to challenges they faced in wishing to protect the marine environment, urgently and permanently.[10] To do this, they had to get together, strengthen ties and be recognized as an identifiable unit, therefore becoming more powerful in pushing for marine conservation action and reinforcing their own capacity and that of the network which was to grow from then on. This group of nodes, if considered as a single node within the wider social 'ecosystem', could be qualified as influent, and if its centrality value or coefficient had been calculated at the time, it would certainly have been very high.

Following its revival in 2008, MedPAN had nine founding members: seven management nodes (including NGOs that manage MPAs) and two NGO nodes that work in support of MPAs. These were: Kornati National Park (Croatia), Egyptian Environmental Affairs Agency, *Parc National de Port-Cros* (France), WWF-France, ADENA (Agde MPA, France), Zakynthos National Park (Greece), Miramare Marine Natural Reserve/WWF-Italy, *Zavod RS zavarstvonarave* (Slovenia) and WWF-Turkey. All of these are considered charismatic nodes bearing centrality, but they do have different scales. Some nodes enjoy popularity primarily at a localized level, either because of their propinquity, or because they filter information flow as semi-passive receptors and create mutual or multi-directional ties only concerning subjects of personal or local interest. Other nodes have a regional/international reach with high multiplexity in their multi-directional ties or the reciprocal flows of information they maintain. With additional support of more peripheral nodes, the establishment of the Secretariat and the constitution of the Board of Directors, MedPAN expanded, welcoming new member-nodes and eventually partner-nodes.

Within the 13 members of the BoD, which still includes some of the founding members, all representatives are respected for their various attributes, such as strong personality/charisma, strategic regional representativity and/or accountability (among other qualities). Each member of the BoD has centrality and multiplexity well anchored within the network and beyond. Many BoD members play a key 'bridging' role in the network whereby they connect the MedPAN core to various regional and international entities. Some BoD nodes also have highly multi-directional ties and multiplexity within sub-sections of the wider MedPAN membership. This is the case with AdriaPAN, the Adriatic Protected Areas Network. This bottom-up initiative was launched by two Italian MPAs (Miramare and *Torre del Cerrano*, both members of MedPAN) and aims to facilitate contact between protected areas in the Adriatic Sea and improve their partnership effectiveness both in management and in planning activities. While this social network currently has about 40 riparian Adriatic members and some 30 associated organizations, not all of them necessarily have reciprocal connections to the MedPAN nodes. It goes without saying that such sub-network sets are highly desirable for the dynamics of the greater whole due to their specificity at addressing ecoregional needs among MPA managers. Furthermore, several nodes within AdriaPAN contribute to bridging connectivity gaps between field nodes and the European Commission policies involving MPAs, as does MedPAN via other node ties.

Returning to the BoD, if it is considered as a single node, it gains in esteem principally for its intensive support of the MedPAN Secretariat, and the cohesive image it projects for those that become aware of its role.

With respect to centrality, the Secretariat plays the principal role within the MedPAN network. Anticipation, organizational skills, multidisciplinary understanding, driving a defined clear purpose, putting into practice a kind of altruism by seeking the best interest for the commons, team complementarity and cohesion, along with great responsiveness are some traits that possibly enhance interactions between all the nodes of the network at large. This imperceptibly becomes a 'balanced network' as defined in social network analysis whereby all triads of nodes have some kind of connection, whether passive or active. The Secretariat, while maintaining and stimulating multi-directional ties, also plays the bridging role with regional and international institutions, recognizing that working hand in hand with legally enshrined organizations benefits common interests, filling structural holes. In this regard, it is clear that the most representative dyad of such interactions is that between the MedPAN Secretariat and RAC/SPA: this positive reciprocal tie and its edges means that RAC/SPA brings legitimacy to MedPAN while MedPAN brings flexibility to RAC/SPA. Although this is the most salient feature of the collaboration, it goes without saying that there are also many other advantages as well.

The Secretariat also benefits from multi-directional ties and edges with all the nodes of the Advisory Committee and Scientific Committee, thus creating multidisciplinary interactions to contribute to a faster achievement of concerted conservation objectives. It continually strengthens its interactions with other networks of MPA managers, territorial social groups, sea users' federations and consortia, and groups of decision makers worldwide to share experience and join forces in contributing to a common purpose.

The interactions with and among members and partners of the network are much more complex yet represent the *raison d'être* of MedPAN. A hundred individual nodes (Figure 9.2) lay the foundation for the network's existence, its dynamics and connections to about 7000 other nodes. If a social network analysis representational map was to be derived from all the interactivity among the MedPAN nodes, the result would surely be astonishing.

Now the question of why the network keeps strong has mostly been answered, but the one of why it expanded so fast has only been hinted at. While being a function of the characteristics of central nodes within the network, 'adhesion' and intensified interest also depends on typical reasons common to all social networks.

Possible Reasons for Joining a Niche Social Network such as MedPAN

- **Identity** and a common sense of the 'self' while sharing common characteristics is arguably one of the main reasons. In the Mediterranean, despite geopolitical upheavals, religious differences and the array of languages, there is a strong feeling of cultural identity. People relate to one another as 'Mediterraneans', which, incidentally, is often recognized as a distinct race. Although controversial, Sergi (1901) considered the Mediterranean race as the 'greatest race of the world', singularly responsible for the most accomplished civilizations of antiquity. Seligman (1924) supports this view, inviting recognition 'that the Mediterranean race has actually more achievement to its credit than any other, since it is responsible for by far the greater part of the Mediterranean civilisation ... and so shaped not only the Aegean cultures, but those of Western and greater part of Eastern Mediterranean lands'. Coon (1958) also acclaimed the race, stating that 'The Mediterraneans occupy the centre of

the stage; their areas of greatest concentration are precisely those where civilization is the oldest'.

- **Belonging** to an identifiable community and to a scheme that enables the self to be part of something important on a grand scale, or in other words a **purpose**, are arguably two further important reasons. For members to belong to the network subconsciously implies that they can gather around a grand cause, share their experiences and provide mutual support (MacMillan and Chavis, 1986). It also lifts them out from a certain state of isolation, and a feeling of having to face problems on one's own, with the offer of a solution-orientated community (Baumeister and Leary, 1995). The efficiency of the supporting unit in providing the needed tools, and orchestrating information flow within the network and beyond, is essential for maintaining **engagement** through the sense of belonging and adhering to a clear purpose over time.

- **Objectives**, when set clearly, attract memberships, another types of 'adhesion', by individual nodes or groups of nodes. Indeed, tangible goals set within a network provide nodes with frames of reference and yardsticks in order to find win–win scenarios for their own benefit and that of their close connections, or the 'edges' they have established (Lardbucket project, 2012).

- **Interest** (and passion) naturally strongly federates nodes in a network while bringing in an assortment of emotions linked to positivism, 'feel-good' reactions and gratification. It is a chief motivational element to the good functioning of a network and maintaining engagement within the community (Rheingold, 1993). To some extent, this goes with ensuring access to **information** that can nourish the interest and passion and that can be conveniently accessible at all times. Further to this the same information needs to be accessible in different formats or levels of detail to remain accessible to all nodes. In fact, storytelling remains a winner for the majority of nodes, especially when obvious replicable elements are underscored.

- **Behaviour betterment**, whether desired for oneself or others, is enabled by a social network and becomes attractive for nodes to buy in when accountability is showcased and when positive peer reinforcement is made available (Borg, 2012). Guidance and incentives are part of this process. Nodes can seek referrals, advice and support/encouragement (together with positive image reinforcement) and also remodel a positive image of themselves by sharing attempts, progress or successes, along with advice, with other nodes.

- **Individual recognition** within and beyond a given social network is arguably the last key reason why the MedPAN community grew so fast. Recognition of individuals is deeply rooted in each of us from the early stages of infancy. Self-construction and self-esteem begin with realizing one exists in the eyes of another individual when recognition by the mother (usually) is identified by the infant via social cues (Jenkins, 2014; Schore, 2015). The quality of information flow in this initial dyad is unsurprisingly of importance to the self. Silent social cues as in facial expressions of admiration versus disapproval set a precedent for self-development and for seeking different types and strengths of recognition in later years. An interactive and highly connected network calls back to this sense of recognition as what one does and says gets appreciated by other members of the community, sometimes even being showcased to all other nodes of the network, together with positive reinforcement. How rewarding! Here, it is to be noted that individual expression of opinion is freely allowed beyond what the node would feel free to express within his/her institution or affiliated organization, which to a great extent contributes to the success of social networks.

Other peripheral facilitators to the growth of the MedPAN network include: the ease of joining the network as a member and financial support to meet and take action regarding planned objectives; how other nodes are connected to one another and how popular or answerable they are perceived to be; a sense of idealistic proximity and ancillary occasions to meet new people or catch up with old friends. There may also be appreciation of talent recognition (with identified champions and opportunities for companionship) and a sense of pride for the community achievements. Or perhaps simply a handful of nodes have a need to 'hang out' live and online or interact out of pure curiosity?

One thing is certain however: with the current international social context of interconnectivity craving, belonging to a social network has become fashionable. Such networks are beginning to supplant institutions, with power shifting back to people because they want to interact with other people and because the opportunity to share an identity within a community that reflects one's self and personal interests is far more enticing than being locked behind a brand and having to stoically convey the identity of an institution. This is obviously a caricature but the process is a reality, especially in niche networks that knit enthusiasts: networks allow humanness to be brought back into a professional community.

Incidentally, this humanness also means that all personality types will be allowed to interact freely, whether they are predominantly of the Openness Type (artistic, curious, imaginative...), the Conscientious Type (efficient, organized...), the Extravert Type (energetic, active, assertive...), the Agreeable Type (compassionate, cooperative...) or the Neurotic Type (anxious, tense, self-pitying...) (Costa and McCrae, 1992). Yet, the social network principles provide a self-regulating process to all these connections and it is rarely necessary for a node of great centrality to have to intervene, especially when nodes know that it can (i.e. exclusion from a social community is not a desirable outcome). Finally it is to be noted that personality traits partly influence responsiveness and reactivity in the interactions between nodes.

How a Network of MPA Managers Strengthens the Management of MPA Networks

The intent here is not to look into the management effectiveness of MPAs (i.e. the positive impact on natural marine features of an MPA via its enforced regulations and operational management measures compared to outside the MPA). Rather, it is to look into the connection between the social dimension of a network and (i) the field management activities required for the actual management of sites by people (and thereby expected derived positive impact for the natural environment), and (ii) the requirements for constructing ecologically coherent networks of MPAs that are actually managed. At this stage, it is to be noted that the term 'ecological coherence' finds no grounding in science and was a sound product from policy makers that today leads to challenging implementation within the policy–science–management interface.

Nonetheless, to address the first question regarding the connection between the social dimension of a network and field management activities required for good management, the dominant term may be 'mutualization' (or a pooling effect). All the reasons and contextual factors described in previous sections underline the propensity of people to better manage an MPA, or at least intend to do so, because of the pooling of experience and knowledge inferred via a dynamic niche social network, and because of socio-psychological individual human traits. Summarizing, it indeed means

sharing problems and solutions, access to processed relevant information, tools and guidance, accompaniment and support, extra funding, a recognized identity, and clarified objectives to attain. The evidence of the benefits brought to an individual MPA managing person or team of belonging to the MedPAN network can be identified through the many results and testimonies from, and information flow between, the nodes of the network.

If looking at this through the prism of thematics, the central node (MedPAN Secretariat) acts as a catalyst and matchmaker between different social communities that need to address a particular challenge. For example, the 2013 experience-sharing workshop entitled 'Surveillance and enforcement of regulations in MPAs: how to maximize the efficiency and sustainability of actions' (Hyères, France) took place near the *Parc National de Port-Cros*. They took the opportunity to share their strategies on strengthening cooperation with competent authorities, even when the legal framework may not offer the adequate tools for compliance and prosecution (such as had been the case in Croatia). It was the opportunity for many MPA managers to hear the perspectives of Maritime Affairs and Public Prosecutor. Typically in a number of MPAs, maintaining contact and good cooperation with these entities, along with inviting them to properly discover the MPA and its management, had strengthened the involvement and role of competent authorities in managing the MPA (thus strengthening efficiency).

Beyond sharing good practice, the workshop was also conducive to exploring problems, such as the enforcement of no-take zones (i.e. no-fishing areas) which are not always accepted let alone complied with by small-scale fishermen. Following the workshop, an exchange visit was organized for Libyan MPA managers and fishers to visit the Gökova MPA in Turkey and share experience with their counterparts.

Gökova has six no-fishing zones yet the fishermen have quickly grasped the additional economic benefits these have brought about and are much involved in the management. Through many multidirectional interactions, information that no-take zones can be positive has gradually been disseminated throughout the network. Indeed, communities of nodes that were initially not involved with MPAs (i.e. groups of small-scale fishermen) have sprouted around the Mediterranean and become linked to MedPAN: MedArtNet (the Mediterranean Platform of Artisanal Fishers), the Libyan Association for the Development of Fishing Resources, the Algerian Network of Artisanal Fishermen's Associations, the Moroccan Confederation of Artisanal Fisheries, and the Tunisian platform of Artisanal Fisheries, among others (now all together under a larger umbrella network of the North African Network for Sustainable Fisheries). What is interesting is that their social power has eclipsed the lack of institutional drive and inadequate legal frameworks in many locations. Indeed, several of these networks currently implement, strictly control and manage defined areas out of their own initiative. To conclude, the shift to social power via networks of organized and identifiable communities on topics of joint interest shows benefits of individual MPA management in the majority of cases.

Concerning the role a social network like MedPAN can play, via collaborations, to strengthen ecological network management, the path bears only fresh footprints in the Mediterranean. To date, despite great efforts by littoral states, the EU, UNEP-MAP RAC/SPA and the Barcelona Convention (Figure 9.1) no true ecologically representative network of MPAs has yet been fully implemented for protecting marine biodiversity. It is therefore early days to draw specific conclusions about how interactions between all manager-nodes of MedPAN can help with this goal, but some key actions underway seem promising.

In 2012, MedPAN and RAC/SPA, in collaboration with many regional partners, conducted a joint analysis of the status of MPAs in the Mediterranean (Gabrié *et al.*, 2012). A number of recommendations were made and incorporated into a roadmap. The latter also represented a consultative initiative which was included in the Antalya Declaration of the 1st Forum of MPAs in the Mediterranean (Antalya, Turkey, 2012), approved by over 300 stakeholders from about 30 countries. In 2015, the roadmap underwent some amendments and was subsequently adopted by the parties to the Barcelona Convention. This demonstrates how a bottom-up initiative from a social network can contribute to engaging top players more strongly in reaching the conservation treaties' objectives, namely attaining ecologically coherent networks of well-managed MPAs. Such initiatives also show how the shift of power towards social networks can address the classic obstacles in conservation matters: lack of political will or concern, poor funding, legal vacuum and loopholes, institutional 'contortionism', discrepancy in perspectives and rare multidisciplinary acuity, spatial heterogeneity of scientific biological data, and 'access to accessible' information— all against the backdrop of ego.

On the subject of data acquisition, a number of internationally funded projects (such as Coast to Coast Networks of Marine Protected Areas – EU FP7) have recently filled many gaps in biological knowledge. A collaborative approach with networks of MPA managers also allowed researchers to acquire data to answer management needs and contribute to expanding the current system of MPAs in the Mediterranean. Nevertheless, because knowledge is still being acquired and the current system of Mediterranean MPAs cannot yet be considered a true 'ecological network', the precautionary principle should still be applied to underpin the implementation of marine conservation measures (which can be adaptive over time as knowledge is accumulated). Here, key nodes within social networks such as MedPAN play the hinging role in reminding decision makers about fundamental field realities and whether knowledge is there or not. The reason is a cautionary one and goes beyond the environmental justification: it is linked to the social consciousness for survival in the face of natural capital destruction, and that early action can prevent costly restitution later.

Concrete examples of how collaborations between MPA managers strengthen network/system management within the above-described context usually seem to apply best to the sub-regional dimension and by topic, largely linked to threats. These factors also often trigger the creation of sub-networks of interest groups. For instance, MPAs noting the presence of the *Betanodavirus*, which can cause encephalopathy and viral retinopathy in groupers and some other fish, share information via the Grouper Study Group towards which the MedPAN Secretariat routinely re-directs managers' enquiries. This contributes to better managing this threat at a larger network level than individual MPA units. Similarly, MPAs affected by intrusion from non-indigenous (and sometimes invasive) species have been supported by the network, via enhanced collaborations led by IUCN-Mediterranean. This has allowed managers to cooperate in acquiring knowledge and capacity to adopt suitable management options when possible and devise a warning system for neighbouring sites, namely through using a mobile application MedMIS (http://www.iucn-medmis.org/?c=Map/show).

Another example is the use of common tools and guidelines shared via the network, such as how to establish underwater trails. Experienced managers have visited other sites to help establish similar trails, in order

to channel tourists or enhance environmental awareness and education. Finally, collaborations could eventually lead to better management of threats to highly mobile, charismatic and endangered species such as turtles, whales and dolphins. Harmonized data collection and storage would permit a deeper understanding of both biological variables and threats so that appropriate management measures can be selected following a harmonization process, covering wider areas. Although scientific data are being collected by many institutions, it is often done in such a way that doesn't necessarily answer a management question, and often in a heterogeneous fashion for a number of reasons including conflicting scientific schools of thought regarding protocols or methodologies. Yet, as indicated here, there is a manifest need for 'science for management' which MedPAN has begun to consolidate synergistically with WWF-Mediterranean and several other strategic partner nodes.

The strategy of the MedPAN network mentioned previously includes several fields of action which may be unfamiliar for nodes. This shortcoming was identified early after the revival of the network in 2008, and so contact with other existing similar networks in the world was sought.

Lessons from Other MPA Social Networks Worldwide

MedPAN is not the only example of a social network for MPA managers. Beyond the Mediterranean sub-regional AdriaPAN mentioned earlier, close to a dozen other similar initiatives linked to MPAs have been established worldwide, namely:

- **Big Ocean**: A peer-learning network created 'by managers for managers' (and managers in the making) of large-scale marine areas. It currently comprises about 10 large MPAs worldwide.

- **NAMPAN** (North American Marine Protected Areas Network): Set up in 1999 under the auspices of the Commission for Environmental Cooperation (CEC). It brings together resource agencies, MPA managers and experts covering Canada, Mexico and the United States.
- **CMAR** (or Corredor Marino): The Tropical Eastern Pacific marine corridor network, which was created in 2004 and involves five core MPAs and brings together managers, civil society, governmental agencies, cooperation agencies and NGOs.
- **MAIA**: The network of MPAs in the Atlantic Arc which was prompted under a European cooperation project in 2010. While the results of the project have been absorbed within OSPAR activities, the social network manages to continue some limited activity.
- **PANACHE** (Protected Area Network Across the CHannel Ecosystem): A project which started in 2012 and ended in 2015. Ongoing human ties are however palpable.
- **Yellow Sea Network**: Similar to other Large Marine Ecosystem initiatives, a network has been built up, closely linked to the UNDP/GEF regional sea approach.
- **NEASPEC and NEA MPA**: The North-Eastern Asian MPA network was established in 2012 involving 171 national and provincial marine nature reserves and 40 marine special protected areas from five countries.
- **LMMA** (Locally Managed Marine Areas): An international network of natural resource management practitioners working in Asia and the Pacific, who have joined together to share best practices and lessons learned, and to amplify their community voices nationally and internationally. Born in 2000, it has grown exponentially.
- **Te ME UM**: Since 2009, some 17 organizations have joined forces in the French overseas territories to assist natural area management professionals via the Te Me Um programme (Terres Mers Ultra-Marines, Lands Seas Overseas). The programme benefits about 150 protected natural areas.
- **WIOMSA** (Western Indian Ocean Marine Science Association): Established in 1993 to promote the educational, scientific and technological development of all aspects of marine sciences throughout the Western Indian Ocean region (consisting of 10 countries: Somalia, Kenya, Tanzania, Mozambique, South Africa, Comoros, Madagascar, Seychelles, Mauritius, Réunion (France)), and aimed at sustaining the use and conservation of its marine resources. Over time they have diversified their range of activities and incorporated links from science to management.

The focus in this section, however, is on two other networks: RAMPAO (*Réseau régional d'Aires Marines Protégées en Afrique de l'Ouest*) and CaMPAM (Caribbean Marine Protected Area Management Network and Forum).

RAMPAO was created in 2007 and brings together MPA managers and NGOs at the initiative of the Regional Coastal Zone and Marine Conservation Programme for West Africa (PRCM) who voluntarily joined and respect the RAMPAO Charter. To date there are 22 MPAs in five countries in the subregion. PRCM has no legal personality: it is a forum of partners who have signed a simple memorandum of understanding and cooperation. PRCM has an MPA component that initiated the creation of RAMPAO; so in effect, RAMPAO is one of the activities of the MPA component of the PRCM. Funding and programme activities are implemented entirely by the International Foundation for the Banc d'Arguin (FIBA). Although RAMPAO does not have a legal entity nor a secretariat, it holds a general assembly with its members. A team of four staff runs the programme of action. The network secured strong legitimacy and endorsement from ministerial entities early on and thus its actions can be supported. Many commonalities are found with MedPAN. These include a similar context, whereby RAMPAO supports a Regional Sea Convention work plan (Abidjan Convention and its protocols) and has developed a strategy to contribute to reaching the CBD objectives. Its mission, strategic axes and activities are also very similar. What is most interesting is that each weakness identified in one of these two networks finds the start of a solution by sharing experience with the other network.

CaMPAM, the Caribbean network of MPA managers, was created in 1997 under the Protocol for Specially Protected Areas and Wildlife of the Cartagena Convention (SPAW). It is a partnership with UNEP, the Fisheries Institute of the Caribbean and Gulf, and governments and NGOs from all Caribbean countries. SPAW covers 37 states and territories and is the equivalent of the Barcelona Convention Protocol on Specially Protected Areas and Biodiversity in the Mediterranean. The SPAW RAC, hosted by the Guadeloupe National Park, is the equivalent of RAC/SPA of the UNEP-MAP in the Mediterranean. CaMPAM is an initiative that brings together researchers working on MPAs, administrations and managers, and has received support from several international financing institutions. It is not a formal structure and, like RAMPAO, offers great flexibility while maintaining a disciplined approach focused on exchange of experience and transfer of science to management. CaMPAM's objectives and activities are similar to those of MedPAN and RAMPAO, although it suffers from under-staffing, as well as low and irregular financial support. For these reasons and the sheer geographical scope of the area (in terms of surface area, political and cultural diversity, and languages) the central node of this network is possibly a little more distant from site managers. However, many inspiring experience-sharing occasions have been seized between CaMPAM and MedPAN, and together with RAMPAO, ties have been strengthened between the three networks since 2013. Among the actions they take, these networks jointly deliver practical recommendations for achieving the CBD Aichi targets.

Common findings by the three social networks on what support they bring to the management of single or systems of MPAs can be summarized as follows:

- Advocacy at the regional and international levels
- Networking of different actors in charge of creating new MPAs and extending existing ones
- Enhancing common understanding and scientific knowledge at the regional level through sharing and collaboration
- Developing regional databases and conducting analysis of MPAs to provide an overview of their status and measure gaps towards reaching ecological coherence via a network of MPAs
- Formulating recommendations for a better integration of the ecological systems of MPAs into MSP and integrated coastal zone management frameworks
- Improving MPA management effectiveness through capacity building, exchanges among MPA managers, and promoting lessons learned and best practices
- Promoting participative governance
- Developing management guidelines and tools
- Playing a key role in mainstreaming the environment into development activities and advocating for governance at multiple levels in multiple sectors and varying cultural contexts.

What Could Come Next?

This chapter has set out a number of elements that show how social networks of MPA managers can successfully support the management of an ecological system of MPAs. To conclude on why such social networks sprout up all around the blue planet, it emerges that the need to overcome institutional bureaucratic slowness, while maintaining a positive reciprocal dyad relation with these very institutions, is possibly the overarching reason that ties together all the others discussed in this chapter.

The Gaia hypothesis has it that all organisms interact with their inorganic surroundings within the biosphere to form a self-regulating complex system that helps to maintain life on the planet in a preferred homeostasis (Lovelock and Margulis, 1974). Human impacts have increasingly aggregated to unsettle this engaging principle. While patching up the damage done has already been urgent for over 50 years, and preventing new gashes to Mother Earth set as a static priority for about the same time, environmental social networks have budded all over her. Their aim is to shake up limping institutions, provide them with crutches and tell them that, yes, they can walk and reach the binding conservation targets in time.

The main objective for MPA social networks is to achieve the CBD target that 10% of the oceans will be protected and well managed by 2020 (CBD, 2008). In effect, that is what should come next, albeit it is unrealistic to think that all contracting parties will achieve both the 'well managed' part along with the 'coverage' part. However, what ought to come next is for at least the coverage part to be established via governmental processes, with sound legal frameworks as well as adequate and sustainable financing. Once the race for coverage is over and done with, institutional focus should then be targeted at strengthening management and seeing that it is effective on the environment; to do so, adequate resources will be required as part of the sustainable development duty to preserve the natural capital.

Because a degree of power has shifted from institutions to niche social networks, a more balanced approach has been adopted that incorporates bottom-up processes, in which social networks have gained legitimacy to work collaboratively with institutions. In the case of social networks of MPA managers, such as MedPAN, CaMPAM and RAMPAO, this means that the social dynamics generated within the network can bring much support to strengthening the management effectiveness of MPAs and thus complement the slower pace of top-down institutional processes.

What will probably also come next is the strengthening of all the different MPA managers' social networks themselves, together with other connected interest groups and communities. This should foster joint action despite sometimes diverging stances, and lead to better integrated MPA management configurations.

Acknowledgements

With thanks to: Bruno Meola, MAPAMED database officer, MedPAN Secretariat, who provided the maps; Marie Romani, MedPAN Secretariat Executive Secretary, for being a central node and principal driver for the network; Paul Goriup, NatureBureau, for his patience; Purificació Canals, Chairwoman of MedPAN, for her inspiration and insight.

Notes

1 This figure does not include: Areas of Special Importance for Cetaceans (Agreement on the Conservation of Cetaceans in the Black Sea, Mediterranean Sea and Contiguous Atlantic Area – ACCOBAMS), described Ecologically and Biologically Significant Areas (EBSAs – Convention on Biological

Diversity), Important Bird Areas
(IBAs – BirdLife International), No
berthing zones/No entry zones except for
Fisheries (Malta), National Fisheries
Reserves, and proposed MPAs. It is also
important to consider MAPAMED figures
as in constant evolution due to regular
updates of the database.

2 A gift economy is a mode of exchange
where valuables are neither traded nor
sold but given without expectation for
immediate or future rewards. This
contrasts with barter or market
economy and with exchanges based on
money or other commodities
(Cheal, 1988).

3 The 'commons' refers to the cultural and
natural resources accessible to all members
of a society, including natural materials and
living resources which are not formally
regulated. These assets are held in common
and not owned privately. They are meant to
be preserved regardless of their return of
capital. Received as a shared right, we have a
duty to pass them on to future generations
in at least the same condition as we received
them and must neither degrade nor destroy
them. The tragedy of the commons is where
individuals acting independently and
rationally according to each's self-interest
behave contrary to the best interests of the
whole group by depleting some common
resource (Lloyd, 1833; Hardin, 1968).

4 In economics the 'economic man' (or *homo
economicus*) is the concept whereby humans
act rationally and self-interestedly to reach
their subjectively-defined ends in an optimal
way. In other words, the economic man acts
to maximize his benefits as a consumer
(utility) and his profits as a producer
(Pareto, 2014). This theory stands in
contrast with that of behavioural
economics, *homo reciprocans* and *homo
sociologicus* (looking at cognition,
cooperation and fulfilling social roles).

5 In the international legal sense of 'treaty',
which is an agreement under international

law entered into by actors in international
law, namely sovereign states and
international organizations. A treaty may
also be known as an agreement, protocol,
covenant, pact and convention…

6 Bosnia and Herzegovina is party to the
Barcelona Convention but has not yet
established an MPA and is therefore not
referred to in this analysis. The European
Union is also party, which brings the
number of parties to 21. Four countries
have not ratified the SPA/DB Protocol of
the Barcelona Convention, yet all aside
one have established MPAs.

7 This figure does not include: sites with no
recognized juridical status (e.g. IBAs and
areas identified by ACCOBAMS), or
linked to direct conservation objectives
(e.g. National Fisheries Reserves), areas
that have solely been identified as
important biologically (e.g. Described
EBSAs), and MPAs proposed or in
process. They are, however, included in
MAPAMED which currently has 1462
entries.

8 It is to be noted that some of these
national, regional and international
designations overlap (i.e. one site may
have several designations which overlap
exactly or partly in terms of surface
coverage and perimeter).

9 Internet: http://www.medpan.org |
Facebook: MedPAN-network (https://
www.facebook.com/MedPAN.network/).

10 The need for cognitive closure in social
psychological terms is the motivation of
an individual to obtain an answer to a
question, seek out information, and
aversion of that individual towards
ambiguity. This process is characterized
by the urgency tendency and the
permanence tendency (Webster and
Kruglanski, 1994). When looking at nodes
in social networks, this psychological trait
can be taken into account for a group,
namely for transitivity (i.e. completeness
of relational triads).

References

Baumeister, R.F. and Leary, M.R. (1995) The need to belong: desire for interpersonal attachments as a fundamental human motivation. *Psychological Bulletin*, **117** (3), 497–529.

Borg, S. (2012) Social networks and health: models, methods, and applications. *Journal of the American Medical Association*, **307** (11), 1203. doi:10.1001/jama.2012.309

Cheal, D.J. (1988) *The Gift Economy*. Routledge. 228 pp.

Claudet, J., Notarbartolo di Sciara, G. and Rais, C. (2011) *Database of Mediterranean Marine Protected Areas: Site Identification Criteria*. MedPAN and UNEP-MAP RAC/SPA. 13 pp.

Clauzon, G., Suc, J-P., Gautier, F. *et al.* (1996) Alternate interpretation of the Messinian salinity crisis: controversy resolved? *Geology*, **24** (4), 363–366.

Convention on Biological Diversity (2008) *COP Decision IX/20 Marine and Coastal Biodiversity*. https://www.cbd.int/decision/cop/default.shtml?id=11663

Coon, C. (1958) *The Story of the Middle East*. Harper and Row, New York. pp. 154–157.

Costa Jr., P.T. and McCrae, R.R. (1992) *Revised NEO Personality Inventory (NEO-PI-R) and NEO Five-Factor Inventory (NEO-FFI) Manual*. Psychological Assessment Resources, Odessa, Florida.

Gabrié, C., Lagabrielle, E., Bissery, C. *et al.* (2012) *The Status of Marine Protected Areas in the Mediterranean Sea*. MedPAN, Marseilles and RAC/SPA, Tunis. MedPAN Collection. 256 pp.

Garcia-Castellanos, D., Estrada, F., Jiménez-Munt, I. *et al.* (2009) Catastrophic flood of the Mediterranean after the Messinian salinity crisis. *Nature*, **462**, 778–781.

Gautier, F., Clauzon, G., Suc, J.P. *et al.* (1994) Age and duration of the Messinian salinity crisis. *Comptes Rendus de l'Académie des Sciences (IIA)*, **318**, 1103–1109.

Grotius, H. (1609) *Mare Liberum, sive de jure quod Batavis competit ad indicana commercia dissertatio*. Elzevir, Lodewijk.

Hardin, G. (1968) The tragedy of the commons. *Science*, **162**, 1243–1248.

Jenkins, R. (2014) *Social Identity*. Routledge.

Kaplan, S. (2000) Human nature and environmentally responsible behavior. *Journal of Social Issues*, **56** (3), 491–508.

Krijgsman, W.G., Langereis, M., Daams, C.G. *et al.* (1996) A new chronology for the middle to late Miocene continental record in Spain. *Earth and Planetary Science Letters*, **142** (3–4), 367–380.

Lardbucket project (2012) *Management Principles V.* http://2012books.lardbucket.org/pdfs/management-principles-v1.1.pdf

Lloyd, W.F. (1833) *Two Lectures on the Checks to Population*. University of Oxford, England.

Lovelock, J.E. and Margulis, L. (1974) Atmospheric homeostasis by and for the biosphere: the Gaia hypothesis. *Tellus*, **26** (1–2), 2–10.

MacMillan, D.W. and Chavis, D.M. (1986) Sense of community: a definition and theory. *Journal of Community Psychology*, **14**, 6–23.

MAPAMED (2015) *MAPAMED, the database on sites of interest for the conservation of Mediterranean marine environment.* MedPAN, Marseilles and UNEP-MAP RAC/SPA, Tunis. December 2015 release.

MAPAMED (2016) *MAPAMED, the database on sites of interest for the conservation of Mediterranean marine environment.* MedPAN, Marseilles and UNEP-MAP RAC/SPA, Tunis. March 2016 release.

Pareto, V. (2014) *Manual of Political Economy: A Critical and Variorum Edition* (eds A. Montesano, A. Zanni, L. Bruni, J. Chipman and M. McLure). Oxford University Press. 664 pp.

Piliavin, J.A. and Hong-Wen, C. (1990) Altruism: a review of recent theory and research. *Annual Review of Sociology,* **16,**27–65.

Rheingold, H. (1993) *The Virtual Community: Finding Connection in a Computerized World.* Addison-Wesley Longman.

Schore, A.N. (2015) *Affect Regulation and the Origin of the Self: The Neurobiology of Emotional Development.* Routledge.

Seligman, C.G. (1924) Presidential address: Anthropology and psychology – a study of some points of contact. *The Journal of the Royal Anthropological Institute of Great Britain and Ireland,* **54,** 13–46. doi:10.2307/2843660

Sergi, G. (1901). *The Mediterranean Race: A Study of the Origin of European Peoples.* W. Scott, London.

Simpson, B. and Willer, R. (2015) Beyond altruism: sociological foundations of cooperation and prosocial behaviour. *Annual Review of Sociology,* **41,** 43–63.

Webster, C., Meola, B., Sostres, M. *et al.* (in press) *The Status of Marine Protected Areas in the Mediterranean Sea.* MedPAN, Marseilles and RAC/SPA, Tunis. MedPAN Collection.

Webster, D. and Kruglanski, A. (1994) Individual differences in need for cognitive closure. *Journal of Personality and Social Psychology,* **67,** 1049–1062.

10

Eyes Wide Shut: Managing Bio-Invasions in Mediterranean Marine Protected Areas

Bella Galil

The Steinhardt Museum of Natural History, Israel National Center for Biodiversity Studies, Tel Aviv University, Tel Aviv, Israel

This [bio-invasion] is one of the most worrying forms of human impact, because it usually does not decrease with distance and time: it is irreversible at human scale ... introduced species undermine in an irreversible way everything that has been done to protect biodiversity, whether through the protection of species or the protection of habitats. For example, there would no longer be any point in setting up MPAs if it were merely to protect uniform meadows of introduced species, e.g. *Caulerpa taxifolia*, *C. racemosa*, *Acrothamnion preissii* and/or *Womersleyella setacea*.

(Boudouresque and Verlaque, 2005)

Introduction

The Mediterranean Sea is one of the world's prime marine biodiversity hotspots (Coll *et al.*, 2010). Intensification of anthropogenic activities, coupled with growth of littoral resident and transient recreational populations, are driving unprecedented changes in the Mediterranean Sea (Micheli *et al.*, 2013; EEA, 2015). Symptoms of complex and fundamental alterations to the sea's ecosystems proliferate, including increases in invasive alien species, which affect the functioning of marine ecosystems and the consequent provision of goods and services. (It should be noted that 'invasive alien species' refers to species introduced by human activities; species undergoing climate-shifted population distributions, without human-assisted spread, are not considered to be alien.)

The rapid degradation of the Mediterranean has prompted calls for more effective approaches to protect, maintain and restore its ecosystems, including calls to increase the number and spatial extent of its 'protected areas' (Gabrié *et al.*, 2012; EEA, 2015). Marine Protected Areas (MPAs) have an inherent appeal, particularly to a public weaned on terrestrial reserves and parks, who would like similar protection extended to the marine environment. Marine Protected Areas, created in part to conserve natural diversity of native species in their habitats, are meant to offer an ecosystem-based approach to conservation, and provide protection to habitats, biodiversity and ecosystem services, and insurance against environmental or management uncertainty (Lubchenco *et al.*, 2003).

It is, however, questionable whether MPAs, or even networks of MPAs, provide

Management of Marine Protected Areas: A Network Perspective, First Edition. Edited by Paul D. Goriup.
© 2017 John Wiley & Sons Ltd. Published 2017 by John Wiley & Sons Ltd.

adequate protection from the suite of current and future threats that include global change impacts (e.g. increase in sea surface temperature, sea-level rise and more severe storm events, acidification); unsustainable extractive practices (e.g. overfishing, oil drilling and sand mining, now proscribed); impacts from marine and terrestrial pollution; widespread habitat destruction and degradation; and the spread of alien species. These threats are exacerbated in the relatively small, landlocked sea which is hemmed in by fast-increasing populations (73 million in 1950, 244 million in 2000, and 590 million forecast in 2050; Zlotnik, 2003; Tosun, 2011).

The reigning theory holds that MPAs, owing to their high species diversity and putative abundance of predators/competitors/parasites of alien species, are resistant to invasion. This chapter examines whether MPAs are effective in protecting native biodiversity under a high load of alien species, and underlines the risk posed to MPAs as management tools for the conservation of Mediterranean marine biodiversity.

Marine Alien Species in the Mediterranean Sea

The number of recorded species introductions (whether accidental or deliberate) into the Mediterranean Sea is far higher than in other European Seas: nearly triple the number of records known from the Western European margin stretching from Norway to Portugal (Galil *et al.*, 2014a). The present number of multicellular alien species stands at 728; of these 138 are recorded in five or more Mediterranean countries (Galil *et al.*, 2016b). 'Erythraean aliens' – species introduced into the sea through the Suez Canal – number 450. It is assumed this is only a partial inventory, as our ignorance of marine biota leads to

'massive under-reporting and thus understatement of ... the altered distributions of non indigenous species' (Carlton, 2000).

The Mediterranean Sea has a long history of bio-invasions: in the first decade of the 20th century, 13 of the 14 alien species recorded for the first time in the Mediterranean entered through the Suez Canal (Galil, 2009). The number of alien species recorded in the Mediterranean more than doubled (213%) between 1970 and 2014, with the greatest increase recorded in the past two decades (Galil *et al.*, 2014b).

The number of recorded alien species differs among the Mediterranean countries, and is substantially greater in the Levant than in the western Mediterranean (Figure 10.1). The Levantine countries (Egypt, Israel, Lebanon, Syria, Turkey, Cyprus, Greece), and to lesser degree Tunisia, saw an increase in the number of introductions through the Suez Canal; in Italy and Spain the majority of recent records stem from shipping, whereas in France shipping and mariculture introduced nearly equal numbers of species. Fish, molluscs and crustaceans are the major taxa introduced to the Levant, whereas macroalgae account for the highest number in Italy, France and Spain. Vectors determine the geographical origin and taxonomic identity of the introduced taxa: in the eastern Mediterranean, where the Suez Canal serves as the main pathway, most alien species are of tropical Indo-West Pacific origin and comprise molluscs, fish and crustaceans; in the western basin, where vessels and aquaculture are the prevailing vectors, the taxonomic composition and native ranges of alien species are diverse and depend on shipping routes and mariculture trade (Galil, 2009).

Alien macrophytes, invertebrates and fish are prominent in many coastal habitats in the Mediterranean, especially affecting the composition of the biota of the southeastern Mediterranean Sea (Steinitz, 1970; Por, 1978; Galil, 2007). Their impacts are determined, in part, by their demographic

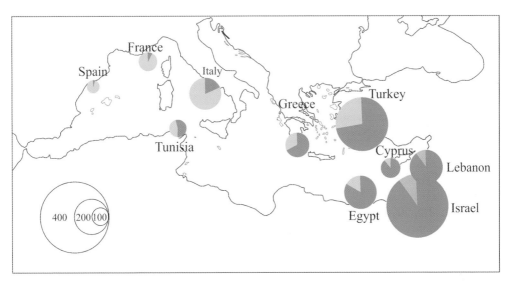

Figure 10.1 The number of non-indigenous species (NIS) in some Mediterranean countries. In dark grey is the fraction of species probably introduced through the Suez Canal. The circle sizes are proportionate to the total number of NIS recorded in the country by 2016.

success (abundance and spread) (Parker *et al.*, 1999). With few exceptions, however, the ecological impact of invasive alien species on the native Mediterranean biota is poorly known (Zibrowius, 1992; Boudouresque, 2004; Katsanevakis *et al.*, 2014), though it is believed that some species have caused major shifts in community composition. Yet, it is difficult to disentangle confounding factors in evaluating the impacts of most bio-invasions: where populations of native Mediterranean species appear to have been outcompeted or displaced by an alien, these could be part of a profound anthropogenic alteration of the marine environment. Nevertheless, a number of Mediterranean invasive aliens have drawn the attention of scientists, management agencies and the media, for the conspicuous impacts on the native biota attributed to them.

Perhaps the most notorious and best-studied invasive alien species in the Mediterranean are a pair of coenocytic chlorophytes: *Caulerpa taxifolia* (known as the 'killer alga'; Meinesz *et al.*, 2002), and *C. cylindracea* (Verlaque *et al.*, 2004).

An invasive strain of a tropical green alga, *C. taxifolia* was unintentionally introduced into the Mediterranean in 1984 (Jousson *et al.*, 1998) and is considered 'a real threat for the balance of the marine coastal biodiversity' (Longpierre *et al.*, 2005). This alga's rapid spread and high growth rate (up to 14 000 blades per m^2) on diverse infralittoral bottom types has led to the formation of homogenized microhabitats. Its presence is associated with replacement of native algal species (Verlaque and Fritayre, 1994; Boudouresque *et al.*, 1995; Harmelin-Vivien *et al.*, 1999), and reduction of species richness, density and biomass of fish assemblages (Francour *et al.*, 1995; Harmelin-Vivien *et al.*, 1999; Levi and Francour, 2004). *Caulerpa cylindracea* was discovered in the Mediterranean in 1990 and has since spread from Cyprus to Spain and the Canary Islands (Verlaque *et al.*, 2004). It overgrows other macroalgae and curtails species number, extent and diversity of the macroalgal community, even in highly diverse, native macroalgal assemblages with dense coverage (Piazzi *et al.*, 2001, 2003). Off Cyprus, within six years it replaced the previously dominant *Posidonia oceanica* community (Argyrou *et al.*, 1999).

Two species of rabbitfish, *Siganus rivulatus* (Figure 10.2) and *S. luridus*, which entered the Mediterranean from the Red Sea through the Suez Canal, were first recorded off the coast of Israel in 1924 and 1955 respectively (Steinitz, 1927; Ben-Tuvia, 1964). The species have now been found as far west as France and Tunisia (Ktari-Chakroun and Bahloul, 1971; Ktari and Ktari, 1974; Daniel *et al.*, 2009). These schooling, herbivorous fish form thriving populations in the Levant Sea where 'millions of young abound over rocky outcrops, grazing on the relatively abundant early summer algal cover' (George and Athanassiou, 1967).

The siganids comprise one-third of the fish biomass in rocky habitats in Israel (Goren and Galil, 2001), 80% of the abundance of herbivorous fish in shallow coastal sites in Lebanon (Bariche *et al.*, 2004),

83–95% of the biomass of herbivorous fish at sites on the Mediterranean coast of Turkey (Sala *et al.*, 2011), and have replaced native herbivorous fish (Papaconstantinou, 1987; Bariche *et al.*, 2004). Their diet has a significant impact on the structure of the algal community: by selective feeding the siganids have nearly extirpated some of their favourite algae locally (Lundberg *et al.*, 2004); 'once flourishing algal forests have disappeared to leave space to sponges and wide areas of bare substratum… The shift from well-developed native algal assemblages to "barrens" implies a dramatic decline in biogenic habitat complexity, biodiversity and biomass … with effects that may move up the food chain to the local fisheries' (Sala *et al.*, 2011). A survey along a thousand kilometres of Greek and Turkish coasts found that in regions with abundant siganids canopy algae were 65% less abundant, benthic biomass was reduced by 60% and species richness reduced by 40% (Vergés *et al.*, 2014).

Figure 10.2 Rabbitfish *Siganus rivulatus* in a heavily grazed rocky reef, Akhziv MPA, Israel. Photo: Bat Sheva Rothman.

The small mytilid mussel *Brachidontes pharaonis* in the early 1970s was '250 times rarer' than the native mytilid *Mytilaster minimus*, that formed dense '*Mytilaster* beds' on intertidal rocky ledges along the Israeli coastline (Safriel *et al.*, 1980). More recently 'the same rocks are … completely covered with the Erythrean *B. pharaonis*, while *M. minimus* is only rarely encountered' (Mienis, 2003: 15). *Brachidontes pharaonis* has spread westwards to southern Italy, where it forms dense populations with over 25 000 specimens per m^2 (Sarà *et al.*, 2006), and to Corsica, France (Merella *et al.*, 1994).

Many more 'replacements' have been noted in the south-eastern Mediterranean. The spiny oyster, *Spondylus spinosus*, has supplanted the native congener *S. gaederopus* within a decade of its first record in Israel (Mienis *et al.*, 1993). Similarly, the jewel box oyster *Chama pacifica* succeeded its native congener, *C. gryphoides* (Mienis, 2003). The native Mediterranean cerithiid gastropods *Cerithium vulgatum* and *C. lividulum*, respectively common and abundant in shallow water along the coast of Israel until the 1970s, were supplanted by the cerithiids *C. scabridum* and *Rhinoclavis kochi* (Mienis, 2003). The dragonet, *Callionymus filamentosus*, has replaced the native callionymids *C. pusillus* and *C. risso* along the Levantine upper shelf (Golani, 1998). The snapping shrimps *Alpheus inopinatus* and *A. audouini* are more common in the Levantine rocky littoral than the native *A. dentipes*, and on the muddy bottoms *A. rapacida* is much more common than the native *A. glaber* (Lewinsohn and Galil, 1982; Galil, 1986).

Moreover, there is the unprecedented number of alien jellyfish. Periodic outbreaks of indigenous scyphozoan jellyfish have long been noted in the Mediterranean (UNEP, 1991; CIESM, 2001). Various anthropogenic perturbations including eutrophication, overfishing, global warming and the increase of littoral man-made hard substrates have been suggested as contributing to their proliferation (CIESM, 2001; Boero *et al.*, 2008; Richardson *et al.*, 2009).

Yet the Mediterranean is unique in hosting six alien scyphozoan jellyfish, in addition to two alien ctenophores: *Cassiopea andromeda*, *Phyllorhiza punctata*, *Rhopilema nomadica* (Galil *et al.*, 1990), *Marivagia stellata* (Galil *et al.*, 2010), *Pelagia benovici* (Piraino *et al.*, 2014), and the recently recorded *Cotylorhiza erythraea* (Galil *et al.*, 2016a). Though occasionally forming large aggregations (e.g. in 2009 swarms of *Mnemiopsis leidyi* appeared along the Ligurian, Tyrrhenian and Ionian shores of Italy, and the Mediterranean coast of Spain and Israel; Boero *et al.*, 2009; Fuentes *et al.*, 2009; Galil *et al.*, 2009), the Mediterranean populations of these alien jellyfish have remained small.

In contrast, the Erythraean *Rhopilema nomadica*, first recorded in the Mediterranean in the 1970s, is notorious for the large swarms it has formed each summer since the early 1980s along the south-east Levantine coast (Galil *et al.*, 1990). As gelatinous plankton plays a pivotal role in marine food-webs and nutrient fluxes (Purcell and Arai, 2001; D'Ambra *et al.*, 2013; Fleming *et al.*, 2015), outbreaks of alien jellyfish may affect production cycles in plankton and benthos. Invasive alien scyphozoans and ctenophorans may impact the ecosystem in ways we neither expect nor understand, and which are more significant than their obvious impacts in economic and human health terms.

Porous Borders: Alien Species in MPAs

The most important generalisation is that all nature reserves, except those in Antarctica, appear to have invasive species.

(Usher, 1988)

The Southern Ocean around Antarctica is no longer free from invasive marine species... Isolated for at least 25 million years, the endemic Antarctic Southern Ocean marine fauna is now being exposed to human-mediated influx of exotic species. Invasive species and polar warming combined can foster the probability of arrival and colonization by non-indigenous species.

(Tavares and De Melo, 2004)

The introduction of the venomous Indo-Pacific lionfish, *Pterois volitans* and *P. miles* (Figure 10.3), from a beachside aquarium during Hurricane Andrew into the western Atlantic is one of the most closely followed marine invasions to date (Whitfield *et al.*, 2002). Their spread across the tropical western Atlantic and Caribbean Sea was swift, the first record dating from only 1985 (Schofield, 2009). Since then, the lionfish have established dense populations with up

to 1320 individuals per hectare (Trégarot *et al.*, 2015). Their predation caused a 95% decrease in the abundance of small native reef fish at some invaded sites and a 65% decline in native fish biomass over two years on heavily invaded reefs in the Bahamas (Green *et al.*, 2012; Albins, 2013). The altered marine ecosystems led to cascading effects on coral reef food-webs and benthic community structure (Albins and Hixon, 2011; Lesser and Slattery, 2011). A census of lionfish in the Florida Keys National Marine Sanctuary (FKNMS) using multiple independent monitoring data sets revealed a rapid rise in their frequency of occurrence, abundance and biomass, with a three- to six-fold increase between 2010 and 2011 alone (Ruttenberg *et al.*, 2012). A monitoring programme in Martinique documented the fast invasion of the island's littoral hard bottom habitats (Trégarot *et al.*, 2015). Despite increasing control efforts, in three years (February 2011 to December 2013) the

Figure 10.3 *Pterois miles*, an Erythraean lionfish, at Akhziv MPA, Israel. Photo: Oren Klein.

lionfish colonized the west coast of Martinique, most of it designated as MPAs (http://campam.gcfi.org/CaribbeanMPA/mapview.php).

The Parque Nacional Arrecife Alacranes (420 km²), located 135 km off the northern coast of the Yucatan Peninsula, Mexico, was established in 1994. The isolated reef has restricted access, save for artisanal Mayan fishermen during the lobster fishing season. A survey conducted during the 2010/11 lobster fishing season suggested that based on the criteria for lionfish invasion established for FKNMS (Morris and Whitfield, 2009), the invasion in Alacranes reef was intermediate to advanced (López-Gómez *et al.*, 2014). Recent surveillance showed a more abundant population with larger-bodied individuals: the largest ever lionfish (total length 44 cm) recorded in the Southern Gulf of Mexico was captured in 2014 (Alfonso Aguilar-Perera, pers. comm., 10 August 2015). The lionfish have even reached the remote coral reefs of the Flower Garden Banks National Marine Sanctuary on the outer continental shelf in the Gulf of Mexico (193 and 172 km offshore from Galveston, Texas) which support diverse and abundant fish populations (Johnston *et al.*, 2013).

A survey of lionfish in 71 Caribbean reefs (Hackerott *et al.*, 2013) established they were present at high densities on reefs with depauperate native predator assemblages, as well as on reefs with high diversity and biomass of native predators; no evidence was found for an effect of native predators on their invasion success across the region. The authors noted that lionfish densities were lower in marine reserves, since they are regularly removed from most Caribbean reserves by managers, dive operations, and tourists in efforts to preserve the biota of the protected reefs. A re-examination of the same data concluded that 'lionfish abundance is reduced in marine protected areas due to some factor other than predator abundance. The negative effect of protection status on lionfish abundance and lack of effect of grouper or other predator biomass on lionfish abundance indicate that culling within protected areas most likely explains the observed pattern' (Valdivia *et al.*, 2014).

Twenty years after the first specimen of *P. miles* was recorded from the Levant, a spate of records from Israel, Lebanon, Cyprus, the Turkish Mediterranean coast and Rhodes, Greece (Bariche *et al.*, 2013; Turan *et al.*, 2014; Corsini-Foka and Kondylatos, 2015; Oray *et al.*, 2015; M.B. Yoke, pers. comm.) attest to the presence of an established population in the Levant. Whether denizens of the Mediterranean rocky reefs will suffer the fate of the Caribbean reef biota remains to be seen (Johnston and Purkis, 2014).

Extensive diving surveys conducted in 2005–2006 in Gökova Specially Protected Area, south-west Turkey, documented, among scores of Erythraean invasive alien species, a huge proliferation of the Indian Ocean seagrass *Halophila stipulacea*, the strombid gastropod *Conomurex persicus*, the Indo-Pacific holothuroid *Synaptula reciporans*, and the Western Indian Ocean rabbitfish *Siganus rivulatus* and *S. luridus*.

A study of rabbitfish populations at three sites on the Mediterranean coast of Turkey – Kaş-Kekova MPA, Fethiye-Göcek Special Environmental Protection Area and Bodrum – found that the percentage of their biomass was only slightly lower in the protected areas (83%) than in Bodrum (95%) (Sala *et al.*, 2011). Similarly, the four MPAs – Karpathos, Fethiye, Kaş and Adrasan – examined by Vergés *et al.* (2014) were all characterized by high abundance of the invasive rabbitfish. Indeed, the highest abundance of alien fish (*Fistularia commersoni*, *Pteragogus pelycus*, *Sargocentron rubrum*, *Siganus luridus*, *S. rivulatus*) among rocky reef fish assemblages in the Mediterranean was observed at Gökova, Fethiye, Kaş and Adrasan, on the southern coast of Turkey, prone to the Erythraean invasion through the Suez Canal (Guidetti *et al.*, 2014, Figure 8).

Since its introduction in 1984 to the Mediterranean, the so-called killer alga *Caulerpa taxifolia* (a popular ornamental plant in aquaria) has spread steadily and established populations in many MPAs (Otero *et al.*, 2013, Figure 4). In 1994 it was first recorded in Port Cros National Park, an island off the coast of Provence, France (Cottalorda *et al.*, 1996). Since then annual campaigns of manual uprooting have kept *C. taxifolia* at bay in the park's littoral. By the mid-2000s its populations had naturally declined (Montefalcone *et al.*, 2015) and the last recorded colonies in Port Cros were found in 2012 (Jaubert *et al.*, 2014).

Unfortunately, the closely allied *C. cylindracea* shows no sign of following this 'boom and bust' model. Since its first record in the Mediterranean (Nizamuddin, 1991), it has spread to 13 other littoral countries. Its impact on native ecosystems is significant (Piazzi *et al.*, 2001; Klein and Verlaque, 2008; Piazzi and Balata, 2008). It was noted in Port Cros National Park (Robert, 2001; Ruitton *et al.*, 2005), in the Nature Reserve 'Bouches de Bonifacio' (Meinesz *et al.*, 2010) and in the Ligurian MPAs of Portofino and Cinque Terre (Montefalcone *et al.*, 2015). A study carried out in the Archipelago of Cabrera National Park, which lies 10 km off Mallorca, Balearic Islands, Spain, recorded it growing in almost 'all habitats present between 0 and 65 m depth' with 'Biomass ... of similar magnitude' to other Mediterranean populations (Cebrian and Ballesteros, 2009: 470, 471). In Datça-Bozburun, a Specially Protected Area in south-west Turkey, *C. cylindracea* meadows were most dense at depths of 40–75 m, as well as occupying gaps within *P. oceanica* meadows in shallower waters (Okuş *et al.*, 2007).

The invasive red alga *Womersleyella setacea*, which was first observed in the Mediterranean in 1987 in the Var region, France (Verlaque, 1989), has since spread to eight countries. This filamentous alga forms thick mats that shroud sublittoral rocky substrata and modify benthic assemblages, and outcompetes key species (Antoniadou and Chintiroglou, 2007; Cebrian and Rodríguez-Prieto, 2012). In the Scandola Natural Reserve, Corsica, France, where it was first noted in 1989, its biomass is 'of the same order of magnitude as those recorded in other Mediterranean areas' (Cebrian and Rodríguez-Prieto, 2012). In 1996 it was first noticed in the Mesco Reef, Cinque Terre Marine Protected Area, Italy, and become dominant in 2008 (Gatti *et al.*, 2015).

A survey of alien opisthobranch species along the Mediterranean coast of Turkey from Bodrum to the Gulf of Iskenderun revealed that the largest number (13 of 18 species) were recorded in the Kaş-Kekova Specially Protected Area (Yokes *et al.*, 2012). A shellgrit sample from Dor marine nature reserve, Israel, 'turned out to be remarkably rich' in invasive Erythraean alien species (Mienis, 1985). An inventory of the mollusc fauna from Akhziv-Rosh HaNiqra nature reserve, Israel, compiled from the literature and samples deposited in the National Collections, private collections and the Zoological Museum of Copenhagen, comprised 283 species, including 38 invasive alien species. Yet, most of the material in the National Collections was collected between 1950 and 1975, whereas most of the invasive alien species records are recent (Mienis and Ben-David-Zaslow, 2004).

Along the Lebanese coast (Enfeh Peninsula, Ras Chekaa cliffs, Raoucheh, Saida, Tyre and Nakoura) surveys were undertaken in 2012 and 2013 in the framework of the 'Regional Project for the Development of a Mediterranean Marine and Coastal Protected Area' (MedMPAnet Project). The resulting report (Ramos-Espla *et al.*, 2014) highlights the prevalence of invasive alien species at all sites, with 31% and 21% respectively of the recorded mollusc and fish species identified as Erythraean aliens. Erythraean species established large populations and are frequently the most common species encountered: the mytilid *Brachidontes pharaonis* 'dominates the abrasion platform and it forms a marked belt in the lower part of the midlittoral'; 'the rocky substrata is bare and empty of erected soft macroalgae; this overgrazing is due to the herbivorous pressure of the fishes *Siganus rivulatus* and *S. luridus*'; the stinging feathery hydroid *Macrorynchia philippina* 'is distributed in all of the observed areas, between 0–7 m depths'; and on 'vertical rock between 0–6 m depth ... *Chama-Spondylus* reefs create a complex habitat ... develop original facies, without comparison along the whole Mediterranean', together with another Erythraean bivalve *Malleus regula*. Similarly, *Synaptula reciporans* is the most common by far of the echinoderms (see report Annex II: Inventory of the taxa observed in the 2012 and 2013 missions).

An Erythraean symbiont-bearing larger foraminifera, *Sorites orbiculus*, is very abundant as well in the Kaş-Kekova Specially Protected Area, as along the south-western coast of Antalya between Kalkan and Kemer (Meriç *et al.*, 2008). This species is found in high numbers in marine nature reserves off Israel, namely Shikmona and Akhziv-Rosh HaNiqra (Lazar, 2007; Merkado *et al.*, 2013).

Boudouresque and Verlaque (2005) decried 'introduced species undermine in an irreversible way everything that has been done to protect biodiversity, whether through the protection of species or the protection of habitats... Frontiers, whether administrative (MPAs) or political (countries), do not exist for invasive species ... there would no longer be any point in setting up MPAs if it were merely to protect uniform meadows of introduced species'. At the time it was a solitary 'cri de coeur' – the reigning hypotheses assumed

that MPAs, owing to their high species diversity and putative abundance of predators/competitors/parasites, are resistant to invasion. Yet, these expectations claimed more than scientific evidence justified: 'The scant empirical research addressing these hypotheses suggests that local facilitation of invasives can indeed occur, though no assessment exists of whether these local effects result in enhanced regional invasion' (Claudet *et al.*, 2011).

An exhaustive search of peer-reviewed literature (Burfeind *et al.*, 2013) disclosed only 13 cases with quantitative data on alien species inside and outside marine ('no-take') reserves. In no case did reserves resist alien species: of the seven reserves established prior to the arrival of aliens, five had no effect and two enhanced alien species; of the six reserves established in areas with pre-existing aliens, two had no effect and four enhanced alien species. Guidetti *et al.* (2014) surveyed rocky reef fish assemblages at 30 sites across the northern Mediterranean Sea comprising 13 protected and 17 unprotected areas. No observable effects of MPAs on alien fish densities were observed, leading the authors to suggest that 'the mechanisms of invasion are not affected by protection', and that MPAs may indeed enhance the number and densities of alien species.

See No Evil

Recognition that species had been introduced into the Mediterranean from other parts of the world dates back to the 19th century. Even before the Suez Canal was fully excavated the French malacologist Vaillant (1865: 97) argued that 'Le percement de l'isthme de Suez ... offrira ... une occasion précieuse de constater les phénomènes que doivent amener l'émigration des espèces et le mélange des faunes'. Vaillant advocated what today would be considered a 'baseline study', and raised provocative and prescient questions:

> La mer Rouge et la mer Méditerranée montrent, quant à leurs mollusques, ... des différences considérables, mais il serait nécessaire de chercher à bien fixer d'avance quelles elles sont maintenant pour pouvoir mieux juger plus tard des changements qui pourront survenir. Sans aucun doute il va y avoir transport des espèces, celles-ci, en changeant de milieu, vont-elles conserver tous leurs caractères ou subir quelques modifications.

A century later communities along the upper shelf of the Levant differed notably from communities elsewhere in the Mediterranean because of the thriving populations of species introduced through the Suez Canal (Galil and Lewinsohn, 1981). While thermophilic species were pouring into the Levantine Basin, vessel-borne species were introduced into ports and commercially valuable species into lagoons, estuaries and bays in the Mediterranean. In the late 1990s the 'Mediterranean Science Commission' (CIESM) undertook the task of assembling and validating many thousands of records of 'exotic' species in the Mediterranean Sea, and presenting them in a standardized, scientifically robust and user-friendly format. This pioneering endeavour resulted in the publication of the first 'Atlas' of 'exotic' species in the Mediterranean (http://www.ciesm.org/online/atlas/) – in fact, the first of its kind worldwide. The 'Atlas' galvanized attention and garnered recognition of the unique situation concerning bio-invasions in the Mediterranean Sea, promoted documentation of the full extent of the diversity of 'exotic' species, and provided information for management and conservation policies.

Yet, despite the clearly visible devastating impacts of increasing numbers of invasive alien species, some scientists have been so intent on 'building a heritage of reserve networks that will safeguard marine communities', couched in quasi-religious terms, such as 'the holy grail of conservation', as to ignore bio-invasions in MPAs (Lubchenco *et al.*, 2003: S6). Some scientists considered invasive alien species 'welcome guests' in the 'impoverished, subtropical cul-de-sac' [the Levantine Basin, Mediterranean Sea], that have 'biologically enriched' the sea (Tortonese, 1973: 327; Por, 1978: 123). Others hoped that 'MPAs, as oases of biodiversity, serve as the last rampart against these invasive species' (Francour *et al.*, 2010). They suggested that 'marine protected areas could be effective tools in limiting invasive species from spreading... The parasite burden of a MPA could be an excellent regulator of invasive species by exercising a control similar to the predator top-down control', and that within long-established MPAs 'a new non-indigenous fish will be controlled by predation'.

However, these authors frankly admit that 'not enough studies are currently available' and that 'the conclusions ... are mainly theoretical and not yet tested throughout the Mediterranean Sea' (Francour *et al.*, 2010). Meanwhile, some scientists are seemingly oblivious of the problem. Mouillot *et al.* (2011: 1) set out to determine 'the spatial overlap – if any – between the present system of Mediterranean MPAs, the hot spots of fish biodiversity ... and the hot spots of anthropogenic stresses' based on a database comprising 282 native coastal and continental shelf teleosts, leaving out even the most abundant and widespread of the over 100 alien fish species. In a subsequent article the authors sought to assess gaps in the representation of taxonomic, phylogenetic and functional diversity among coastal fishes in MPAs, using distribution data for 340 native species, again excluding alien fish species recorded in the sea (Guilhaumon *et al.*, 2015). Similarly, Micheli *et al.* (2013) chart out 'Priorities for Regional Conservation Planning in the Mediterranean Sea', with barely a mention of bio-invasion among the anthropogenic threats besetting the sea. These authors identified 10 areas for immediate conservation action, including the highly invaded south-east Aegean Sea (see *Porous Borders: Alien Species in MPAs*, above).

Specialists, representing the Caribbean Marine Protected Area Management Network and Forum (CaMPAM), Network of Managers of Marine Protected Areas in the Mediterranean (MedPAN), World Wildlife Fund (WWF), Mediterranean Programme Office, Regional Activity Centre for the Specially Protected Areas and Wildlife Protocol, and the United Nations Environment Programme – Caribbean Environment Programme reviewed regional case studies in the Mediterranean and the Caribbean, aiming to 'transform paper parks into functional and ecologically effective marine protected areas' (Bustamante *et al.*, 2014). Two of the three sites chosen (Kaş-Kekova, Turkey, and Hol Chan, Belize) are notable for proliferation of invasive alien species, yet the authors failed to address bio-invasions among their recommendations to achieve effective management.

A report on 'The Status of Marine Protected Areas in the Mediterranean Sea 2012' prepared by specialists representing MedPAN in collaboration with the Regional Activity Centre for Specially Protected Areas (RAC/SPA), barely touched on the issue. Though briefly noting that 'The introduction of non-native species appears to be one of the most important ecological and economic threats to the Mediterranean' (Gabrié *et al.*, 2012: 135), and that '25% of surveyed MPA managers reported invasive species among key pressures on habitats and species' (Gabrié *et al.*, 2012: 135), no mention was made of invasive alien species in the nine pages of 'Recommendations' (Gabrié *et al.*, 2012: 157–165). A recent Marine Board Position Paper entitled 'Achieving Ecologically Coherent MPA Networks in Europe: Science Needs and Priorities' omitted the subject altogether, save for a single mention when considering

resilience to climate change: 'Species in European waters might move northward out of protected areas, while some southern species, as well as exotics, could thrive'(Olsen *et al.*, 2013: 51).

A collaborative study by IUCN, WWF and MedPAN (Abdulla *et al.*, 2008: 58) found 'Uncertainty and lack of information regarding marine introduced species was high in the MPAs we surveyed as in average half the MPA managers (54.8%) did not know the status of the introduced species reported in the MPA'.

Management: Eyes Wide Shut

The first assessment of Europe's seas at an EU-wide scale by the European Environment Agency (EEA, 2015) acknowledges that alien species are a growing threat to the environment, that climate-driven changes may factor importantly in expanding their spread, and that the Mediterranean Sea is particularly exposed to introductions: 'a growing number of non-indigenous species have been entering Europe's seas since the 1950s, with the highest rate of introductions being observed in the 2000s. These species are mostly brought in through shipping and the Suez Canal in the Mediterranean Sea' (EEA, 2015: 188). However, the assessment fails to recognize the significance of bio-invasion on MPAs and revels instead in their number and size: 'Europe is doing well in coverage of coastal waters, with more than 16% of coastal marine areas now inside an MPA'. MedPAN recognized that

Marine Protected Areas in the Mediterranean have not escaped this general trend [of bio-invasion] and most of them have been affected by the introduction of alien invasive species for a long time, threatening marine biodiversity... MPAs across the MedPAN Network face common challenges, among them, the lack of awareness and understanding of the impacts of invasive species, the scarcity of information on best practices for management as well as the insufficient baseline information, guidelines and trained local staff to identify and gather knowledge on species introductions and impacts... At a regional level ... there is still a weak networking, coordination and collaboration on this issue.

(Otero et al., 2013: 10)

MedPAN focuses its attention on the internal governance, strategies, and management effectiveness of MPAs (Otero *et al.*, 2013). These authors highlighted the risk posed by alien species to MPAs, introduced a management strategy and actions, provided a priority list of invasive species with the greatest potential impact, presented invasive alien species monitoring and data recording/recording protocols and offered well-illustrated fact sheets for priority Mediterranean invasive alien species. This is a worthwhile initiative: there are far too few region-wide targeted efforts to survey alien species in MPAs (see Guidetti *et al.*, 2014).

The European Union's ecosystem-based Marine Strategy Framework Directive (MSFD; see Braun, this volume) acknowledged that alien species represent one of the main threats to marine biodiversity and related ecosystem services, placing them among the 11 qualitative descriptors for determining 'Good Environmental Status' (GES). Member States were required to establish a monitoring programme for the ongoing assessment and the regular update of targets (by 15 July 2014), and develop (by 2015) a programme of measures designed to achieve or maintain GES (including 'Non-indigenous species introduced by human activities are at levels that do not adversely alter ecosystems') by 2020 – highly unlikely in the case of bio-invasions in the Mediterranean Sea. Indeed, the European Commission's report on 'The first phase of

implementation of the Marine Strategy Framework Directive' (EC, 2014) notes that 'the Commission's assessment of Member States' reports gives rise to concern: Member States' definition of good environmental status and the path they set out to achieve it shows overall limited ambition, often fails to take into account existing obligations and standards and lacks coherence across the Union, even between neighbouring countries within the same marine region'.

The crucial elements for an effective strategy for slowing the influx of invasive alien species are a scientifically sound policy and coordination among all Mediterranean countries to ensure consistency in legal rules and standards to address all major vectors/pathways. In point of fact, the documents adopted in the recent meeting (February 2016) of the Contracting Parties to the 'Barcelona Convention' (Convention for the Protection of the Marine Environment and the Coastal Region of the Mediterranean and its Protocols) are rich in pious expressions of concern for the well-being of the Mediterranean marine environment. Alas, the adopted 'Updated Action Plan concerning Species Introductions and Invasive Species in the Mediterranean Sea' (UNEP(DEPI)/MEDIG.22/L.3/Add.12 Annex III) and 'Monitoring and Assessment Programme' (UNEP(DEPI)/MED IG.22/L.3/Add.7), that ostensibly deal with non-indigenous and invasive species, avoid the most significant pathway – the enlarged Suez Canal (Galil *et al.*, 2014b). It may seem an expedient compromise, but this bureaucratic act of denialism does not change the actuality that introductions through the Suez Canal contribute the largest number of invasive alien propagules in the Mediterranean, affecting the well-being of the Mediterranean Sea and its MPAs.

Considering the highly connected nature of the Mediterranean Sea (Boero, this volume), an MPA, even less networks of MPAs, except for the very large and isolated (an impossibility in the Mediterranean Sea), will not be free of alien species unless embedded in an integrated ecosystem management regime that reduces alien propagule load. The success of a basin-wide ecosystem-based policy on bio-invasions is the key to achieving the long-term objectives of MPAs.

So What Is To Be Done?

- Science-based stewardship has to replace wilful blindness and unrealistic targets. Scientists, stakeholders, policy makers and management should face the reality of the anthropogenically impacted sea.
- Management is hampered by political, economic and societal fragmentation: only seven of the Mediterranean riparian countries are EU Member States. The option of implementing European environmental policies in those states alone may seem pragmatic, but piecemeal protection is futile.
- The crucial element of an effective strategy for slowing the influx of marine alien species into MPAs in the Mediterranean Sea is policy coordination with the Regional Sea Convention ('Barcelona Convention') to ensure consistency in legal rules, standards and implementation.
- Given the near impossibility of eradication of established marine alien species, the precautionary approach for their management is prevention of introductions (primary and secondary) through control of invasion vectors and pathways.
- New MPAs should be located away from the regional hubs of vectors and pathways (i.e. ports, marinas, fish and shellfish farming), and down-current from the major pathway of invasion in the Mediterranean, namely the Suez Canal.
- MPAs located in areas with high alien load, or near invasion hubs, should be surveyed and risk assessment conducted

concerning secondary spread. Cost-effective options for long-term control of alien populations should be identified. If alien populations are at levels that adversely affect native natural diversity, alter native natural habitats and risk secondary spread, changes to protection status should be considered.
- Stakeholders should be informed of the scope and status of bio-invasions in MPAs, and consulted as to the management actions and commitment of resources for their control, as well as possible changes to protection status.

If not acted upon with alacrity, protected effectively and managed efficiently, MPAs under high propagule pressure of alien species, such as those located in the Levantine Basin, may serve as 'seed banks' for invasions inducing 'spillover effect' to adjacent areas, rather than valued tools for native biodiversity conservation.

References

Abdulla, A., Gomei, M., Maison, E. and Piante, C. (2008) *Status of Marine Protected Areas in the Mediterranean Sea*. IUCN, Malaga and WWF-France. 152 pp.

Albins, M.A. (2013) Effects of invasive Pacific red lionfish *Pterois volitans* vs. a native predator on Bahamian coral reef fish communities. *Biological Invasions*, **15**, 29–43.

Albins, M. and Hixon, M. (2011) Worst case scenario: potential long-term effects of invasive predatory lionfish (*Pterois volitans*) on Atlantic and Caribbean coral-reef communities. *Environmental Biology of Fishes*, **96** (10–11), 1151–1157.

Antoniadou, C. and Chintiroglou, C. (2007) Zoobenthos associated with the invasive red alga *Womersleyella setacea* (Rhodomelacea) in the northern Aegean Sea. *Journal of the Marine Biological Association*, **87**, 629–641.

Argyrou, M., Demetropoulous, M. and Hadjichristophorou, M. (1999) The impact of *Caulerpa racemosa* on the macrobenthic communities in the coastal waters of Cyprus, in 'Proceedings of the Workshop on Invasive *Caulerpa* in the Mediterranean', Heraklion, Crete, Greece, 18–20 March 1998. *MAP Technical Report Series*, **125**, 139–158.

Bariche, M., Letourneur, Y. and Harmelin-Vivien, M. (2004) Temporal fluctuations and settlement patterns of native and Lessepsian herbivorous fishes on the Lebanese coast (eastern Mediterranean). *Environmental Biology of Fishes*, **70**, 81–90.

Bariche, M., Torres, M. and Azzurro, E. (2013) The presence of the invasive lionfish *Pterois miles* in the Mediterranean Sea. *Mediterranean Marine Science*, **14** (2), 292–294.

Ben-Tuvia, A. (1964) Two siganid fishes of Red Sea origin in the eastern Mediterranean. *Bulletin of the Sea Fisheries Research Station, Haifa*, **37**, 1–9.

Boero, F., Bouillon, J., Gravili, C. *et al.* (2008) Gelatinous plankton: irregularities rule the world (sometimes). *Marine Ecology Progress Series*, **356**, 299–310.

Boero, F., Putti, M., Trainito, E. *et al.* (2009) First records of *Mnemiopsis leidyi* (Ctenophora) from the Ligurian, Thyrrhenian and Ionian Seas (Western Mediterranean) and first record of *Phyllorhiza punctata* (Cnidaria) from the Western Mediterranean. *Aquatic Invasions*, **4**, 675–680.

Boudouresque, C.F. (2004) Marine biodiversity in the Mediterranean: status of species, populations and communities. *Scientific Reports of the Port-Cros National Park*, **20**, 97–146.

Boudouresque, C.F. and Verlaque, M. (2005) Nature conservation, Marine Protected Areas, sustainable development and the flow of invasive species to the Mediterranean Sea. *Scientific Reports of the Port-Cros National Park*, **21**, 29–54.

Boudouresque, C.F., Meinesz, A., Ribera, M.A. and Ballesteros, E. (1995) Spread of the green alga *Caulerpa taxifolia* (Caulerpales, Chlorophyta) in the Mediterranean: possible consequences of a major ecological event. *Science*, **59** (Supplement), 21–29.

Burfeind, D.D., Pitt, K.A., Connolly, R.M. and Byers, J.E. (2013) Performance of non-native species within marine reserves. *Biological Invasions*, **15**,17–28.

Bustamante, G., Canals, P., Di Carlo, G. *et al.* (2014) Marine protected areas management in the Caribbean and Mediterranean seas: making them more than paper parks. *Aquatic Conservation: Marine and Freshwater Ecosystems*, **24** (Supplement 2), 153–165.

Carlton, J.T. (2000) Global change and biological invasions in the oceans, in *Invasions in a Changing World* (eds H.R. Mooney and R.J. Hobbs). Island Press, Covelo. pp. 31–53.

Cebrian, E. and Ballesteros, E. (2009) Temporal and spatial variability in shallow- and deep-water populations of the invasive *Caulerpa racemosa* var. *cylindracea* in the Western Mediterranean. *Estuarine, Coastal and Shelf Science*, **83**, 469–474.

Cebrian, E. and Rodríguez-Prieto, C. (2012) Marine invasion in the Mediterranean Sea: the role of abiotic factors when there is no biological resistance. *PLoS ONE*, **7** (2), e31135.

Claudet, J., Guidetti, P., Mouillot, D. *et al.* (2011) Ecological effects of marine protected areas: conservation, restoration and functioning, in *Marine Protected Areas: A Multidisciplinary Approach* (ed. J. Claudet). Cambridge University Press, Cambridge. pp. 37–71.

Coll, M., Piroddi, C., Steenbeek, J. *et al.* (2010) The biodiversity of the Mediterranean Sea: estimates, patterns, and threats. *PLoS ONE*, **5** (8), e11842. doi:10.1371/journal.pone.0011842

Commission Internationale pour l'Exploration Scientifique de la Méditerranée (CIESM)

(2001) Gelatinous zooplankton outbreaks: theory and practice. *CIESM Workshop Series*, **14**, 1–112. Monaco. http://www. ciesm.org/publications/Naples

Corsini-Foka, M. and Kondylatos, G. (2015) First occurrence of the invasive lionfish *Pterois miles* in Greece and the Aegean Sea, in 'New Mediterranean Biodiversity Records (October 2015) Collective Article A'. *Mediterranean Marine Science*, **16** (3), 692.

Cottalorda, J.M., Charbonnel, R.P., Dimeet, E. *et al.* (1996) Eradication de la colonie de *Caulerpa taxifolia* découverte en 1994 dans les eaux du parc National de Port-Cros (Var, France), in *Second International Workshop on Caulerpa taxifolia* (eds M.A. Ribera, E. Ballesteros, C.F. Boudouresque *et al.*), University of Barcelona. pp. 149–155.

D'Ambra, I.G.W., Carmichael, R.H., Malej, A. and Onofri, V. (2013) Predation patterns and prey quality of medusae in a semi-enclosed marine lake: implications for food web energy transfer in coastal marine ecosystems. *Journal of Plankton Research*, **35** (6), 1305–1312.

Daniel, B., Piro, S., Charbonnel, E. *et al.* (2009) Lessepsian rabbitfish *Siganus luridus* reached the French Mediterranean coasts. *Cybium*, **33** (2), 163–164.

European Commission (2014) *Report from the Commission to the Council and the European Parliament: The First Phase of Implementation of the Marine Strategy Framework Directive (2008/56/EC). The European Commission's Assessment and Guidance.* Brussels, 10 pp. http://eur-lex. europa.eu/legal-content/EN/TXT/ PDF/?uri= CELEX:52014DC0097 andfrom=EN

European Environment Agency (EEA) (2015) *State of Europe's Seas.* European Environment Agency Report No. 2/2015. 220 pp. http://www.eea.europa.eu/ publications/state-of-europes-seas

Fleming, N.E.C., Harrod, C., Newton, J. and Houghton, J.D.R. (2015) Not all jellyfish are equal: isotopic evidence for inter- and intraspecific variation in jellyfish trophic ecology. *PeerJ*, **3**, e1110.

Francour, P., Harmelin-Vivien, M., Harmelin, J-G. and Duclerc, J. (1995) Impact of *Caulerpa taxifolia* colonization on the littoral ichthyfauna of north-western Mediterranean: preliminary results. *Hydrobiologia*, **300**/301, 345–353.

Francour, P., Mangialajo, L. and Pastor, J. (2010) Mediterranean marine protected areas and non-indigenous fish spreading, in *Fish Invasions of the Mediterranean Sea: Change and Renewal* (eds D. Golani and B. Appelbaum Golani). Pensoft Publishers, Moscow. pp. 127–144.

Fuentes, V.L., Atienza, D., Gili, J.M. and Purcell, J.E. (2009) First records of *Mnemiopsis leidyi* A. Agassiz 1865 off the NW Mediterranean coast of Spain. *Aquatic Invasions*, **4**, 671–674.

Gabrié, C., Lagabrielle, E., Bissery, C. *et al.* (2012) *The Status of Marine Protected Areas in the Mediterranean Sea.* MedPAN, Marseilles and RAC/SPA, Tunis. MedPAN Collection. 256 pp.

Galil, B.S. (1986) Red Sea decapods along the Mediterranean coast of Israel: ecology and distribution, in *Environmental Quality and Ecosystem Stability* (eds Z. Dubinsky and Y. Steinberger), volume IIIA/B. Ramat Gan, Bar-Ilan University Press. pp. 179–183.

Galil, B.S. (2007) Seeing red: alien species along the Mediterranean coast of Israel. *Aquatic Invasions*, **2** (4), 281–312.

Galil, B.S. (2009) Taking stock: inventory of alien species in the Mediterranean Sea. *Biological Invasions*, **11** (2), 359–372.

Galil, B.S. and Lewinsohn, C. (1981) Macrobenthic communities of the eastern Mediterranean continental shelf. *Marine Ecology*, **2** (4), 343–352.

Galil, B.S., Spanier, E. and Ferguson, W.W. (1990) The Scyphomedusae of the Mediterranean coast of Israel, including two Lessepsian migrants new to the Mediterranean. *Zoologische Mededelingen*, **64**, 95–105.

Galil, B.S., Shoval, L. and Goren, M. (2009) *Phyllorhiza punctata* (Scyphozoa: Rhizostomeae: Mastigiidae) reappeared off the Mediterranean coast of Israel. *Aquatic Invasions*, **4**, 481–483.

Galil, B.S., Gershwin, L.A., Douek, J. and Rinkevich, B. (2010) *Marivagia stellata gen. et sp. nov.* (Scyphozoa: Rhizostomeae: Cepheidae), another alien jellyfish from the Mediterranean coast of Israel. *Aquatic Invasions*, **5** (4), 331–340.

Galil, B.S., Marchini, A., Occhipinti-Ambrogi, A. *et al.* (2014a) International arrivals: widespread bio-invasions in European Seas. *Ethology, Ecology and Evolution*, **26** (2–3), 152–171. doi:10.1080/03949370.2014.897651

Galil, B.S., Boero, F., Campbell, M.L. *et al.* (2014b) 'Double trouble': the expansion of the Suez Canal and marine bio-invasions in the Mediterranean Sea. *Biological Invasions*, **17** (4), 973–976.

Galil, B.S., Gershwin, L.A., Zorea, M. *et al.* (2016a) *Cotylorhiza erythraea* Stiasny, 1920 (Scyphozoa: Rhizostomeae: Cepheidae), yet another Erythraean jellyfish from the Mediterranean coast of Israel. *Marine Biodiversity*, doi:10.1007/s12526-016-0449-6

Galil, B.S., Marchini, A. and Occhipinti-Ambrogi, A. (2016b) East is East and West is West? Management of marine bioinvasions in the Mediterranean Sea. *Estuarine, Coastal and Shelf Science*, doi:10.1016/j.ecss.2015.12.021

Gatti, G., Bianchi, C.N., Parravicini, V. *et al.* (2015) Ecological change, sliding baselines and the importance of historical data: lessons from combining observational and quantitative data on a temperate reef over 70 years. *PLoS ONE*, **10** (2), e0118581. doi:10.1371/journal.pone.0118581

George, C.J. and Athanassiou, V. (1967) A two-year study of the fishes appearing in the seine fishery of St George Bay, Lebanon. *Annali del Museo Civico di Storia Naturale di Genova*, **79**, 32–44.

Golani, D. (1998) Impact of Red Sea fish migrants through the Suez Canal on the aquatic environment of the Eastern Mediterranean, in 'Transformations of Middle Eastern Natural Environments: Legacies and Lessons' (eds J. Albert, M. Bernhardsson and R. Kenna). Yale University Conference. *Yale University School of Forestry and Environmental Studies Bulletin*, **103**, 375–387.

Goren, M. and Galil, B.S. (2001) Fish biodiversity and dynamics in the vermetid reef of Shiqmona (Israel). *Marine Ecology*, **22**, 369–378.

Green, S.J., Akins, J.L., Maljkovic, A. and Cote, I.M. (2012) Invasive lionfish drive Atlantic coral reef fish declines. *PLoS ONE*, **7** (3), e32596. doi:10.1371/journal.pone.0106229

Guidetti, P., Baiata, P., Ballesteros, E. *et al.* (2014) Large-scale assessment of Mediterranean marine protected areas effects on fish assemblages. *PLoS ONE*, **9** (4), e91841. doi:10.1371/journal.pone.0091841

Guilhaumon, F., Albouy, C., Claudet, J. *et al.* (2015) Representing taxonomic, phylogenetic and functional diversity: new challenges for Mediterranean marine-protected areas. *Diversity and Distributions*, **21**, 175–187.

Hackerott, S., Valdivia, A., Green, S.J. *et al.* (2013) Native predators do not influence invasion success of Pacific lionfish on Caribbean reefs. *PLoS ONE*, **8** (7), e68259. doi:10.1371/journal.pone.0068259

Harmelin-Vivien, M., Francour, P. and Harmelin, J-G. (1999) Impact of *Caulerpa taxifolia* on Mediterranean fish assemblages: a six-year study, in 'Proceedings of the Workshop on Invasive *Caulerpa* in the Mediterranean', Heraklion, Crete, Greece, 18–20 March 1998. *MAP Technical Report Series*, **125**, 127–138.

Jaubert, R., Cottalorda, J-M., Barcelo, A. *et al.* (2014) Résultats des campagnes 2012 et 2013 de recherche et d'éradication du

Chlorobionte invasif *Caulerpa taxifolia* (Vahl) C. Agardh dans les eaux de l'île de Port-Cros, coeur du Parc national de Port-Cros (Var, France). *Scientific Reports of the Port-Cros National Park*, **28**, 189–194.

Johnston, M.A., Nuttall, M.F., Eckert, R.J. *et al.* (2013) *Long-term Monitoring at the East and West Flower Garden Banks National Marine Sanctuary, 2009–2010, volume 1: Technical Report.* US Department of Interior, Bureau of Ocean Energy Management, Gulf of Mexico OCS Region, New Orleans, Louisiana. OCS Study BOEM 2013–2014. 202 pp.

Johnston, M.W. and Purkis, S.J. (2014) Are lionfish set for a Mediterranean invasion? Modelling explains why this is unlikely to occur. *Marine Pollution Bulletin*, **88**, 138–147.

Jousson, O., Pawlowski, J., Zaninetti, L. *et al.* (1998) Molecular evidence for the aquarium origin of the green alga *Caulerpa taxifolia* introduced to the Mediterranean Sea. *Marine Ecology Progress Series*, **172**, 275–280.

Katsanevakis, S., Wallentinus, I., Zenetos, A. *et al.* (2014) Impacts of marine invasive alien species on ecosystem services and biodiversity: a pan-European review. *Aquatic Invasions*, **9** (4), 391–423.

Klein, J. and Verlaque, M. (2008) The *Caulerpa racemosa* invasion: a critical review. *Marine Pollution Bulletin*, **56**, 205–225.

Ktari, F. and Ktari, M.H. (1974) Présence dans le golfe de Gabes de *Siganus luridus* (Rüppell, 1829) et de *Siganus rivulatus* (Forsskal, 1775) (Poissons, siganides) parasites par *Pseudohaliotrematoides polymorphus*. *Bulletin de l'Institut National Scientifique et Technique d'Oceanographie et de Peche de Salammbo*, **3**, 95–98.

Ktari-Chakroun, F. and Bahloul, M. (1971) Capture de *Siganus luridus* (Rüppell) dans le golfe de Tunis. *Bulletin de l'Institut National Scientifique et Technique d'Oceanographie et de Peche de Salammbo*, **2**, 49–52.

Lazar, S. (2007) Recent and late Pleistocene carbonate-rich sediments in the Mediterranean shelf of Israel: sedimentary, biogenic and genetic analysis. *Geological Survey of Israel Report GSI/08/2007* (in Hebrew, English abstract). 105 pp.

Lesser, M.P. and Slattery, M. (2011) Phase shift to algal dominated communities at mesophotic depths associated with lionfish (*Pterois volitans*) invasion on a Bahamian coral reef. *Biological Invasions*, **13**, 1855–1868.

Levi, F. and Francour, P. (2004) Behavioural response of *Mullus surmuletus* to habitat modification by the invasive macroalga *Caulerpa taxifolia*. *Journal of Fish Biology*, **64**, 55–64.

Lewinsohn, C. and Galil, B.S. (1982) Notes on species of *Alpheus* (Crustacea, Decapoda) from the Mediterranean coast of Israel. *Quaderni del Laboratorio di Tecnologia della Pesca*, **3**, 207–210.

Longpierre, S.R.A., Levi, F. and Francour, P. (2005) How an invasive alga species (*Caulerpa taxifolia*) induces changes in foraging strategies of the benthivorous fish *Mullus surmuletus* in coastal Mediterranean ecosystems. *Biodiversity and Conservation*, **14**, 365–376.

López-Gómez, M.J., Aguilar-Perera, A. and Perera-Chan, L. (2014) Mayan diver-fishers as citizen scientists: detection and monitoring of the invasive red lionfish in the Parque Nacional Arrecife Alacranes, southern Gulf of Mexico. *Biological Invasions*, **16**, 1351–1357.

Lubchenco, J., Palumbi, S.R., Gaines, S.D. and Andelman, S. (2003) Plugging a hole in the ocean: the emerging science of marine reserves. *Ecological Applications*, **13** (1) (Supplement), S3–S7.

Lundberg, B., Ogorek, R., Galil, B.S. and Goren, M. (2004) Dietary choices of siganid fish at Shiqmona reef, Israel. *Israel Journal of Zoology*, **50**, 39–53.

Meinesz, A., Simberloff, D. and Quammen, D. (2002) *Killer Algae.* University of Chicago Press. 360 pp.

Meinesz, A., Chancollon, O. and Cottalorda, J-M. (2010) *Observatoire sur l'expansion de Caulerpa taxifolia et Caulerpa racemosa en Méditerranée: campagne Janvier 2008 – Juin 2010.* Université Nice Sophia Antipolis, E.A. 4228 ECOMERS. 50 pp.

Merella, P., Porcheddu, A. and Casu, S. (1994) La malacofauna della riserva naturale di Scandola (Corsica Nord-occidentale). *Bollettino Malacologico*, **30** (5–9), 111–128.

Meriç, E., Avşar, N. and Yokes, M.B. (2008) Some alien foraminifers along the Aegean and southwestern coasts of Turkey. *Micropaleontology*, **54** (3–4), 307–349.

Merkado, G., Holzmann, M., Apotheloz-Perret-Gentil, L. *et al.* (2013) Molecular evidence for Lessepsian invasion of soritids (larger symbiont bearing benthic foraminifera). *PLoS ONE*, **8** (10), e77725. doi:10.1371/journal.pone.0077725

Micheli, F., Halpern, B.S., Walbridge, S. *et al.* (2013) Cumulative human impacts on Mediterranean and Black Sea marine ecosystems: assessing current pressures and opportunities. *PLoS ONE*, **8** (12), e79889. doi:10.1371/journal.pone.0079889

Mienis, H.K. (1985) *Metaxia bacilla* and *Kleinella amoena*: two other Indo-Pacific species from the Mediterranean coast of Israel. *Levantina*, **54**, 619–620.

Mienis, H.K. (2003) Native marine molluscs replaced by Lessepsian migrants. *Tentacle*, **11**, 15–16.

Mienis, H.K. and Ben-David-Zaslow, R. (2004) A preliminary list of the marine molluscs of the National Park and Nature Reserve of Akhziv-Rosh Haniqra. *Triton*, **10**, 13–37.

Mienis, H.K., Galili, E. and Rapoport, J. (1993) The spiny oyster, *Spondylus spinosus*, a well-established Indo-Pacific bivalve in the Eastern Mediterranean off Israel (Mollusca, Bivalvia, Spondylidae). *Zoology in the Middle East*, **9**, 83–91.

Montefalcone, M., Morri, C., Parravicini, V. and Bianchi, C.N. (2015) A tale of two invaders: divergent spreading kinetics of the alien green algae *Caulerpa taxifolia* and *Caulerpa cylindracea*. *Biological Invasions*, **17**, 2717–2728.

Morris, J.A. and Whitfield, P.E. (2009) *Biology, Ecology, Control and Management of the Invasive Indo-Pacific Lionfish: An Updated Integrated Assessment.* NOAA Technical Memorandum NOS NCCOS 99.

Mouillot, D., Albouy, C., Guilhaumon, F. *et al.* (2011) Protected and threatened components of fish biodiversity in the Mediterranean Sea. *Current Biology*, **21**, 1–7.

Nizamuddin, M. (1991) *The Green Marine Algae of Libya.* Elga Publishing, Bern. 227 pp.

Okuş, E., Yüksek, A., Yilmaz, I.N. *et al.* (2007) Marine biodiversity of Datça-Bozburun specially protected area (Southeastern Aegean Sea, Turkey). *Journal of the Black Sea/Mediterranean Environment*, **13** (1), 39–49.

Olsen, E.M., Johnson, D., Weaver, P. *et al.* (2013) *Achieving Ecologically Coherent MPA Networks in Europe: Science Needs and Priorities* (eds K.E. Larkin and N. McDonough). Marine Board Position Paper 18. European Marine Board, Ostend, Belgium. 88 pp.

Oray, I.K., Sınay, E., Saadet Karakulak, F. and Yıldız, T. (2015) An expected marine alien fish caught at the coast of Northern Cyprus: *Pterois miles* (Bennett, 1828). *Journal of Applied Ichthyology*, **31**, 733–735.

Otero, M., Cebrian, E., Francour, P. *et al.* (2013) *Monitoring Marine Invasive Species in Mediterranean Marine Protected Areas (MPAs): A Strategy and Practical Guide for Managers.* MedPAN North project. IUCN, Malaga, Spain. 136 pp.

Papaconstantinou, C. (1987) Distribution of the Lessepsian fish migrants in the Aegean Sea. *Biologia Gallo-Hellenica*, **13**, 15–20.

Parker, I.M., Simberloff, D., Lonsdale, W.M. *et al.* (1999) Impact: toward a framework for understanding the ecological effects of invaders. *Biological Invasions*, **1**, 3–19.

Piazzi, L. and Balata, D. (2008) The spread of *Caulerpa racemosa* var. *cylindracea* in the Mediterranean Sea: an example of how biological invasions can influence beta diversity. *Marine Environmental Research*, **65**, 50–61.

Piazzi, L., Ceccherelli, G. and Cinelli, F. (2001) Threat to macroalgal diversity: effects of the introduced alga *Caulerpa racemosa* in the Mediterranean. *Marine Ecology Progress Series*, **210**, 149–159.

Piazzi, L., Balata, D., Cecchi, E. and Cinelli, F. (2003) Co-occurrence of *Caulerpa taxifolia* and *C. racemosa* in the Mediterranean Sea: interspecific interactions and influence on native macroalgal assemblages. *Cryptogamie Algologie*, **24**, 233–243.

Piraino, S., Aglieri, G., Martell, L. *et al.* (2014) *Pelagia benovici sp. nov.* (Cnidaria, Scyphozoa): a new jellyfish in the Mediterranean Sea. *Zootaxa*, **3794** (3), 455–468.

Por, F.D. (1978) *Lessepsian Migration: The Influx of Red Sea Biota into the Mediterranean By Way of the Suez Canal.* Ecological Studies 23. Springer-Verlag, Berlin-Heidelberg-New York. 228 pp.

Purcell, J.E. and Arai, M.N. (2001) Interactions of pelagic cnidarians and ctenophores with fish: a review. *Hydrobiologia*, **451**, 27–44.

Ramos-Espla, A.A., Bitar, G., Khalaf, G. *et al.* (eds) (2014) *Ecological Characterization of Sites of Interest for Conservation in Lebanon: Enfeh Peninsula, Ras Chekaa cliffs, Raoucheh, Saida, Tyre and Nakoura.* RAC/SPA – MedMPAnet Project, Tunis. 168 pp + annexes.

Richardson, A.J., Bakun, A., Hays, G.C. and Gibbons, M.J. (2009) The jellyfish joyride: causes, consequences and management responses to a more gelatinous future. *Trends in Ecology and Evolution*, **24**, 312–322.

Robert, P. (2001) *Mission d'éradication localisée de l'algue Caulerpa racemosa dans les eaux du Parc national de Port-Cros.*

Août–Septembre 2001. Parc national de Port-Cros publications, Hyères, France. 17 pp.

Ruitton, S., Javel, F., Culioli, J.-M. *et al.* (2005) First assessment of the *Caulerpa racemosa* (Caulerpales, Chlorophyta) invasion along the French Mediterranean coast. *Marine Pollution Bulletin*, **50**, 1061–1068.

Ruttenberg, B.I., Schofield, P.J., Akins, J.L. *et al.* (2012) Rapid invasion of Indo-Pacific lionfishes (*Pterois volitans* and *Pterois miles*) in the Florida Keys, USA: evidence from multiple pre- and post-invasion data sets. *Bulletin of Marine Science*, **88** (4), 1051–1059.

Safriel, U.N., Gilboa, A. and Felsenburg, T. (1980) Distribution of rocky intertidal mussels in the Red Sea coasts of Sinai, the Suez Canal, and the Mediterranean coast of Israel, with special reference to a recent colonizer. *Journal of Biogeography*, **7**, 39–62.

Sala, E., Kizilkaya, Z., Yildirim, D. and Ballesteros, E. (2011) Alien marine fishes deplete algal biomass in the Eastern Mediterranean. *PLoS ONE*, **6** (2), e17356. doi:10.1371/journal.pone.0017356

Sarà, G., Romano, C. and Mazzola, A. (2006) A new Lessepsian species in the western Mediterranean (*Brachidontes pharaonis* Bivalvia: Mytilidae): density, resource allocation and biomass. *Marine Biodiversity Records*, **5087**, 1–7. doi:10.1017/S175526720600087X

Schofield, P. (2009) Geographic extent and chronology of the invasion of nonnative lionfish (*Pterois volitans* [Linnaeus 1758] and *P. miles* [Bennett 1828]) in the Western North Atlantic and Caribbean Sea. *Aquatic Invasions*, **4**, 473–479.

Steinitz, H. (1970) A critical list of immigrants via the Suez Canal. *Biota of the Red Sea and Eastern Mediterranean*, pp. 59–63.

Steinitz, W. (1927) Beiträge zur Kenntnis der Küstenfauna Palästinas. I. *Pubblicazioni della Stazione Zoologica di Napoli*, **8**, 331–353.

Tavares, M. and De Melo, G.A.S. (2004) Discovery of the first known benthic invasive species in the Southern Ocean: the North Atlantic spider crab *Hyas araneus* found in the Antarctic Peninsula. *Antarctic Science*, **16** (2), 129–131.

Tortonese, E. (1973) Facts and perspectives related to the spreading of Red Sea organisms into the eastern Mediterranean. *Annali del Museo Civico di Storia Naturale 'Giacomo Doria'*, **79**, 322–329.

Tosun, M.S. (2011) *Demographic Divide and Labor Migration in the Euro-Mediterranean Region*. IZA Discussion Paper No. 6188. http://ssrn.com/abstract=1973917

Trégarot, E., Fumaroli, M., Arqué, A. *et al.* (2015) First records of the red lionfish (*Pterois volitans*) in Martinique, French West Indies: monitoring invasion status through visual surveys. *Marine Biodiversity Records*, **8**, e1. doi:10.1017/ S1755267214001341

Turan, C., Erguden, D., Gurlek, M. *et al.* (2014) First record of the Indo-Pacific lionfish *Pterois miles* (Bennett, 1828) (Osteichthyes: Scorpaenidae) for the Turkish marine waters. *Journal of Black Sea/Mediterranean Environment*, **20** (2), 158–163.

United Nations Environment Programme (UNEP) (1991) Jellyfish blooms in the Mediterranean. *MAP Technical Reports Series*, **47**, 1–320.

Usher, M.B. (1988) Biological invasions of nature reserves: a search for generalizations. *Biological Conservation*, **44**, 119–135.

Vaillant, L. (1865) Recherches sur la faune malacologique de la baie de Suez. *Journal de Conchyliologie*, **13**, 97–127.

Valdivia, A., Bruno, J.F., Cox, C.E. *et al.* (2014) Re-examining the relationship between invasive lionfish and native grouper in the Caribbean. *PeerJ*, **2**, e348. https://dx.doi. org/10.7717/peerj.348

Vergés, F.T., Cebrian, E., Ballesteros, E. *et al.* (2014) Tropical rabbitfish and the deforestation of a warming temperate sea. *Journal of Ecology*, **102** (6), 1518–1527.

Verlaque, M. (1989) Contribution à la flore des algues de la Méditeraneé: espèces rares ou nouvelles pour les côtes françaises. *Botanica Marina*, **32**, 101–113.

Verlaque, M. and Fritayre, P. (1994) Modification des communautés algales méditerranéennes en presence de l'algue envahissante *Caulerpa taxifolia* (Vahl) C. Agardh. *Oceanologia Acta*, **17**, 659–672.

Verlaque, M., Afonso-Carrillo, J., Candelaria Gil-Rodríguez, M. *et al.* (2004) Blitzkrieg in a marine invasion: *Caulerpa racemosa* var. *cylindracea* (Bryopsidales, Chlorophyta) reaches the Canary Islands. *Biological Invasions*, **6**, 269–281.

Whitfield, P.E., Gardner, T., Vives, S.P. *et al.* (2002) Biological invasion of the Indo-Pacific lionfish *Pterois volitans* along the Atlantic coast of north America. *Marine Ecology Progress Series*, **235**, 289–297.

Yokes, M.B., Dalyan, C., Karhan, S.U. *et al.* (2012) Alien opisthobranchs from Turkish coasts: first record of *Plocamopherus tilesii* Bergh, 1877 from the Mediterranean. *Triton*, **25** (Supplement 1), 1–9.

Zibrowius, H. (1992) Ongoing modification of the Mediterranean marine fauna and flora by the establishment of exotic species. *Mésogée*, **51**, 83–107.

Zlotnik, H. (2003) The population of the Mediterranean region during 1950–2000, in *Security and Environment in the Mediterranean: Conceptualizing Security and Environmental Conflicts* (eds H.G. Brauch, L.A. Marqrina, P.F. Rogers and M. El-Sayed Selim). Springer, Germany. pp. 593–614.

11

Marine Protected Areas and Marine Spatial Planning, with Special Reference to the Black Sea

Eva Schachtner

Leibniz Institute of Ecological Urban and Regional Development, Dresden, Germany

Marine Spatial Planning and Marine Protected Areas: Compatible or Conflicting Concepts?

In its broadest sense, Marine Spatial Planning (MSP) is 'a public process of analyzing and allocating the spatial and temporal distribution of human activities in marine areas to achieve ecological, economic and social objectives that have been specified through a political process' (UNESCO and IOC).[1]

This definition of MSP often leads to the assumption that ecological, economic and social objectives are of equal importance and should be balanced equally in the MSP process (e.g. Schäfer, 2009). This view seems to correspond to the concept of the three pillars of sustainability: economic development, social development and environmental protection.[2]

In contrast, Marine Protected Areas (MPAs) are primarily selected on the basis of ecological and/or geomorphological criteria and focus on the protection of those features. They have a specified object of protection, for example marine mammals, or they aim at tackling environmental threats from a particular source, like shipping or fisheries. As a positive side effect, protected areas can nevertheless contribute to the non-environmental objectives of MSP, for example by conserving nursery areas for fisheries production or by enhancing tourism revenues. Thus, MSP and protected area programmes are in many cases mutually beneficial (Clark, 1992). But even in multiple-purpose MPAs, a holistic, cross-sectoral approach is often not truly implemented in practice. Marine Protected Areas therefore cannot be considered as a small-scale 'predecessor' for MSP (Drankier, 2012).

In recent times, long-standing sea uses have become more intense and new forms of use have emerged. The negative effects include over-fishing, loss and destruction of habitats, pollution and climate change (Douvere, 2008). It could thus be worth considering shifting the orientation of MSP and using it as a tool to redress the balance in favour of the marine environment.

Restrictions on economic activities do, however, often seem less acceptable than stresses on the environment, since negative effects on the environment are often felt only with a time lag, whereas economic downturn immediately threatens livelihoods. Especially in countries with fast-growing maritime industries and still-developing economies, it is viewed as problematic to overly prioritize

Management of Marine Protected Areas: A Network Perspective, First Edition. Edited by Paul D. Goriup.
© 2017 John Wiley & Sons Ltd. Published 2017 by John Wiley & Sons Ltd.

ecosystem conservation (Qiu and Jones, 2013). To gain public acceptance for MSP concepts in the Black Sea region, it might therefore seem necessary to provide more leeway for development than in other European Seas.

Yet, since functional ecosystems are an essential precondition for social and economic development, the balance between economic, social and environmental interests can be found only within the framework of environmental compatibility (ARL, 2000). The carrying capacity of the sea has to be respected, not only to preserve the intrinsic value of nature, but also to secure future prosperity. By destroying their environment, countries deprive themselves of development chances that, for example, genetic resources might offer in the future and whose value cannot yet be estimated. The protection of the environment should therefore not be considered a 'luxury' problem that only rich countries can afford to tackle.

It is therefore crucial to set the right course today by assigning to MSP not only a coordinating role between the different interests, but also a steering role towards ecosystem-based management.

Protection of the Sea

There are basically two concepts of area protection. The segregation approach is based on the dichotomy of 'protection area' and 'pollution area'. Thus, nature protection areas and areas for economic activities are spatially separated. However, because of the highly connected nature of the sea, MPAs are vulnerable to natural resource exploitation and other activities even if they occur far outside the protected areas. For example, pollution does not respect the boundaries of MPAs and therefore endangers habitats and species within those areas. Also, the state of the neighbouring ecosystems can influence the health and productivity of the MPA ecosystem. Protected areas should therefore

not be managed in isolation, as 'islands of protection' (Salm *et al.*, 2000).

The integration approach, on the other hand, aims to overcome the aforementioned dichotomy by combining environmental protection and economic use (Mose and Weixlbaumer, 2007). Nature protection is thus instituted across 100% of the area by regulating the type and intensity of the anthropogenic use of space (Spektrum, 2001). Marine Spatial Planning can unite the advantages of both concepts by integrating MPAs into a comprehensive spatial development strategy.

Protection of Open Space

At sea, intensive use can have a similar negative effect for species and their habitats as the sealing of the soil on land (Janssen *et al.*, 2008). Moreover, due to the absence of land prices and the seemingly endless expanse, space is often too generously used (Buchholz, 2004). The viability of ecosystems, however, depends on sufficient open space and unspoiled nature (Ritter, 2005).

Protection of open space is ideally quantitative, structural and qualitative. Quantitative protection means there is an adequate amount of open space; structural protection means the conservation of sufficiently large continuous areas of open space is ensured; and qualitative protection means ecological connectivity is respected (Ritter, 2005).

To effectively implement protection of open space, MSP should not only define 'where' and 'how' a use takes place, but also decide 'if' a use is really necessary. This also means that uses undesired on land are not simply relocated in the sea. The sea should rather be reserved for uses for which it provides a particular locational advantage.

Surface recycling can help to further reduce claims on areas so far undisturbed by human activities. For instance, spatial planning can ensure that new generations of offshore wind farms or other installations are built over decommissioned and dismantled plants (Köppel *et al.*, 2006). Also, the pre-definition of minimum capacities, especially for power plants, can help to reduce space requirements (BBSR, 2011).

Moreover, uses should occupy as little space as possible, taking account of all three dimensions of the sea (seabed, water column and water surface): to use the available space efficiently, uses ought to be concentrated, and installations bundled (BfN, 2006). Marine Spatial Planning can also promote synergies and facilitate co-use. Offshore wind farms, for example, can be combined with aquaculture. The advantages of the concentration of uses, however, have to be balanced against the then locally multiplied environmental impact.

Cumulative Effects and Interactions

Environmental pressures result from the individual or various activities of one or more users, which may occur simultaneously or at different times, independently or interrelated. There are additive effects, such as the accumulation of similar effects, and synergetic effects from the combined effects of various pressures. The severity of these effects on the environment depends mainly on the quantity, type and intensity of the impacts, their spatial distribution and their sequence in time, but also on the vulnerability and adaptability of the affected ecosystems. 'Time-crowding' and 'space-crowding' constitute the biggest threats to the environment, but gradual processes also need to be considered (Siedentop, 2003).

Often, as a result of a series of small, apparently independent and environmentally compatible decisions, a far-reaching process can be set in motion without ever consciously addressing the issue (Odum, 1982). For example, through the cumulative effects of small decisions, the sea gradually becomes more and more eutrophic, or acidic, or laden with plastics, each of which can significantly alter ecosystem functions. Marine Spatial Planning offers a framework suitable for the implementation of a holistic perspective and the consideration of all possible pressures within the planning area. To identify incremental effects, indicators can be used, for example the cumulative loss of habitats, the cumulative level of noise pollution or the cumulative fragmentation of an area (Hanusch *et al.*, 2007).

Through MSP, reasonable placement alternatives can be considered in the planning process and their respective impacts on the environment compared: uses can then be sited where they cause the least environmental impacts. Fragmentation effects can thus be minimized, while migration routes and retreat areas are protected. Similarly, buffer zones can be placed around sensitive areas, for example to reduce exposure of marine mammals to harmful levels of noise emissions. Temporal coordination can further help to alleviate the impacts of uses, since adequate periods of low use or no use are crucial for the regeneration of the environment. For example, construction activities can be planned on a staggered basis to reduce their cumulative impact (Janssen *et al.*, 2008).

Unanticipated results can also occur when interaction webs are overlooked or manipulated. For example, removing top predators including marine mammals, sharks and other large fishes can generate cascade effects for the whole food chain. In the MSP process, the most important ecological features of an ecosystem and possible indirect effects can be identified (Crowder and Norse, 2008) and, consequently, these effects can as far as possible be avoided.

Prevention of and Compensation for Negative Effects

Because the marine environment is particularly sensitive and because there is a significant knowledge deficit about the functioning of its ecosystems (see Boero, this volume), the observance of the precautionary principle is essential. Where scientific understanding is still incomplete, recourse to this principle helps to avoid possible risks. This means, for example, that if there are indications of special vulnerability of an area, its protection must be ensured by appropriate spatial planning measures, even if a definite assessment is not yet possible. Marine Spatial Planning can thus play a proactive role, and not just react to problems after they have occurred.

Furthermore, MSP can help to ensure compensation and replacement for interference in the natural environment. If stipulations in a marine spatial plan are likely to have unavoidable negative consequences for the environment, corresponding stipulations can provide for commensurate compensation. Possible measures are restrictions on other, less important uses, the requirement to dismantle out-of-date installations before new installations are constructed, or even measures onshore that contribute to the regeneration of the sea. The limits of the planned compensation possibilities then also set a limit for impacts on the environment and thereby ensure sustainable development (ARL, 2000).

Compensation, however, always implies that the pre-existing natural conditions have been seriously damaged or even destroyed. Care must thus be taken that the overriding principle of avoidance of environmental damage does not get undermined.

Flexible and Proportionate Planning

Marine spatial plans reflect the state of knowledge at the time of their adoption and therefore tend to perpetuate errors (Beaucamp, 2002). Planning should therefore be understood to be a continuous adjustment process and plans regularly reviewed and adapted. Stipulations are ideally not definite, but keep planning possibilities open by

ensuring a certain spatial disposability. For example, the sea should not be used as a space for permanent fixed installations, and the dismantling of decommissioned installations should generally be required (Wende *et al.*, 2007). Similarly, all other activities should only be granted permission for a manageable period of time. Otherwise, the implementation of later decisions on the establishment of protected areas or on other protective measures that may become necessary because of increased knowledge of the marine environment will be considerably more complicated.

Conversely, to enhance acceptance of protective measures by users, activities should not be excessively restricted. Some species only need protection in one of the three dimensions of the sea (water surface, water column and sea bed). For example, some benthic communities only need protection from impacts on the seabed, like bottom-trawl fisheries. Moreover, since the need for protection of species and the vulnerability of areas can vary over time, temporal aspects can be taken into consideration as a fourth dimension of planning. Marine Spatial Planning can consequently provide for proportionate spatial management by placing only certain areas under protection and, if appropriate, only at certain times.

Creation of an Efficient Network of MPAs

Even though the sea is characterized by great permeability and therefore ecosystems are better connected than onshore, the guarantee of an undisturbed exchange of organisms and nutrients between MPAs through a protected network can considerably multiply their effectiveness (Boero, this volume). Furthermore, well-designed networks of MPAs are more resilient and better suited to mitigating the effects of dynamic natural processes, or imposed processes such as climate change, than unconnected MPAs. Networks of no-take and partially protected

MPAs are thus increasingly considered as an essential element of ecosystem-based MSP (Jones *et al.*, 2016).

A network of MPAs should be designed as a synergistic system, based on cells of ecosystem functioning (Boero, this volume) where the 'whole is greater than the sum of the parts'. Networks ideally reflect the migration paths of certain species to connect their sub-habitats or scattered populations, or they connect similar habitats to reinforce the respective protection effect (Beal *et al.*, this volume). The degree of protection of the connecting areas has to be at least commensurate with the function they need to fulfil. Migration corridors or stepping stones can be established to ensure connection, or MPAs can be optimally positioned in relation to each other, for example to ensure exchange through currents. The MPAs might also be established as dynamic MPAs that protect dynamic ocean features (like eddies or fronts) or the seasonal migration of protected species (Crowder and Norse, 2008).

However, by creating networks of protected areas, it is important not to lose sight of the goal of a comprehensive protection of the sea. Environmental protection must not be relegated to the spatial sidelines, such as narrow migration corridors (Leibenath, 2009).

Towards Implementation of the Ecosystem Approach

As early as 1992, at the UN Earth Summit in Rio de Janeiro, it was recognized that the traditional sectoral approach to natural resource and environmental management did not adequately address human impacts on the environment (Laffoley *et al.*, 2004). In consequence, management has shifted towards a more holistic approach, 'mainstreaming' the environment into economic sectors. However, even this was soon recognized as flawed. Accordingly, at the fifth Conference of the Parties of the Convention on Biological Diversity in 2000, it was recommended that the ecosystem approach be applied, and 12 principles have been developed for its implementation[3] that also seem to be relevant for MSP. For example, Principle 7 states that the ecosystem approach should be undertaken at the appropriate spatial and temporal scales.

Within the EU, the Marine Strategy Framework Directive (MSFD, 2008/56/EC) and the Maritime Spatial Planning Directive (MSPD, 2014/89/EU) now require the application of an ecosystem approach. Furthermore, the Baltic Marine Environment Protection Commission (HELCOM) and Visions and Strategies around the Baltic Sea (VASAB) have adopted the ecosystem approach as an overarching principle for Maritime Spatial Planning[4] and agreed on a 'Guideline for the implementation of ecosystem-based approach in Maritime Spatial Planning (MSP) in the Baltic Sea area.'[5]

Ecosystems can be defined as 'subdivisions of the Earth's surface, including marine areas, and lower atmosphere within which natural processes operate and biological communities perpetuate themselves' (Ehler and Douvere, 2007). Humans, with their cultural diversity, are regarded as an integral component of ecosystems.[6] The ecosystem approach is, according to one definition, 'a strategy for the integrated management of land, water and living resources that promotes conservation and sustainable use in an equitable way' (Convention on Biological Diversity).[7] By taking the full array of interactions among ecosystem components and human users into consideration, the ecosystem approach can help to arbitrate between the increasing diversity and intensity of human activities and the carrying capacity of the sea.

The spatial component is a key characteristic of the ecosystem approach to management, since in most cases ecosystems are fixed in space for long periods of time. And, since MSP addresses inter-sectoral conflicts and user–environment conflicts, taking account of temporal aspects, it is an ideal tool to implement the holistic ecosystem approach.

Limits of MSP

Establishing maritime spatial plans does not yet guarantee the achievement of environmental objectives. Therefore, the establishment of marine spatial plans should not be considered as the ultimate goal. The goal should rather be to achieve real outcomes such as sustainable energy supplies, reduced conflicts among human activities, or the conservation of marine ecosystems (Ehler, 2012). Moreover, while through MSP space can be allocated, conflicts reduced and synergies maximized, the quality of uses and the concrete impacts of individual projects cannot be controlled (Schultz-Zehden *et al.*, 2008). Other instruments like environmental impact assessment therefore need to be employed alongside MSP.

The Law of the Sea: A Hindrance to MSP?

There is no international convention exclusively dedicated to spatial planning at sea. Some relevant regulations, however, can be found in the United Nations Convention on the Law of the Sea (UNCLOS). Apart from Turkey, all states of the Black Sea area have signed and ratified this convention.

UNCLOS sets out different zones of the sea and defines the rights and obligations of its contracting parties in each of them. Article 2(1) of UNCLOS states that the sovereignty of a coastal state covers its land territory and internal waters. The coastal state is thus free to make laws, to regulate any use, to use any resource and, therefore, to submit its internal waters to MSP. According to Art. 2(1) of UNCLOS, the sovereignty of the coastal state comprises its territorial sea, extending up to 12 nautical miles from the baseline (Art. 3). That sovereignty derives from the sovereignty over the land territory. Consequently, the coastal state can undertake spatial planning activities in that part of the sea. Ships of all states, however, enjoy the right of innocent passage through the territorial sea (Art. 17).

Beyond its territorial sea, a coastal state may claim an exclusive economic zone (EEZ) that extends up to 200 nautical miles from the baseline (Art. 55, 57). Since the Black Sea is quite small and all the riparian states have declared EEZs (Oral, n.d.), it is completely divided between them (Black Sea Commission). Thus, there are no areas that lie beyond national jurisdiction (high seas/the Area).

UNCLOS provides coastal states with certain functional rights in their EEZ for the purpose of exploring and exploiting, conserving and managing natural resources and with regard to other activities for the economic exploitation and exploration of the zone, such as the production of energy from the water, currents and winds and with regard to the establishment and use of artificial islands, installations and structures (Art. 56). The exercise of these rights is subject to various conditions, such as the respect of the right of any state to lay submarine pipelines and cables, and the freedom of navigation of other states' vessels (Art. 58). Concerning the seabed and subsoil, the rights of the coastal state in the EEZ shall be exercised in accordance with Part VI of UNCLOS on the continental shelf (Art. 56(3)).

Article 56(1) of UNCLOS does not expressly assign to the coastal state a sovereign right or jurisdiction to undertake planning activities in the EEZ. This, however, does not necessarily mean that MSP there is unlawful. Under Art. 60(1) of UNCLOS, for example, the coastal state has the exclusive right to construct, to authorize and to regulate the construction, operation and use of artificial islands, installations and structures.

It is left to the coastal state to determine if and how these rights are to be executed (Proelß, 2009). Therefore, it seems justified to conclude that MSP is allowed if planning activities are directly linked to the rights expressly assigned to the coastal state by Part V of UNCLOS.

In enclosed or semi-enclosed seas like the Black Sea, contracts between all riparian states could allow MSP measures that go beyond the scope of measures allowed by UNCLOS. Of course, in this case, only the contracting states are bound by the contract and only the rights of those states can be affected by its provisions.

EU Instruments: A Fresh Impetus to MSP

Recommendation on Integrated Coastal Zone Management

The European Parliament and the Council adopted on 30 May 2002 the Recommendation 2002/413/EC on Integrated Coastal Zone Management (ICZM) that outlines the steps that the Member States should take to promote ICZM along their shorelines and defines the principles of sound coastal planning and management. Those principles include the need to base planning on in-depth knowledge, to take a long-term and cross-sectoral perspective, to involve stakeholders, and to take into account both the terrestrial and the marine component of the coastal zone. The recommendation, however, lacks binding force.

Item 5.9 of the Roadmap for Maritime Spatial Planning of the Commission (COM (2008) 791 final) concerns the relation between MSP and ICZM and says 'coastal zones are the "hinge" between maritime and terrestrial development. Drainage areas or land-based impacts from activities such as agriculture and urban growth are relevant in the context of MSP. This is why terrestrial spatial planning should be coordinated with MSP. Furthermore, according to a Commission Staff Working Paper of 2013, 'MSP and ICZM connect in their geographical coverage (transition area from land to sea) and in their overall objective (to manage human uses in their respective areas of application)' (EC, 2013b).

Consequently, the Commission has decided to develop these two tools together, an approach that is reflected in the new MSPD: Art. 6 No. 2 lit. (c) encourages Member States to promote coherence between MSP and the resulting plan or plans and other processes, such as integrated coastal management.

The Example of the Mediterranean Sea

The Convention for the Protection of the Marine Environment and the Coastal Region of the Mediterranean (Barcelona Convention) entered into force on 12 February 1978. The European Community as well as all the EU Mediterranean Member States are Contracting Parties to the Convention. Within its framework, a draft protocol on ICZM was prepared, and, after a lengthy negotiation process, adopted on 21 January 2008.

The protocol aims to minimize the impact of economic activities on the environment and to guarantee a sustainable use of resources (Art. 9), to protect coastal ecosystems, landscapes, islands and cultural heritage (Art. 10–13), and to ensure participation and raise awareness (Art. 14–15). In order to ensure that corresponding measures are adopted in a coherent way, the text requires that they are made part of a broader planning system. Article 18(1) says that 'each Party shall further strengthen or formulate a national strategy for integrated coastal zone management and coastal implementation plans and programmes'.

Since it has, in contrast to the ICZM Recommendation of the EU, binding power, the protocol significantly advances the ICZM process. However, even if the protocol is binding, the wording of some of its provisions resembles recommendations rather than strict obligations.

Marine Strategy Framework Directive

The most recent policy driver for the protection of the marine environment is the MSFD. The objective of the MSFD is to achieve a Good Environmental Status (GES) of the EU's marine waters by 2020 by applying an ecosystem approach towards marine management and governance.

Each Member State is required to assess the current state of its marine environment, to define the desirable 'good environmental status' of its region and to establish detailed environmental targets as well as monitoring programmes.

The MSFD can be interpreted as applying the 'hard' sustainability approach, of which ecosystem conservation is the basis. The taking into account of all relevant impacts constitutes a novel, holistic approach to environmental protection at the EU level, through which many of the sectoral efforts of the past can be complemented or even replaced (ARCADIS, 2011). Together with the Water Framework Directive (WFD, 2000/60/EC), the MSFD provides for an integrated environmental management system that stretches from the basin catchment area through the coast to the open sea (Qiu and Jones, 2013).

The MSFD does not explicitly require the Member States to implement MSP, but they are required to take management measures into consideration that 'influence where and when an activity is allowed to occur' (Spatial and temporal distribution controls/Art. 13(1) in conjunction with Annex VI(3)).

Furthermore, the MSFD promotes spatial protection measures, contributing to coherent and representative networks of MPAs, adequately covering the diversity of the constituent ecosystems (Art. 13(4)). The establishment of such a coherent and representative network of MPAs requires a level of protection that goes beyond the level of protection guaranteed by Natura 2000 sites (Braun, this volume). The Birds Directive (2009/147/EC) and the Habitats Directive (92/43/EEC), which form the basis for the protection of those sites, do not reflect the modern ecosystem approach. They were only designed to protect certain species and habitats, not to create a coherent and fully representative network of MPAs across Europe (Qiu and Jones, 2013). To form an effective network, the Natura 2000 sites have to be complemented, for example by national MPAs, by protection corridors or by 'stepping stones'.

Maritime Spatial Planning Directive

From the Birds Directive to the MSFD, a clear trend of mainstreaming environmental concerns into wider planning and development programmes can be recognized in European legislation (Qiu and Jones, 2013). Right in line with that trend, the MSPD has recently been adopted, constituting a milestone in European legislation with regard to spatial planning. The EU for the first time includes not only individual spatial planning elements in environmental regulations (Schubert, 2015). In particular because of the increasing and uncoordinated use of coastal and maritime areas that leads to an inefficient and unsustainable use of marine and coastal resources, the Directive rather aims to cover all policy areas with an impact on coasts, seas and oceans (EC, 2013a).

The Directive, however, does not set new sectoral policy targets. Through maritime spatial plans, the objectives defined by national or regional sectoral policies are to be integrated and linked, and steps taken to prevent or alleviate conflicts between different sectors and to achieve the Union's objectives in marine and coastal related sectoral policies (EC, 2013a). The operational objectives of the Directive are thus procedural in nature. It supports ongoing implementation of sea-related policies in Member States through more efficient coordination and increased transparency (EC, 2013a).

Consequently, the Directive only establishes a 'framework' for maritime spatial planning (Art. 1(1)). The EU has opted for such a 'framework-type' Directive to provide flexibility and to allow the Member States to develop their own national policies. The Directive is deliberately not aimed at assigning a new planning task to the EU or at reshaping the different national spatial planning systems (Schubert, 2015).

According to the Directive, 'when establishing and implementing maritime spatial planning, Member States shall consider economic, social and environmental aspects to support sustainable development and growth in the maritime sector, applying an ecosystem-based approach, and to promote the coexistence of relevant activities and uses' (Art. 5(1)). The definition of the objectives of the ecosystem-based approach corresponds to the definition in Art. 1(3) of the MSFD and so requires that 'the collective pressure of all activities is kept within levels compatible with the achievement of good environmental status and that the capacity of marine ecosystems to respond to human-induced changes is not compromised, while contributing to the sustainable use of marine goods and services by present and future generations' (Preamble, Recital 14).

The ecosystem-based approach is considered a basic principle of MSP within the EU and links the MSPD clearly to the MSFD. In reality, however, the two Directives seem to function more on an antagonistic than synergistic basis. By often prioritizing 'blue growth' over environmental protection towards the achievement of GES, Member States undermine the closer coupling that has been called for (Jones *et al.*, 2016).

Moreover, the appropriate balance between ecological, economic and social objectives of MSP and the respect of the carrying capacity of the sea, required by the ecosystem-based approach, seems to be difficult to strike. It could be argued that, at least if the sea is affected by planning decisions to such an extent that its ecosystems cannot recover in the foreseeable future, insufficient weight has been given to the protection of the environment (Schubert, 2015). Such an interpretation ensures that the ecosystem-based approach does not conflict with the requirement to consider also economic and social interests, but just prevents manifest errors of consideration.

The Black Sea: Evaluation of Progress on MSP at a Regional Level

The Black Sea is surrounded by six countries. The countries of the west coast, Bulgaria and Romania, form part of the European Union. Turkey, located on the south coast, is an EU candidate country. The states on the north and east coasts (Ukraine, the Russian Federation and Georgia) arose following the break-up of the Soviet Union, which still influences their legal system, although both Ukraine and Georgia signed Association Agreements with the EU in 2014 which implies increasing harmonization of their legislation with the acquis communautaire.

Despite its anoxic zone below 300 m, the Black Sea is relatively rich in biological resources (Alexandrov *et al.*, this volume). The sea and its coastal wetlands provide spawning grounds for various fish species and breeding and resting places for many endangered birds. Also, three species of marine mammals live in the Black Sea. Eutrophication, pollution and irresponsible fishing, however, brought the environment of the Black Sea to the edge of collapse.

The Bucharest Convention on the Protection of the Black Sea Against Pollution

The Convention on the Protection of the Black Sea Against Pollution (also referred to as the Bucharest Convention) was signed in Bucharest in April 1992, and ratified by all legislative assemblies of the six Black Sea riparian states in early 1994. Acting on the mandate of the Black Sea countries, the Commission on the Protection of the Black Sea Against Pollution (the Black Sea Commission) implements the provisions of the Convention, its four Protocols and the Black Sea Strategic Action Plan (BSC, 2007). The Commission is assisted by its Permanent Secretariat located in Istanbul, Turkey.

Efforts towards ICZM and MSP

The original Odessa Declaration of 1993 (Ministerial Declaration on the Protection of the Black Sea) calls on coastal states 'to elaborate and implement national coastal zone management policies, including legislative measures and economic instruments, in order to ensure the sustainable development in the spirit of Agenda 21' (point 15).

A Regional Activity Center on the Development of Common Methodologies for Integrated Coastal Zone Management (AC ICZM) was established in Krasnodar (Russia).

In the Sofia Declaration of 2009 on 'Strengthening the Cooperation for the Rehabilitation of the Black Sea Environment', the Ministers of Environment of the Contracting Parties to the Convention have, under point 9, agreed to 'incorporate up-to-date environmental management approaches, practices and technologies, with particular attention to integrated coastal zone management, introduction of green technologies, sustainable human development and ecosystem based management of human activities'.

The Protocols to the Convention also deal with ICZM. Particularly relevant is Art. 7 of the Black Sea Biodiversity and Landscape Conservation Protocol (2002) that says that 'the Contracting Parties shall encourage introduction of intersectoral interaction on regional and national levels through the introduction of the principles and development of legal instruments of integrated coastal zone management seeking the ways for sustainable use of natural resources and promotion of environmentally friendly human activities in the coastal zone'.

In addition, the Protocol on the Protection of the Marine Environment of the Black Sea from Land-Based Sources and Activities (2009; entry into force pending) requires the Contracting Parties, in order to achieve the purpose of the Protocol, to 'endeavour to apply the integrated management of coastal zones and watersheds' (Art. 4(2) lit. f).

Within the Bucharest Convention system, Strategic Action Plans are adopted at regular intervals. The Strategic Action Plan for the Environmental Protection and Rehabilitation of the Black Sea of 2009 lists, as key environmental management approaches under 3.1, Integrated Coastal Zone Management (ICZM), the Ecosystem Approach and Integrated River Basin Management (IRBM).

A binding ICZM/MSP Protocol for the Black Sea could be the logical next step to advance those concepts within the Bucharest Convention system.

The question could be raised, however, if the EU membership of two Black Sea countries and the ongoing process of approximation of three other Black Sea countries towards the EU renders such a regional cooperation superfluous. As EU Member States, Bulgaria and Romania have to respect the MSPD. Turkey is a candidate country to the EU and Ukraine and Georgia have signed Association Agreements. According to those agreements, the Parties shall promote maritime spatial planning (Art. 411 lit. b of the Association Agreement between the EU and Ukraine signed on 27 June 2014/Art. 339 lit. b of the Association Agreement between the EU and Georgia signed on 27 June 2014).

But since the MSPD, due to a lack of EU competences for the comprehensive regulation of MSP (Schubert, 2015), only sets a general framework, it explicitly requires further cooperation among Member States and with third countries in Art. 11 and 12, inter alia within regional institutional cooperation structures such as Regional Sea Conventions. The aim is to ensure that maritime spatial plans are coherent and coordinated across the marine region concerned. Thus, even the implementation of the relevant EU Directives in the Black Sea region could not be considered a substitute for a more detailed regulation of ICZM/MSP within the Bucharest Convention system.

Steps have been taken to advance MSP for other regional seas also. For example, the members of HELCOM (Baltic Marine Environment Protection Commission – Helsinki Commission) and VASAB (Vision and Strategies around the Baltic Sea – intergovernmental multilateral cooperation of 11 countries of the Baltic Sea Region in spatial planning) have agreed on a Regional Baltic Maritime Spatial Planning Roadmap (2013–2020) to fulfil the goal of drawing up and applying maritime spatial plans throughout the Baltic Sea region by 2020 which are coherent across borders and apply the ecosystem approach.

Readiness of the Region for a Binding Instrument

Even though the importance of ICZM has been recognized by the Contracting Parties to the Bucharest Convention, their approach to the concept still seems piecemeal and unsystematic. Several pilot projects for ICZM and spatial planning have been implemented in the Black Sea area, for example in the resorts of Malaya Yalta (Ukraine) and Gelendzhik (Russia), in Akçakoca (Turkey) and in Tskhaltsminda village (Georgia) (Pegaso Project, 2014). The beneficial effects of pilot projects, however, do not often last beyond the duration of the project. To establish only such temporary management measures in localized areas results, in the best case, in an 'oasis in the desert' (Billé and Rochette, 2010). That project-orientated approach thus goes against the basic principle of sustainable development 'which requires not that "exceptions" be created, but that the "rule" (legal framework) and the routine (the way the coast is actually managed), be changed' (Billé and Rochette, 2010).

So far, the management of coastal and marine zones through legislation that is specifically dedicated to such areas is still exceptional in the Black Sea area. In addition, in many cases, there is a lack of consistency between sector-specific policies with regard to environmental protection, as well as a lack of coordination between decision-makers. Very likely, steps towards MSP and ICZM would be considerably more efficient within a strong implementation framework. A legally binding ICZM/MSP protocol for all Black Sea countries could help to fill the gaps in the existing national legal frameworks, to coordinate efforts and to thereby reconcile the development of coastal and marine zones with the protection of the environment in the whole Black Sea region (Rochette and Billé, 2012).

Since there are no national regulations on MSP in the Black Sea region yet and few regulations on ICZM, now seems to be an opportune time to advance those concepts. A binding protocol would not conflict with existing national regulations and would largely influence the content of new ones, facilitating a consistent planning concept for the whole Black Sea. Moreover, there are not many permanent structures in the Black Sea yet (e.g. there are no offshore wind farms).

Thus, planning and regulation possibilities are not severely restricted by hardly reversible decisions.

The legally binding nature of a protocol can, however, also be regarded as a disadvantage, especially if there is a need for a fast and efficient response to environmental problems. Until a protocol enters into force, there is usually a lengthy process of drafting and negotiating the text. The Protocol on the Protection of the Marine Environment of the Black Sea from Land-Based Sources and Activities of 2009, which updates the corresponding Protocol on Protection of the Black Sea Marine Environment Against Pollution from Land-Based Sources of 1992, for example, still needs to be ratified. As a consequence, there is a long regulatory vacuum. In addition, in the Black Sea there is not yet an effective 'soft law' instrument on ICZM or MSP that could bridge the time gap (Vinogradov, 2007).

Moreover, a protocol is usually less detailed than 'soft law' instruments. States are often reluctant to commit themselves to detailed legal obligations. The regulation of issues that are typically a matter of national competence (e.g. urban planning) at a regional level often meets with particular resistance (Rochette and Billé, 2012). This results in very general and vague provisions and in 'framework' type protocols that have, in the end, a similarly weak effect as 'soft law' instruments. It is thus questionable if such a protocol is worth the complicated adoption process (Vinogradov, 2007).

The problem is aggravated by the fact that, especially compared to other European Seas (the Mediterranean, the Baltic and the North Sea), cooperation in environmental matters in the Black Sea seems still to be at an early stage (Vinogradov, 2007). The activities under the Bucharest Convention already allowed a significant increase in public involvement in environmental protection and the efficient addressing of transboundary environmental issues. To achieve all the objectives of the Convention, however, progress still needs to be made, especially with regard to financing and enforcement. Notably, the Convention does not contain any instruments to ensure compliance with its provisions.

Moreover, the Black Sea Commission has yet to achieve the level of efficiency of HELCOM or the OSPAR Commission (Protection of the Environment of the North-East Atlantic). The current organizational structure of the Black Sea Commission is too complex and there is too little accountability for environmental performance. In the past, missed deadlines have often simply been replaced by new ones or activities have been postponed to the next working period (BSC, 2007). Furthermore, the Commission does not seem to be adequately staffed and funded to draft and implement an additional protocol (Vinogradov, 2007).

To conclude: a binding protocol is not always the magic bullet for establishing efficient ICZM and MSP structures (Rochette and Billé, 2012).

Quickly Realizable Options

With a 'Code of Practice' or with guidelines, the future course of action of states can be fast and efficiently determined. Even if they do not have the same force as binding instruments, such 'soft law' instruments can help to advance ICZM and MSP by establishing common standards, by helping states to improve their legal and institutional framework, and by further anchoring the ICZM and MSP concept in the region. An important characteristic of such 'soft law' instruments is their flexibility. Because of their non-binding nature, states are more easily convinced to adopt or modify them without lengthy discussions about every detail, which is particularly advantageous in the face of pressing environmental problems (Vinogradov, 2007).

As a first step, the formulation of guidelines therefore seems to be a useful option, perhaps complemented by the establishment of an action plan that determines concrete practical measures (Vinogradov, 2007). These guidelines could also provide the basis for a later development of a binding ICZM and MSP instrument.

Evaluation of Progress at National Level

Bulgaria

Bulgaria is located in south-eastern Europe; its coastline measures 378 km and comprises the provinces of Dobrich, Varna and Burgas (EC, 2009). The Balkan Mountains reach the edge of the Black Sea at Cape Emine, dividing the coastline into a southern and a northern part. Parts of Bulgaria's northern Black Sea coast feature rocky headlands with cliffs up to 70 m high, whereas the southern coast is known for its wide sandy beaches. The two largest cities and main seaports on the Bulgarian coast are Varna in the north and Burgas in the south.

The increasing urbanization of the coast as well as industrial activities, shipping, pollution and wastewater discharge put valuable territories, protected areas, dunes and beaches in danger. Also, the vast beaches along the Bulgarian Black Sea coast and the temperate continental climate favour the tourist industry, which constitutes another risk factor for the ecosystems of the coastal zone (Palazov and Stanchev, 2006).

Bulgaria has only recently become an EU Member State and has also just started the ICZM process. To harmonize its legislation with the acquis communautaire, many laws, plans and programmes have been issued concerning environmental protection, sustainable development and spatial planning (Thetis, 2011). The main policy action undertaken in Bulgaria to protect the coastal zones was the adoption of the Black Sea Coast Spatial Planning Act, promulgated in State Gazette No. 48/2007, with the objective to create conditions for the stable and integrated development and protection of the Black Sea coastline (Art. 2). The law distinguishes two development zones (Zone A and Zone B) for which specific restrictions with regard to the density of buildings, the maximum building height as well as the minimum space for green areas have been stipulated (EC, 2009).

Bulgaria has so far developed neither a strategy nor an action plan for ICZM and there is no authority competent to implement the ICZM principles yet. Among the strategic objectives of the National Concept for Spatial Development for the period 2013–2025 (National Centre for Regional Development, Sofia, 5 November 2012) is, however, 'Integrated management and sustainable development of the Black Sea coastal municipalities, including through cross-border cooperation with neighbouring countries from the Black Sea Region, for introduction of an Integrated Maritime Policy' (Objective 5.1).

Georgia

Georgia's coastline stretches approximately 315 km along the Black Sea, across 12 administrative districts and three port cities, Batumi, Poti and Sokhumi. The coastal zone is dominated by wetland ecosystems. On the north and south end of the coast, there are also steep cliffs and mountains.

Human activities are putting increasing pressure on the ecosystems of the coastal zone (World Bank, 2007). Areas of forest and vegetation have significantly decreased, there is a progressive erosion of the coast, untreated water pollutes the sea and there are many examples of unsustainable developments, like unnecessary infrastructure projects and the illegal construction of dachas.

In October 1998, the State Consultative Commission for Integrated Coastal Zone Management was established by Presidential Decree No. 608 in order to develop the institutional framework for an integrated planning and management of the coastal resources of Georgia. A law on ICZM has been drafted, but has never been adopted. Instead, the Law of Georgia on Spatial Planning and Urban Development was adopted in 2005 and now regulates planning at local, regional and national levels. The draft ICZM law was reworked into non-binding guidelines. The institutional and legal framework for ICZM is thus still in its initial phase (World Bank, 2007).

With regard to the protection of the Black Sea, the National Environmental Action Programme of Georgia 2012–2016 (Approved by the Resolution of the Government No. 127, Tbilisi, 24 January 2012) states that 'Existing national legislation needs to be updated in accordance with modern European practices. Introduction of Integrated Coastal Zone Management (ICZM) approaches and protection of the coastal zone from degradation also requires appropriate legislation to be in place.'

Romania

Romania is located in south-eastern Europe at the lower reaches of the Danube River. Its coast on the Black Sea stretches about 245 km from Ukraine in the north to Bulgaria in the south. The coastal region is called Dobrogea and is subdivided into two regional administrative units, Tulcea in the north and Constanta in the south. The northern part, Tulcea County, is characterized by sandy beaches, low altitudes and gentle submarine slopes. The Danube Delta dominates this area. The southern part features limestone cliffs, small sandy beaches and steep submarine slopes. It is the focal point of Romanian seaside tourism activities. The capital of Constanta County is Constanta, the second biggest city of Romania, with the country's largest port (Demmers *et al.*, 2004).

As one of the more recent EU Member States, Romania is in a process of rapid economic development. The activities in the Romanian coastal and sea area include fishing, shipping, tourism, military activities and oil and gas extraction. These activities are not always compatible (Coman *et al.*, 2008) and for a prosperous development of the country both now and in the future, it is essential not to neglect the protection of the valuable resources of the Black Sea (Varga *et al.*, 2011).

In 2002, the Governmental Emergency Ordinance 202/2002 was issued as the legal basis for ICZM. That Ordinance was updated by the Law No. 280/2003, following the European Parliament and Council Recommendation of 30 May 2002 on Integrated Coastal Zone Management in Europe (2002/413/EC). It regulates the designation of coastal zones, restrictions of certain human activities, management measures, finance, public participation and enforcement. A National Committee of the Coastal Zone (NCCZ) was established in 2004.

Romania is thus the first Black Sea country that has a special legal and institutional framework for ICZM and already more than 70% of the Romanian coastline has protected status, including particularly the Danube Delta Biosphere Reserve (Nicolaev, 2011). However, there is still no single 'planning authority' for the sea, but a specific authority for each activity (Coman *et al.*, 2008). Moreover, sectoral controls are often not able to respond quickly to new pressures (Coman *et al.*, 2008) and to pay due regard to the cumulative impacts of the various sea uses.

Russia

The Krasnodar Region is the southernmost region of Russia and borders the Black Sea and the Sea of Azov. Geographically, the area is split by the Kuban River into two different parts. The western extremity of the Caucasus range lies in the southern third of the region, within the Crimean sub-Mediterranean forest ecoregion.

The Krasnodar Region is one of the most economically developed regions in Russia, with an important port in Novorossiysk. Being the warmest region of the country, the Black Sea coast of Krasnodar has also become the most popular tourist destination of Russia, focused on the resort city of Sochi.

Between 1993 and 1994, a number of Presidential Decrees relevant to ICZM were adopted. Following these, a federal target programme called 'Integrated Coastal Zone Management for the Black and Azov Seas Taking into Account the Task of Rational Use of Natural Resources in the Black Sea and Adjacent Territory' was prepared and approved. However, in 1997, the programme was suspended again (Vlasyuk, 2005).

In the Russian Federation legislation, the coastal zone is not yet regarded as an integral, natural 'land-sea' complex. Instead, there are various sectoral regulations for the protection and management of coastal and marine resources and various government bodies are responsible for their implementation. This situation is not beneficial for the implementation of an integrated management approach,[8] which is listed in the Maritime Doctrine of Russian Federation 2020 (27 July 2001) as one of the principles of the future national maritime policy (an 'integrated approach to maritime activities').

Turkey

Turkey has 1701 km of coastline bordering the Black Sea. The Black Sea region is divided into an eastern and a western part that show very different characteristics. Along the eastern part, mountain ranges run parallel to the coast and severely limit the width of the coastal area, sometimes to a few metres, which renders the area unsuitable for many coastal uses (Ozhan, 2005). On the western Black Sea, there are alluvial plains (e.g. Kizilirmak and Yesilirmak). The coastal area along these alluvial and deltaic shores widens significantly from a handful of kilometres to a few tens of kilometres,

comprising agricultural land of very high productivity (Ozhan, 2005).

Shipping, fishing, urbanization, and the conservation of natural and cultural heritage are the traditional sectors that have featured in the coastal zone. Recently, new sectors such as tourism and mariculture have become increasingly important (Ozhan, 2005).

Even though there have been several efforts since the late 1980s to apply a more integrated approach to the management of coastal zones and to transfer more responsibilities to local administrations, the management of coastal development in Turkey is still centralized and highly sectoral (Ozhan, 2005).

The main aims of the Coastal Law No. 3621 of 1990, amended in 1992 (Ozhan, 2005), are to protect the coasts, to utilize the coastal resources only for public benefit, and to ensure free access of the public to the coast. On the first 50 m of the shore strip, most constructions are forbidden. However, the Coastal Law is not a coastal management law that comprehensively regulates all activities (Unsal, 2013) and establishes a special institutional structure. It is also clearly focused on activities on the shore, not in the sea (Kaya, 2010).

The consequences of this lack of a holistic legal framework for ICZM are overlapping competences of various organizations (more than 20 institutions are responsible for the sea and coastal areas) and gaps in the management of the coast. Therefore, efforts to advance ICZM policies do not go beyond project level (EC, 2011).

Ukraine

The Black Sea coastline of Ukraine (about 1829 km, including the Crimean Peninsula) includes the northern and north-western shores of the Black Sea and the Sea of Azov. The cities of Odessa and Mariupol are located on the Ukrainian coast. The coast of the Black Sea is intersected by rivers, the largest of which are the Danube River, the Dniester River and the Dnieper River.

The land here is relatively flat and there are many sandy beaches.

A major environmental problem in Ukraine is the inefficient treatment of industrial and municipal wastewater, which is causing eutrophication and bacterial and chemical pollution of the country's main rivers and subsequently of the Black Sea (UNECE, 2007). Tourism and industrial activities along the coast also cause stress to the environment.

The development of an ICZM policy in Ukraine started with the Ministerial Declaration on the Protection of the Black Sea in Odessa in 1993, which confirmed the commitment to ICZM and sustainable development of coastal areas and the marine environment under national jurisdiction (Onderstal, 2000). It was afterwards decided to implement national coastal zone policies, including legislative measures and economic instruments. However, even though concepts and guidelines on ICZM have been developed, and a law 'on the coastal zone' (Radchenko, 2012) has been drafted, concrete regulations have not yet been adopted.

Conclusion

An additional protocol would be a great challenge for the Black Sea Commission and the Black Sea states. Therefore, the 'Feasibility Study for the Black Sea ICZM Instrument' of 2007 (Vinogradov, 2007) favours a two-step approach. As a first step, it recommends a combination of 'soft law' instruments. Depending on the success of those instruments, it recommends the adoption of a binding protocol as a second step.

However, in light of the international and especially European progress on ICZM and the wider concept of MSP, 'soft law' can only be an option for a short transitional period. The problems resulting from the different stages of progress in this area of the six Black Sea countries have to be taken into account. The measures that EU Member States are required to take to protect the marine environment by the MSFD and the MSPD might largely run aground if not all Black Sea riparian states, especially including Russia, pull together and regulate uses in their common basin in a binding fashion. A protocol could be adapted to the specific regional situation. It should, however, at least anchor the ecosystem approach as a basic principle of MSP.

This conclusion is also supported by the participants of the 3rd Black Sea and Upgrade Black Sea Scene Joint Scientific Conference BS-OUTLOOK (Odessa, Ukraine, 1–4 November 2011). During Session 4, they agreed on the 'necessity to initiate consultations in support of the development of [an] ICZM legal instrument (protocol) for the Black Sea region'. Furthermore, they agreed to 'introduce and develop in the Black Sea area the new field of maritime spatial planning in a coherent manner and in close integration with ICZM'. One of the overall conclusions of the conference was that 'spatial planning in the Black Sea is mandatory (as part of ecosystem-based management) for a correct management of its resources'.

Notes

1 UNESCO/IOC, Marine Spatial Planning Initiative, http://www.unesco-ioc-marinesp.be/marine_spatial_planning_msp
2 Johannesburg Declaration on Sustainable Development (point 5), World Summit on Sustainable Development, A/CONF. 199/20, Chapter 1, Resolution 1, Johannesburg, September 2002.
3 Convention on Biological Diversity, Ecosystem Approach/Principles, https://www.cbd. int/ecosystem/principles. shtml

4 Baltic Sea Broad-Scale Maritime Spatial Planning (MSP) Principles, adopted by HELCOM HOD 34-2010 at the 54th Meeting of VASAB CSPD/BSR, Principle 2.

5 HELCOM/VASAB, 'Guideline for the implementation of ecosystem-based approach in Maritime Spatial Planning (MSP) in the Baltic Sea area', October 2015, http://www.helcom.fi/Documents/ HELCOM%20at%20work/Groups/MSP/ Guideline%20for%20the%20implementation %20of%20ecosystem-based

%20approach%20in%20MSP%20in %20the%20Baltic%20Sea%20area.pdf

6 Convention on Biological Diversity, The Ecosystem Approach, http://www.cbd. int/ecosystem/

7 Convention on Biological Diversity, The Ecosystem Approach, http://www.cbd. int/ecosystem/

8 'Legal principles of coastal zone management in the Russian Federation', UNESCO, Sustainable Development in Coastal Regions and Small Islands, http://www. unesco. org/csi/act/russia/legalpro7.htm

References

Akademie für Raumforschung und Landesplanung (ARL) (2000) *Nachhaltigkeitsprinzip in der Regionalplanung: Handreichung zur Operationalisierung.* Forschungs- und Sitzungsberichte der ARL 212. Hannover.

ARCADIS (2011) *Comparative Analysis of the OURCOAST Cases: Report on ICZM in Europe.* European Commission, Brussels. http://ec.europa.eu/ourcoast/download. cfm?fileID=1709

Beaucamp, G. (2002) *Das Konzept der zukunftsfähigen Entwicklung im Recht.* Mohr Siebeck, Tübingen.

Billé, R. and Rochette, J. (2010) *Feasibility Assessment of an ICZM Protocol to the Nairobi Convention.* Report for Indian Ocean Commission and the Nairobi Convention Secretariat. 101 pp.

Buchholz, H. (2004) Raumnutzungs- und Raumplanungsstrategien in den deutschen Meereszonen. *Konzeptionelles Entwicklungsziel Nr. 1, Informationen zur Raumentwicklung,* 7/**8**, 485–489.

Bundesamt für Naturschutz (BfN) (2006) *Naturschutzfachlicher Planungsbeitrag zur Aufstellung von Zielen und Grundsätzen der Raumordnung für die deutsche Ausschließliche Wirtschaftszone der Nord- und Ostsee.* BfN, Bonn. 38 pp.

Bundesinstitut für Bau-, Stadt- und Raumforschung (BBSR) (2011) *Beitrag der Raumordnung zur Steigerung des Anteils erneuerbarer Energien: Rechtliche Möglichkeiten der Raumordnung im Bereich 'Repowering', Forschungsprojekt, 2011.* http://www.bbsr.bund.de/BBSR/DE/FP/ MORO/Forschungsfelder/2011/ Repowering/01_Start.html?nn=432564

Clark, J.R. (1992) *Integrated Management of Coastal Zones, Section 2.11: Creating Protected Areas.* Food and Agriculture Organization of the United Nations, Rome. http://www.fao.org/docrep/003/T0708E/ T0708E02.htm

Coman, C., Alexandrov, L., Dumitru, V. and Lucius, I. (2008) *PlanCoast Project in Romania: Extending Coastal Spatial Planning to the Marine Zone.* http://www. nodc.org.ua/ukrncora/index2. php?option=com_docmanandtask=doc_ viewandgid=77andItemid=35

Commission on the Protection of the Black Sea Against Pollution (BSC) (2007) *Black Sea Transboundary Diagnostic Analysis and Strategic Action Plan (TDA).* Istanbul. 269 pp. http://www.blacksea-commission.org/_ tda2008.asp

Crowder, L. and Norse, E. (2008) Essential ecological insights for marine ecosystem-based management and marine spatial planning. *Marine Policy*, **32**, 772–778.

Demmers, I., Keupink, E., Popa, B. and Timmer, R. (2004) *Outline Strategy for the Integrated Management of the Romanian Coastal Zone*. Draft Report, Royal Haskoning, Netherlands. http://ec.europa.eu/environment/iczm/evaluation/iczmdownloads/romania2004.pdf

Douvere, F. (2008) The importance of marine spatial planning in advancing ecosystem-based sea use management. *Marine Policy*, **32**, 762–771.

Drankier, P. (2012) Embedding maritime spatial planning in national legal frameworks. *Journal of Environmental Policy and Planning*, **14**, 7–27.

Ehler, C.N. (2012) Perspective: 13 myths of marine spatial planning. *Marine Ecosystems and Management*, **5** (5), 1–3.

Ehler, C. and Douvere, F. (2007) *Visions for a Sea Change*. Report of the First International Workshop on Marine Spatial Planning, Intergovernmental Oceanographic Commission and Man and the Biosphere Programme. IOC Manual and Guides 46, ICAM Dossier 3. UNESCO, Paris.

European Commission (EC) (2009) *Country Overview and Assessment of Climate Change Adaptation: Bulgaria*. http://ec.europa.eu/maritimeaffairs/documentation/studies/documents/bulgaria_climate_change_en.pdf

European Commission (EC) (2011) *Exploring the Potential of Maritime Spatial Planning in the Mediterranean: Country Report – Turkey*. http://ec.europa.eu/maritimeaffairs/documentation/studies/documents/turkey_01_en.pdf

European Commission (EC) (2013a) *Proposal for a Directive of the European Parliament and of the Council establishing a Framework for Maritime Spatial Planning and Integrated Coastal Management*. COM(2013) 133 final, Explanatory Memorandum.

European Commission (EC) (2013b) *Commission Staff Working Paper/Executive Summary of the Impact Assessment of the Proposal for a Directive of the European Parliament and of the Council establishing a Framework for Maritime Spatial Planning and Integrated Coastal Management*. SWD(2013) 64 final. http://ec.europa.eu/maritimeaffairs/policy/maritime_spatial_planning/documents/swd_2013_64_en.pdf

Hanusch, M., Eberle, D., Jacoby, C. *et al.* (2007) *Umweltprüfung in der Regionalplanung*. E-Paper der ARL Nr. 1, Hannover.

Janssen, G., Sordyl, H., Albrecht, J. *et al.* (2008) *Anforderungen des Umweltschutzes an die Raumordnung in der deutschen Ausschließlichen Wirtschaftszone (AWZ): einschließlich des Nutzungsanspruches Windenergienutzung*. Forschungsprojekt im Auftrag des Umweltbundesamtes FuE-Vorhaben Förderkennzeichen 205 16 101.

Jones, P.J.S., Lieberknecht, L.M. and Qiu, W. (2016) Marine spatial planning in reality: introduction to case studies and discussion of findings. *Marine Policy*, **71**, 256–264.

Kaya, L.G. (2010) Application of collaborative approaches to the integrative environmental planning of Mediterranean coastal zone: case of Turkey. *Journal of Bartin University*, **12** (18), 21–32.

Köppel, J., Wende, W. and Herberg, A. (2006) *Naturschutzfachliche und naturschutzrechtliche Anforderungen im Gefolge der Ausdehnung des Raumordnungsregimes auf die deutsche Ausschließliche Wirtschaftszone*. Technische Universität Berlin.

Laffoley, D., Maltby, E., Vincent, M.A. *et al.* (2004) *The Ecosystem Approach: Coherent Actions for Marine and Coastal Environments*. A Report to the UK Government. English Nature, Peterborough. 65 pp.

Leibenath, M. (2009) Biotopverbund – eine komplexe Koordinationsaufgabe? Chancen,

Risiken und Nebenwirkungen des Metaphernpaares 'Fragmentierung – Biotopverbund'. Kurzfassung des Vortrages/ Wissenschaftliche Plenarsitzung der Akademie für Raumforschung und Landesplanung zum Thema 'Wenn zwei sich streiten… Bessere Planung durch Koordination'.

Mose, I. and Weixlbaumer, N. (2007) A new paradigm for protected areas in Europe?, in *Protected Areas and Regional Development: Towards a New Model for the 21st Century* (ed. Ingo Mose). Ashgate Studies in Environmental Policy and Practice, Aldershot, UK.

Nicolaev, S. (2011) *Integrated Coastal Zone Management in Europe: The Way Forward.* OurCoast Conference Riga. http://www. ourcoastconferenceriga.eu/presentations/ session2/simion_nicolaev.pdf

Odum, W.E. (1982) Environmental degradation and the tyranny of small decisions. *BioScience*, **32** (9), 728–729.

Onderstal, M. (2000) *Coastal Guide: Coastal Management in Ukraine.* EUCC International Secretariat. http://www. coastalguide.org/icm/blacksea/index.html

Oral, N. (n.d.) *Summary of EEZ Zones in the Black Sea.* Commission on the Protection of the Black Sea Against Pollution. http:// www.blacksea-commission.org/_socio-economy-eez.asp

Ozhan, E. (2005) *Coastal Area Management.* UNEP/MAP, Priority Actions Programme Regional Activity Centre. http://www. medcoast.org.tr/publications/cam%20in %20turkey.pdf

Palazov, A. and Stanchev, H. (2006) *Human Population Pressure, Natural and Ecological Hazards along the Bulgarian Black Sea Coast.* SENS'2006, Varna. http://www.space. bas.bg/astro/ses2006/Cd/E23.pdf

Pegaso Project (2014) The Black Sea Basin and ICZM, in *Evaluation Report on CASEs Multi-sector, Multi-administrative and Multi-scale Work, Integrated Approach Method in CASEs.* EU FP7 – ENV.2009.2.2.1.4. http://www. pegasoproject.eu/links-9.html

Proelß, A. (2009) Völkerrechtliche Grenzen eines maritimen Infrastrukturrechts. *EurUP Zeitschrift für Europäisches Umwelt- und Planungsrecht*, **7** (1), 9.

Qiu, W. and Jones, P.J.S. (2013) The emerging policy landscape for marine spatial planning in Europe. *Marine Policy*, **39**, 182–190.

Radchenko, V. (2012) *Marine Spatial Planning: Challenges and Opportunities in Ukraine.* The East Asian Seas Congress. http://eascongress.pemsea.org/sites/ default/files/document-files/presentation-msp-radchenko.pdf

Ritter, E.-H. (2005) Freiraum/Freiraumschutz, in *Handwörterbuch der Raumordnung.* Akademie für Raumforschung und Landesplanung, Hannover.

Rochette, J. and Billé, R. (2012) ICZM Protocols to Regional Seas Conventions: What? Why? How? *Marine Policy*, **36**, 977–984.

Salm, R.V., Clark, J. and Siirila, E.E. (2000) *Marine and Coastal Protected Areas: A Guide for Planners and Managers.* IUCN, Washington, DC.

Schäfer, N. (2009) Maritime spatial planning: about the sustainable management of the use of our seas and oceans, in *Understanding and Strengthening European Union–Canada Relations in Law of the Sea and Ocean Governance.* The Northern Institute for Environmental and Minority Law, Rovaniemi, Finland.

Schubert, M. (2015) Meeresraumordnung und Europarecht: Die Richtlinie (2014/89/EU zur Schaffung eines Rahmens für die maritime Raumplanung), in *Jahrbuch des Umwelt-und Technikrechts 2015.* Erich Schmidt Verlag, Berlin.

Schultz-Zehden, A., Gee, K. and Scibior, K. (2008) *Handbook on Integrated Maritime Spatial Planning.* PlanCoast Project

(2006–2008), EU INTERREG IIIB CADSES. 100 pp. http://www.plancoast.eu/files/ handbook_web.pdf

Siedentop, S. (2003) Prüfung kumulativer Umweltwirkungen in der Plan-UVP, in *Umweltprüfung für Regionalpläne*. Akademie für Raumforschung und Landesplanung, Hannover.

Spektrum (2001) *Lexikon der Geographie*. Spektrum Akademischer Verlag, Heidelberg. http://www.spektrum.de/lexikon/ geographie/integrativer-naturschutz/3815

Thetis (2011) *Analysis of Member States Progress Reports on Integrated Coastal Zone Management (ICZM)*. Final Report to DG Environment, European Commission. http://ec.europa.eu/environment/iczm/pdf/ Final%20Report_progress.pdf

UN Economic Commission for Europe (UNECE) (2007) *Ukraine, Second Review*. Committee on Environmental Policy, Environmental Performance Reviews Series No. 24. New York and Geneva.

Unsal, F. (2013) *Overview of Turkish Coastline Policy and Implementation: Mare Nostrum Project*. http://www.marenostrumproject. eu/PDFs/MareNostrum_KickoffMeeting_ Turkish_Coastline_Policy.pdf

Varga, A.L., Coman, C. and Stanica, A. (2011) Romania: ICZM planning in an initial stage, in *Climate of Coastal Cooperation, 2011*. http://www.coastalcooperation.net/ part-I/I-3-3.pdf

Vinogradov, S. (2007) *A Feasibility Study for the ICZM Instrument to the 1992 Bucharest Convention*. UNDP-GEF Black Sea Ecosystem Recovery Project, Phase II.

Vlasyuk, K. (2005) *Inventory of Policies and Regulatory Acts and Practices of Its Implementation: Development of Proposals in Sphere of Water Protection, Water Management and Integrated Coastal Zone Management*. UNDP-GEF Black Sea Ecosystem Recovery Project, Phase II, Istanbul.

Wende, W., Herberg, A., Köppel, J. *et al.* (2007) Meeresnaturschutz und Raumordnung in der deutschen Ausschließlichen Wirtschaftszone. *Naturschutz und Landschaftsplanung*, **39** (3), 79–85.

World Bank (2007) *Implementation, Completion and Results Report on a Credit to Georgia for an Integrated Coastal Management Project*. Report#: ICR443.

12

Black Sea Network of Marine Protected Areas: European Approaches and Adaptation to Expansion and Monitoring in Ukraine

Boris Alexandrov, Galina Minicheva and Yuvenaliy Zaitsev

Institute of Marine Biology, National Academy of Sciences of Ukraine, Odessa, Ukraine

Introduction

This chapter brings together several strands of current research concerning Marine Protected Areas (MPAs) in the Black Sea in general, and in Ukraine in particular. First, it provides a more accurate assessment of the total area of MPAs of different status within six Black Sea countries. Second, the impact of eutrophication on the features and the development of MPAs in Ukraine is considered. This is followed, thirdly, by a brief overview of the method used for identifying and justifying the designation of new MPAs (or expanding existing MPAs) in Ukraine, based on integrated evaluation of anthropogenic impact, aquatic plant morphological indicators, and determining the ecological value of marine areas. Finally, the opportunity of developing public ecological monitoring for the Black Sea is explored.

Overview of MPAs in the Black Sea

It is well known that the reproduction of most living marine natural resources takes place in the coastal zones (Zaitsev, 2006)

because of the edge effect in which physico-chemical and biological interactions are most intense at the interface between land and water. It is no coincidence that most protected areas are located near coasts. At the same time, this zone suffers the highest human pressure because of urban expansion, transport and other infrastructure development, exploitation of living and non-living resources and steady extension of recreation areas. Around 15 million people live in the 2 km wide coastal zone of the Black Sea, 6 million of them in Ukraine alone (Panchenko, 2009).

Conflict between economic activities and the need to maintain living resources has led to the establishment of MPAs. One of the first Black Sea MPAs, the Black Sea Biosphere Reserve, was established in Ukraine as early as 14 July 1927 to protect coastal and marine communities near the Dnieper River delta.

It is difficult to determine the precise extent of the existing Black Sea MPA network. First, almost all the MPAs comprise not only marine waters but also terrestrial areas, which are generally larger. Second, parts of the aquatic area are lagoons or closed limans, isolated from the sea, which

Management of Marine Protected Areas: A Network Perspective, First Edition. Edited by Paul D. Goriup.
© 2017 John Wiley & Sons Ltd. Published 2017 by John Wiley & Sons Ltd.

cannot be included with the Black Sea by definition. Third, the definition and classification of protected areas in the Black Sea countries differ to a greater or lesser degree from the IUCN classification (Lausche, 2011). For example, where the IUCN has seven categories of protected area, Bulgaria has five, Romania has 10 (Begun *et al.*, 2012), and Ukraine has 11; moreover their classification criteria are different.

Another difficulty in determining the total area of MPAs in different countries is that their areas often include sites with multiple designations. For example, the transnational Danube Delta Biosphere Reserve in Romania and the Danube Biosphere Reserve in Ukraine also include wetlands in the Ramsar list. The Natura 2000 protected area 'Ropotamo' (Ropotamo wetland complex) in Bulgaria contains four natural reserves (Begun *et al.*, 2012), several Ramsar wetlands (Marushevsky, 2003) and the Blato Alepu nature monument. A recent publication on Black Sea MPAs says that there are no protected areas in Turkey apart from Ramsar wetlands in the Kizilirmak River delta (Begun *et al.*, 2012). However, we know about two nature reserves (Igneada Flooded Forest and Sarikum Lake) and a permanent wildlife reserve in Yesilirmak Delta (Marushevsky, 2003; Öztürk *et al.*, this volume).

To consolidate the existing data about the actual area of the existing Black Sea MPAs, they were divided into three groups: (i) protected areas (reserves) of international significance (importance); (ii) Ramsar wetlands; and (iii) areas of national significance. Protected areas of local importance were not taken into account. Map measurement was used to determine the areas of the MPAs connected with the Black Sea in cases where the figures were absent from the available literature (Marushevsky, 2003).

Analysis of the information collected enabled us not only to map the current distribution of MPAs in the Black Sea (Figure 12.1), but also to establish some important quantitative characteristics about them. Thus, the area of water-bodies in the MPAs connected with the Black Sea amounts to a total of 755 840 ha. The Black Sea countries can be ranked by their MPA extent as follows: Ukraine – 82.0%; Romania – 14.7%; Georgia – 2.2%; Turkey – 0.7%; Bulgaria – 0.4%; and Russia – 0.1%.

Ecological Characteristics of the Ukrainian Part of the Black Sea

Geographic Features

The Ukrainian part of the Black Sea coast has a length of some 1829 km. It has special geographical conditions and associated ecosystems that have to be taken into account when planning a network of MPAs. The vast, shallow (15 to 55 m depth) shelf platform in the north-western Black Sea (Öztürk *et al.*, this volume), from the Danube River to Cape Tarchankut, extends over more than 55 000 km^2. It receives the waters from three large nutrient-rich European rivers: the Danube, Dniester and Dnieper. These conditions result in the shelf being the most biologically productive area of the Black Sea (Zaitsev, 2006), contrasting with the Crimean Peninsula coast (acknowledged by IUCN as one of nine centres of European biological diversity) which is less productive but has the highest national level of landscape and biological diversity (Yena *et al.*, 2004).

Biodiversity

According to the Black Sea Transboundary Diagnostic Analysis, Annex 4 (Commission on the Protection of the Black Sea Against Pollution, 2007), the Black Sea hosts 44 distinct habitat types. Of these, 42 are present in the Ukrainian part of the Black Sea,

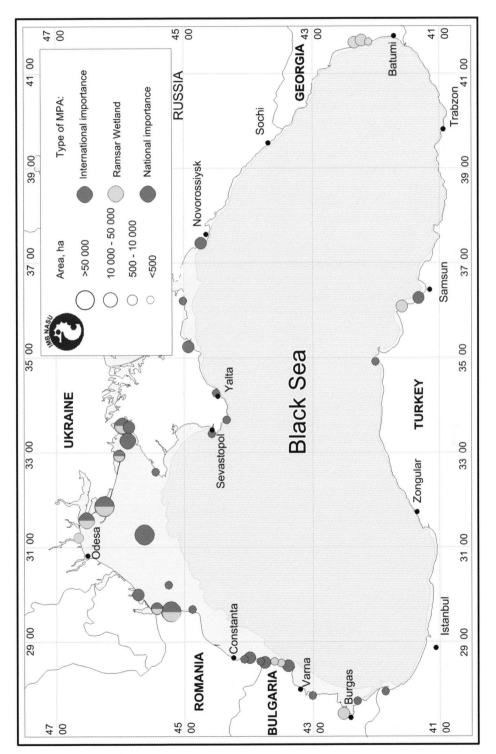

Figure 12.1 The Black Sea MPAs of international and national importance.

with 40 in Bulgaria, 35 in Romania, 28 in Turkey, 25 in Russia and 18 in Georgia. The Red Data Book of Ukraine includes 1368 species. Of these, 10.5% or 88 plant and 57 animal species are Black Sea inhabitants (Black Sea Environment Programme, 2009). This confirms the importance, and responsibility, of Ukraine for conserving marine biodiversity in the Black Sea.

At the same time, the very diversity of the Ukrainian Black Sea area, lying on the intersection of many wildlife migratory paths and human transportation routes, explains why it also has more non-indigenous species than any other Black Sea country. Out of the 261 non-indigenous marine species registered in the database of the Permanent Secretariat of the Black Sea Commission by 2013, some 148 were recorded in Ukraine, with 94 in Turkey, 82 in Romania, 80 in Bulgaria, 51 in Russia and 34 in Georgia. More than 80% of the species originated from the Atlantic Ocean and the Mediterranean Sea (Alexandrov *et al.*, 2013; data available at http://www.corpi.ku.lt/databases/index.php/aquanis). The spread of non-indigenous species common to neighbouring countries follows the counter-clockwise Black Sea coastal cyclonic current. Thus, the highest percentage of common non-indigenous species between neighbouring countries is between Ukraine and Russia (64.0%) and Ukraine and Romania (61.2%), while the lowest percentage is between Bulgaria and Turkey (32%).

Eutrophication of the Black Sea Shelf Area

As mentioned above, the Ukrainian Black Sea shelf is the most biologically productive area of the Black Sea and therefore has the highest level of eutrophication risk connected with nutrient pollution, phytoplankton blooms and hypoxia (Zaitsev, 1992). Analysis of long-term biological changes in response to eutrophication since the 1970s

has shown increases in production of dominant phytoplankton species (by 150%), zooplankton species (by 280%), macrophytes (by 54%) and zoobenthos (by 112%). Among the dominant species, non-indigenous ones generally had the highest levels of production (Alexandrov and Zaitsev, 1998).

Four distinct periods of Black Sea shelf eutrophication have been distinguished using indices derived from morphological parameters of aquatic vegetation associated with the ecosystem's trophic status (Minicheva *et al.*, 2008; see below): *natural state* (before the 1970s), *intensive eutrophication* (early 1980s), *immobility* (mid-1990s) and a *steady trend of de-eutrophication* since the turn of the millennium (Figure 12.2).

However, the recent steady trend of de-eutrophication has sometimes been interrupted by abnormal climatic conditions. In 2010, for example, the Danube River discharge was 45% below its average multi-annual level which, combined with unusually high summer temperatures, created conditions that stimulated primary production processes. As a result, the Ecological Status Class (ESC) of the Ukrainian Black Sea shelf, which had been recorded as 'Good' during the previous decade, had to be revised to 'Poor' (Minicheva, 2013).

The MPA Network of Ukraine

The formation of an ecological network in Ukraine is regulated by national legislation (Verkhovna Rada Ukrainy, 2000, 2004). The main aims of the National Program of Forming a National Ecological Network of Ukraine in 2000–2015 were to determine the network's spatial structure in order to unite natural habitats, and to increase the protected area territory from 4% to 10.4% of the country's total area within 15 years.

There are two marine elements within the structure of the Ukrainian National Ecological Network – the Black Sea natural region (north-west shelf of the Black Sea),

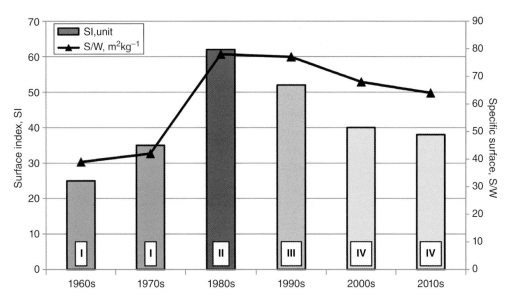

Figure 12.2 Historical stages of eutrophic status in the north-west Black Sea shelf: I – *natural state*;
II – *intensive eutrophication*; III – *immobility*; IV – *steady trend of de-eutrophication* (SI reflects the intensity of primary production processes in marine coastal ecosystem; S/W reflects the ecological activity of bottom vegetation: see text for details).

and the natural coastal corridor along the Sea of Azov and the Black Sea. Ukrainian MPAs of international, national and local levels, as well as marine wetlands of international importance, are located within the boundaries of these two elements of the ecological network, totalling an area of just over $6090\,km^2$ (Table 12.1, Figure 12.3). Most of the MPAs are represented by coastal complexes attached to terrestrial protected areas of different categories and different levels of protection. There are just two MPAs not connected to the coast: Zernov's Phyllophora Field, which is well known and the largest protected area on the north-west shelf, and the Small Phyllophora Field located in the central part of Karkinitskyi Gulf. Thus, practically all the existing accumulations of *Phyllophora* red algae on the north-west shelf, together with their associated communities of invertebrates and fish, are protected by the State.

At present, the Ukrainian Black Sea MPAs (excluding unprotected Ramsar-listed

wetlands) cover almost 11% of the national marine area $(55\,750\,km^2)$, which is much more than in the other Black Sea countries. In this respect, it is fair to say that Ukraine has fulfilled its commitments under the Convention on Biological Diversity Aichi Targets (CBD, 2008), namely to establish MPAs over at least 10% of the ocean by 2020 (in 2010, approximately 6000 MPAs had been declared worldwide, but they covered only 1.17% of the total marine area; Toropova *et al.*, 2010).

A distinctive feature of Ukraine's marine ecological network is the very uneven distribution of sites between the two Black Sea ecoregions, which differ markedly in their biological structures and ecological processes. About 99.8% of the Ukrainian MPA area is situated in the north-west shelf, above the line connecting the Ukrainian part of the Danube Delta and Cape Tarchankut. Accordingly, the coastal ecosystems of the Crimean Peninsula, which are valued for their underwater habitats and

Table 12.1 Black Sea MPAs of international and national level in Ukraine.

No.[a]	MPA	Protected status	General area (ha)	Marine area (ha)
1	Danube	Biosphere Reserve	50 253	6 686
10	Chornomorskyi	Biosphere Reserve	109 255	93 960
25	Karadag	Natural Reserve	2 874	809
14	Lebiazhi Islands	Natural Reserve	9 612	9 612
23	Cape Martian	Natural Reserve	240	120
27	Cape Opuk	Natural Reserve	1 592	62
3	Tuzla liman complex	National Natural Park	27 865	883
16	Tarchankut Cape	National Natural Park	10 900	360
9	Biloberezhia Sviatoslava	National Natural Park	35 223	25 000
11	Dzharylgachskyi	National Natural Park	10 000	2 469
5	Zernov's Phyllophora Field	State Significance Preserve (botanical)	402 500	402 500
12	Small Phyllophora Field	Nationally Important Reserve (botanical)	38 500	38 500
2	Zmiyiny Island	Nationally Important Reserve (zoological)	640	232
13	Karkinitskyi Gulf	Nationally Important Reserve (zoological)	27 646	27 646
19	Kozachia Bay	Nationally Important Reserve (zoological)	23	23
21	Cape Aiya	Nationally Important Reserve (landscape)	1 132	208
Total areas			728 256	609 070

a) Numbers refer to sites shown in Figure 12.3.

high level of marine biodiversity, include only 0.2% of all Ukrainian MPAs of international and national importance.

Approaches to Management and Monitoring of MPAs in Ukraine

Taking Account of Anthropogenic Influence in the Justification of an MPA

As mentioned above, the coastal zone supports high biological diversity and concentration of life due to edge effects.

This zone also experiences significant conflicts between different human economic activities (such as construction, agriculture, industry and recreation). These conflicts adversely affect the state of marine ecosystems to a greater or lesser degree. A matrix comprising 27 human-caused stress factors and 15 types of biota response (Zaitsev, 2006) was proposed for integrated assessment of the anthropogenic impact (AI).

If the intensity of anthropogenic impacts is assessed on a seven-point scale from 'very negative' (1) to 'very positive' (7), it is possible to estimate an overall AI score for a given area. For this purpose, a matrix of expert

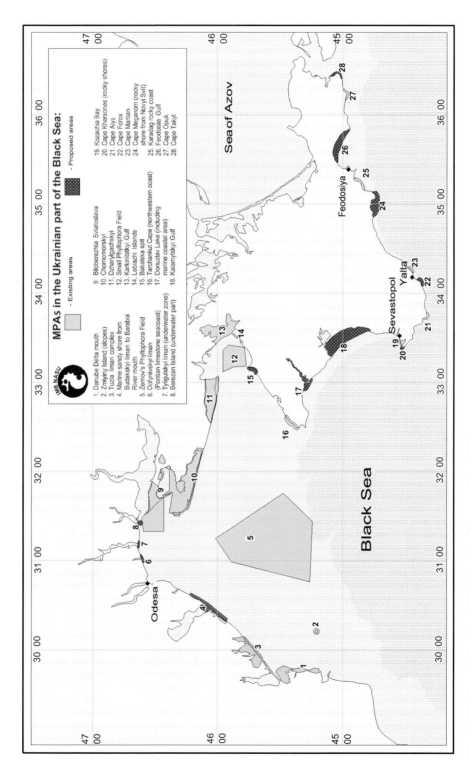

Figure 12.3 Current Ukrainian MPA network and proposed new MPAs.

assessment of stress factors and biota responses can be used (see Table 12.2). For example, the average AI scores for 26 areas of the Black Sea in Ukraine, from the Danube Delta to the Kerch Strait, are given in Table 12.3. The AI scores correspond well with protected areas and can be used as an additional indicator in support of the MPA.

The least number of stress factors (3) influenced the Zernov's Phyllophora Field MPA, while the most (24) affected the Odessa Gulf ecosystem. The AI scores show that Sukhoy liman, which hosts a commercial seaport, had the highest level of anthropogenic impact. In contrast, the marine areas having protected status and situated at some distance from the coastline (Zernov's Phyllophora Field and Zmiyiny Island) had the lowest level of anthropogenic impact.

Plant Morphological Indicators for Rapid Monitoring of MPAs

In 2015, the Commission on the Protection of the Black Sea Against Pollution approved the use of plant morphological indicators (Minicheva *et al.*, 2014) as part of the Black Sea Integrated Monitoring and Assessment Programme standards. These indicators directly reflect the ecological function of the bottom vegetation and therefore have advantages over other structural phytoindicators such as floristic composition, biomass and cover. The simple morphological methods involved allow rapid and accurate assessment of the intensity of autotrophic processes and thus the ESC of the marine ecosystem.

The main aim of the Marine Strategy Framework Directive (MSFD, 2008/56/EC) is to achieve Good Environmental Status (GES) of marine waters, such that they provide ecologically diverse and dynamic oceans and seas which are clean, healthy and productive. Reaching GES is not only the main aim of joint efforts by European states in marine protection and management, but also an important aspect of MPA monitoring and assessment. To interpret what GES means in practice, the MSFD sets out 11 descriptors, which describe what the environment will look like when GES has been achieved. Each descriptor reflects different aspects of the marine environment's resilience to the most widespread and intensive human impacts on it. Quantitative evaluation of the descriptors requires a measuring tool, and different indicators of the ecosystem's state could be used as such a tool. The selection of the most suitable indicators for GES assessment out of the huge number of available hydro-ecological parameters is a vital task. If the indicators selected for monitoring MPA condition only reflect the dynamics of biological features, then the functional state of biological elements and the real ecological status of the protected ecosystem could be obscured. Thus, the GES indicators should reflect the functional properties of biological elements (intensity of production and destruction processes on which high biological diversity depends, branching of food chains, good quality of biological resources and aquatic environment) and at the same time applicable to several descriptors at once.

Indicators based on morphological features of aquatic vegetation, in particular the active surface area to weight ratio, could be a sensitive means for rapid assessment of the ESC as part of MPA monitoring (Minicheva, 1998). The main advantage of such an indicator is that it is based on simple measurement methods of macrophytes (which are permanent and functionally important components of coastal ecosystems). In addition to the assessment of ESC, indicators based on macrophyte morphology can be used for quantitative evaluation of four GES descriptors, namely:

- Descriptor 1: Biodiversity is maintained
- Descriptor 4: Elements of food webs ensure long-term abundance and reproduction
- Descriptor 5: Eutrophication is minimised
- Descriptor 6: The sea floor integrity ensures functioning of the ecosystem.

Table 12.2 Generalized matrix of expert assessments of ecological processes in the Black Sea coastal zone.

Stress	Response	Changes of life conditions									Biological and general changes					
		Salinity	Currents	Transparency	Pollution	Trophicity	Bottom sediments	Oxygen content	Disturbance	Concentration	Biological diversity	Bottom hypoxia	Stocks	Health risks	Marine food quality	Aesthetic qualities
I	Fishing	4	4	3	3	4	1	2	1	4	2	3	1	3	4	3
I	Mining	4	3	2	2	3	1	2	1	4	2	3	1	3	4	3
I	Industrial wastes	1	4	2	1	3	3	2	1	1	1	1	1	1	1	1
II	Pesticides	4	4	2	2	3	2	2	2	1	1	1	1	1	1	1
II	Soil erosion	4	4	1	2	3	1	2	3	3	2	3	1	2	2	2
II	Agricultural runoff	3	4	1	1	3	1	1	2	1	2	1	1	1	1	1
III	Residual foods	4	4	2	1	5	3	2	3	4	4	3	4	3	3	2
III	Genetic degeneration	4	4	4	4	4	4	4	4	4	2	4	5	4	4	4
IV	Ports development	4	2	2	1	3	1	1	1	1	3	3	3	2	3	3
IV	Deepening, dumping	4	2	2	2	2	1	2	1	4	2	3	3	3	4	3
IV	Ballast waters and exotic species	4	4	4	3	4	4	4	4	4	3	4	2	4	4	4
IV	Shipwrecks	4	4	3	2	3	4	4	3	1	1	4	4	3	3	3
V	Urban sewage	3	4	2	1	3	2	2	3	1	1	1	1	1	1	2
V	Rain waters	3	3	2	2	3	3	3	3	1	2	3	3	2	2	2
VI	Addition of sand	4	4	3	4	4	1	4	3	4	1	4	2	2	4	4
VI	Coast protection constructions	4	2	3	2	3	2	2	2	4	6	4	4	3	4	3
VII	Dams	3	3	4	4	4	4	4	2	4	4	4	1	4	4	4
VII	Reservoirs	3	3	4	4	4	4	4	2	4	4	4	1	4	4	4
VIII	Resort development	4	3	3	2	3	3	3	2	3	3	4	4	3	3	4
VIII	Resort sewage	4	3	2	2	3	3	2	2	2	2	3	3	2	2	2
VIII	Recreational activities	4	3	4	3	4	4	3	1	4	4	4	4	4	4	4
IX	Nature conservation	4	5	5	7	6	6	6	7	4	7	6	7	7	7	7
IX	Environmental control	4	5	5	6	6	6	6	7	6	6	6	7	7	7	7
IX	Artificial reefs	4	3	6	6	6	6	7	7	4	7	6	6	6	6	7
X	Environmental education	6	6	6	6	6	6	6	6	6	6	6	6	6	6	7
X	Field trips	6	6	6	6	6	6	6	6	6	6	6	6	6	6	7
X	Books, posters, films	6	6	6	6	6	6	6	6	6	6	6	6	6	6	7
Integrated coastal zone management		7	7	7	7	7	7	7	7	7	7	7	7	7	7	7

Source: After Zaitsev (2006).

Key: *Consequences*: 1, very negative; 2, negative; 3, more negative than positive; 4, uncertain; 5, more positive than negative; 6, positive; 7, very positive. *Uses*: I, industry; II, agriculture; III, pisciculture; IV, sea transport; V, municipal economy; VI, coastal protection; VII, hydro-power engineering; VIII, tourism, resorts; IX, nature conservation; X, environmental education and environmental ethics.

Table 12.3 Average score of anthropogenic impact (AI) on selected marine and coastal sites in Ukraine.[a]

Site name[b]	AI score
Sukhoy liman	3.08
Budakskyi liman	3.10
Kalamytskyi Gulf	3.35
Feodosiia Gulf	3.41
Dniester liman	3.44
Kerch Strait	3.45
Khadjibeyskyi liman	3.45
Sasyk reservoir	3.46
Tuzla liman complex*	3.51
Kuyalnytskyi liman	3.56
Odessa Gulf	3.56
Karkinitskyi Gulf*	3.59
Dnipro and Bug liman*	3.60
Donuzlav Lake	3.66
Dofinovskyi liman	3.68
Sevastopol Bays	3.74
Grigorivskyi liman	3.81
Tyligulskyi liman*	3.82
Danube Delta mouth*	3.87
Berezanskyi liman	3.93
Karadag coast*	4.14
Zhebrianskyi Bay	4.18
Tendrivskyi Bay*	5.52
Yagorlytskyi Bay*	5.52
Zmiyiny Island (slopes)*	5.57
Zernov's Phyllophora Field*	6.07

a) The higher the score, the lower the level of impact (based on expert evaluation of anthropogenic impacts on the sites using the stress factors in Table 12.2).

b) Asterisk (*) indicates Marine Protected Area.

Macroalgae and angiosperms are Biological Quality Elements in the EU Water Framework Directive (WFD, 2000/60/EC), and their exchange processes with water go via the external contour of the thallus. The specific surface (thallus surface/weight ratio − S/W) is the basic parameter reflecting the intensity of water vegetation function, from which a set of indicators can be derived. Depending on the morphological structure and size of the thallus of a particular species, the S/W ratio can vary between several and several hundred square metres of photosynthetic surface area per kilogram of the plant's weight. Thus, the S/W ratio can be used to characterize the ecological function of the species concerned. Under conditions of high rates of biological production–destruction processes, species with high S/W ratios (small filamentary forms with short life cycle and high growth rates) tend to proliferate. Conversely, where production–destruction processes are relatively slow, populations of plants with low S/W ratios (big, perennial, slow-growing, habitat-forming species) tend to increase. The degradation of coastal ecosystems associated with a decline of biological diversity, simplification of food chains, increase of eutrophication level and decrease of benthic communities is accompanied by replacement of species having low S/W ratios (about $8–25\,m^2\,kg^{-1}$) with macroalgae having S/W ratios from 100 to more than $1000\,m^2\,kg^{-1}$). Accordingly, the morphological portrait of coastal and shelf bottom vegetation contains information about the intensity of ecological processes, and hence about the ecological status of protected ecosystems.

The S/W ratios of the most abundant macrophyte species growing in Ukrainian MPAs are available (Minicheva *et al.*, 2003). To use information about the ecological properties of different macrophytes (*r*- and *k*-selected species) for assessment of the ESC of marine ecosystems, ecological evaluation indices (EEI) have been proposed (Orfanidis *et al.*, 2011). As indicators derived from the S/W ratios enable us to go from qualitative to quantitative assessment of marine plants' ecological properties, they also appear to be effective to express the EEI determining the ESC of marine coastal eco-systems. For rapid assessment and monitoring of MPAs, the two simplest indicators are proposed (Minicheva, 2013): Three Dominants Ecological Activity (S/W_{3Dp}) and Phytocoenosis Surface Index (SI_{ph}).

Rapid ESC assessment of a number of existing or proposed protected areas in the Ukrainian part of the Black Sea using the S/W_{3Dp} indicator showed that most of them are in the categories 'High' and 'Good' (Minicheva, 2014). Good Environmental Status corresponding to high ESC is characteristic of Ukrainian MPAs having international and national levels of protection. There are marine areas in Ukraine with high ESC, but which have no conservation status at present, and so are promising for further expansion of the ecological network; these include Donuzlav Lake, Kalamytskyi Gulf and Feodosiia Gulf. At the same time, there are protected areas in the categories 'Moderate' and 'Poor' (Zhebriyanskyi Bay, Danube Delta mouth). This can be largely explained by the fact that these water areas are situated near, and suffer the influence of, big rivers.

Thus, the method of bottom vegetation morphological indicator assessment, which is simple to use, can be very helpful for determining the ecological status of a marine area and determining the need for its protection; it can also be used for routine monitoring of existing MPAs.

Method for Determining the Ecological Value of MPAs

Marine Protected Areas are not only intended to protect and restore endangered flora and fauna; they also serve as reference sites for assessment of GES according to the MSFD descriptors. The main ecological criteria for identifying potential MPAs are: uniqueness, rarity, representativeness, diversity, naturalness, dependency, critical habitats, vulnerability, and connectivity

(Begun *et al.*, 2012). The expansion of MPA coverage should also take into account the creation of ecological corridors, or networks, which should ensure adequate reproduction of wide-ranging species (see Beal *et al.*, this volume).

To justify designating new MPAs and to expand existing ones in Ukraine, a novel integrated indicator of the biological value of a marine water area was developed (Alexandrov *et al.*, 2010; Alexandrov, 2012). This indicator is derived from both the biological diversity of bottom and pelagic communities as well as their productivity. To calculate the integrated indicator of biological value (K_f) of marine water areas, the following formula is used:

$$K_f = (K_i^{a_i})^{0.5 \cdot} \left(K_1^{a_1 \cdot} K_2^{a_2 \cdot} \ldots K_n^{a_n} \right)^{1/2n}$$

where K_1, K_2, ... K_n are the values of seven distinct characteristics reflecting the state of the ecosystem in the area concerned (the so-called metrics; see below); a_1, a_2, ... a_n are weight coefficients of the characteristics reflecting their level of significance; $K_i^{a_i}$ is the minimum value of all metrics (with their weight coefficients) that characterize the area concerned; and n is the total number of characteristics taken into account in accordance with the number of criteria selected. The K_f value thus unites heterogeneous characteristics taking into consideration the level of their significance. Since the parameters considered are not independent, the resulting value of K_f represents the general status of the characteristics it comprises. The selection of the metrics (K_i) and determination of the weights (a_i) of characteristics were done taking into account the following conditions:

$$0 < K_i \leq 1, \quad \text{and} \quad 0 < a_i \leq 1.$$

All the characteristics selected can be divided into two categories: (i) indirect indices of biodiversity such as: primary production of phytoplankton (K_{PP}); ecological activity of macrophytes (see previous section) as an index of primary production of phytobenthos (K_{EAM}); ratio of biomass of plankton to benthos ($K_{P/B}$); and (ii) direct indices of biodiversity such as: number of macrozoobenthic species (K_{MZB}); total number of benthic biocoenoses (K_{BB}); and number of Red Data Book species (K_{RDB}). The numbers of direct and indirect biodiversity indicators in K_f are equal. However, there is a feedback between these indicators: high primary production reduces the species diversity of ecosystems. It was shown above that the value of anthropogenic impact (K_{AI}) is highly correlated with the state of ecosystems in protected areas. Thus K_{AI} can also be treated as an indirect indicator of biological value and included in K_f calculations. All of these metrics reflect the indicative lists of characteristics, pressures and impacts (MSFD Annex III, Table 1; 2008/56/EC): physical and chemical features, habitat types (structure and substrata composition of the seabed), biological features (phytoplankton and zooplankton communities; macroalgae and invertebrate bottom fauna; status of species), and other features (chemicals, sediments contamination, hotspots, health issues).

The weight coefficients of characteristics (a_i) were determined from paired correlation coefficients of the selected metrics value with two of them, K_{RDB} and K_{EAM}, as

these were the most important direct and indirect metrics respectively for assessing the biological significance of a marine area (Table 12.4).

The approach was applied to 26 brackish or marine areas in the Ukrainian part of the Black Sea coast from the Danube Delta to the Kerch Strait: 11 limans, eight bays and gulfs, one island, one delta, one open shelf area, one reservoir, one lake, one coastal cliff and one strait (Table 12.5). The characteristics required for calculating K_f values were taken from Alexandrov *et al.* (2010). Special attention was paid to the fact that values of K_f have to be determined not for the whole area, but for each component ecosystem present (Alexandrov, 2012). To determine the boundary values of K_f for the five classes envisaged by the MSFD (High, Good, Moderate, Poor, Bad), the percentile rule was used (Ohio Environmental Protection Agency, 1987). When a metric tends to decrease with the increase of human pressure, a deviation of more than 25% from the norm is evidence of an aggravated ecological situation.

Applying the method described here (which now incorporates K_{AI} in the K_f calculation originally used by Alexandrov, 2012; values of AI metric normalized similar to direct indices of biodiversity) shows that those marine ecosystems having the highest biological significance (and thus protected

Table 12.4 Matrix of cross-correlation between seven selected biological characteristics of marine ecosystems for determination of their weight coefficients (a_i).

Characteristics (metrics)	RDB	EAM	BB	MZB	P/B	PP	AI
RDB	—	0.24	**0.51**[a]	**0.48**	−0.09	−0.03	0.31
EAM	0.24	—	**0.43**	**0.37**	−0.22	−0.18	**0.40**
Weight coefficients of characteristics (a_i)	0.6	0.6	0.9	0.9	0.5	0.1	0.8

a) Bold values indicate significant coefficients of cross-correlation at <5% confidence level ($k = 32$).
Key: RDB, number of Red Data Book species; EAM, ecological activity of macrophytes; BB, number of benthic biocoenoses; MZB, total number of macrozoobenthic species; P/B, ratio of total plankton to benthos biomass; PP, gross primary production of phytoplankton; AI, integrated anthropogenic impact.

Table 12.5 Status of protection of Black Sea coastal and marine areas in Ukraine and associated integrated index of biological value (K_f).

No.[a]	Assessed area	Protected area		Ramsar Site	K_f	K_f normalized[b]	Biological value
		Status	Aquatic area (%)				
Sediments (sand, mud, clay)							
13	Karkinitskyi Gulf	1, 2	15.17	+	0.68	1.00	H
17	Donuzlav Lake	NP	0		0.50	0.49	G
–	Zhebriyansky Bay	NP	0		0.38	0.14	M
5	Zernov's Phyllophora Field	2	100.00		0.33	0.00	P
–	Odessa Gulf	NP	0		0.33	0.00	P
Rocky coast							
19	Sevastopol Bays (Kozachia)	2, 3	49.02		0.67	1.00	H
2	Zmiyiny Island (slopes)	2	0.50		0.65	0.95	G
25	Karadag coast	1, 2	4.92	+	0.57	0.77	M
18	Kalamytskyi Gulf	3	0.35		0.51	0.64	M
26	Feodosia Gulf	NP	0		0.46	0.52	P
–	Kerch Strait	1, 2, 3	1.82		0.23	0.00	B
Saline lagoons (inlets)							
10	Tendrivsky Bay (Chornomorskyi)	1	59.84	+	0.72	1.00	G
9	Yagorlytsky Bay (Biloberezhia Sviatoslava)	1, 2	96.46	+	0.71	0.98	M
3	Tuzla liman complex	1, 2	8.90	+	0.20	0.00	P

Table 12.5 (Continued)

No.[a]	Assessed area	Protected area		Ramsar Site	K_f	K_f normalized[b]	Biological value
		Status	Aquatic area (%)				
Limans and deltas							
–	Dnipro and Bug liman	1	24.85	+	0.53	1.00	H
–	Berezansky liman	NP	0		0.48	0.89	H
4	Dniester liman	1, 2	30.55	+	0.47	0.87	G
1	Danube Delta mouth	1	95.50	+	0.44	0.80	G
–	Khadzhibeisky liman	NP	0		0.20	0.28	M
–	Grigorivsky liman	NP	0		0.17	0.22	M
–	Tyligulsky liman	1, 3	100.00	+	0.14	0.15	P
–	Sukhyi liman	NP	0		0.11	0.09	P
–	Budaksky liman	NP	0		0.10	0.07	B
–	Dofinovsky liman	NP	0		0.07	0.00	B
Wetlands and salt marshes							
–	Sasyk reservoir	1	16.19	+	0.42	1.00	G
–	Kuyalnytsky liman	NP	0		0.06	0.00	P

a) Numbers refer to sites shown in Figure 12.3.
b) Normalization of K_f values within each habitat carried out using the formula: $(x - min)/(max - min)$.
Key: *Protection level of areas*: 1, international; 2, national; 3, local; NP, not protected. *Biological value*: H, high; G, good; M, moderate; P, poor; B, bad.

status) also typically have high values of K_f. It allows a more accurate ranking of the biological value of 26 coastal and marine areas of Ukraine and thus potential expansion of the number of MPAs, or a change of the protection status of some existing MPAs. The method will also help to work out a better quantitative framework for establishing the boundaries of MPAs and their connections through ecological corridors (see Table 12.5, Figure 12.3).

Using Environmental Sentinels for Public Monitoring of MPAs

The work of Vernadsky (1968) and its further development by Zaitsev (1986, 2015) shows that marine life has a non-uniform (or 'contoured') distribution: the main concentrations of organisms are located on the outer boundary of the pelagic zone while life in the water column is sparse (Vernadsky called it 'dispersed life'). Yet, traditional sampling methods regarding the biology and ecology of the sea largely overlook this phenomenon. This is not to suggest that further study of the water column and great depths is not required, but that more attention should be paid to peripheral biotopes and communities that have been neglected in marine biology and ecology to date.

The external boundaries of the water column, which are in contact and interact with the atmosphere, sandy and rocky coasts or silty bottoms are especially rich in life. Here are found the greatest concentrations of living matter, the effects of external influences are powerful, and the most significant 'hotspots' are located. On the other hand, the ecological conditions in the water column and at lower depths are much more stable.

Peripheral biotopes are inhabited by a large number of diverse organisms adapted to these specific conditions, from bacteria

(Tsyban, 1971), unicellular and multicellular algae, protozoans, fungi, molluscs, crustaceans, worms and other invertebrates to the eggs, larvae, fry and adults of fish such as Gobiidae, Mugilidae and Pleuronectidae (Zaitsev, 2006, 2015). As the result of natural processes, many toxic substances accumulate in the same biotopes causing serious consequences for the communities, especially at the early stages of invertebrate and fish development. Some of these plants and animals are sensitive indicator species, whose presence, abundance or absence is indicative of changes in the biotope. They are the first to signal a change and could be termed 'environmental sentinels' (ES) (Zaitsev, 2015).

The ES from peripheral biotopes provide the clearest evidence of the consequences of anthropogenic eutrophication of the northwest Black Sea shelf, whose waters are strongly affected by discharges from three big rivers: the Danube, Dniester and Dnieper. For example, a particularly sensitive ES is the perennial brown alga *Cystoseira barbata*. Between 1979 and 1981, *C. barbata* that once occurred in dense beds on hard substrates at 1–3 m depth disappeared from the rocky coasts of Odessa Gulf and Zmiyiny Island. Organisms closely associated with *C. barbata*, including the polychaetes *Janua pagenstecheri* and *Spirobranchus triqueter*, which are usually attached to *C. barbata* thalli and surfaces of molluscs and crabs, also disappeared. At present, mussels and crabs are free from these polychaetes, showing that immediate contact with *C. barbata* is essential for maintaining the polychaetes' populations.

During the same period, populations of the polychaete *Ophelia bicornis* and bivalve mollusc *Donacilla cornea* disappeared from sandy coasts. In the 1960s, the abundance of *D. cornea* in the mediolittoral zone of the shelf reached dozens of thousands per square metre (Zakutsky and Vinogradov, 1967) and it was even used as a raw material for local handicrafts. Similarly, in the neustonic biotope, the abundance of the neustonic copepods *Pontella mediterranea* and *Anomalocera patersoni*, decapod larvae, flathead grey mullet *Mugil cephalus* larvae, and fry belonging to the genera *Mugil, Liza, Belone, Solea* and *Callionymus* and other fish developing in the neuston layer also shrank by several orders of magnitude.

The number of grey mullet fry coming to the Black Sea coast in summer is a particularly important indicator of the ecological state of the neuston. This fish hatches from eggs laid on the water surface in the open sea, tens of kilometres away from the coastline. Reaching a body length of 4–5 mm, the fry remain in the neuston while migrating towards the coast to feeding grounds in shallow bays and limans. The quantity of fry reaching the coast between July and September could be used to assess the ecological condition of the sea surface for the period from their hatching until arrival at the coast (Alexandrov and Zaitsev, 1989).

Phytoplankton blooms are easily recognized. On sandy beaches they can clog the interstices between sediment particles with detritus, which decreases the rinsing and drainage of the sand by seawater and reduces its aeration. On rocky coasts a phytoplankton bloom could impede filter feeding by sedentary organisms such as sponges and polychaetes. Furthermore, the production of toxic substances by algal metabolites can occur.

Thus, the most dramatic ecological changes, when entire populations of marine organisms practically disappear, take place only in peripheral biotopes. By contrast, in the water column of the pelagic zone and at great depths, the chemical composition and other properties of the water mass are more stable. This explains why stocks of the commercial pelagic fish species sprat *Sprattus phalericus* and whiting *Merlangius euxinus* hardly changed during the major eutrophication episode from the 1980s to 1990s and retained their socioeconomic value.

Using ES to assess the ecological status of peripheral marine biotopes has a number of advantages compared to traditional methods: it requires no research vessels; it clearly reveals sharp changes in the marine environment; it shows precisely the location of ecological 'hotspots' and time of their emergence; and it encourages the involvement of amateur naturalists (especially young ones), under the leadership of experienced specialists, in ecological monitoring of the coastal zone.

A preliminary list of ES genera comprises: attached brown algae of *Cystoseira* and *Sargassum*; gastropod molluscs of *Littorina* and *Melaraphe*; bivalve molluscs of *Patella, Fissurella* and *Diodora*; polychaetes of *Ophelia, Janua, Spirobranchus* and *Serpula*; mullet fry of *Mugil* and *Liza*; and piscivorous birds hunting for mullet fry: little egret *Egretta garzetta* and grey heron *Ardea cinerea*.

Expansion of the Ukrainian MPA Network

The Ukrainian ecological network to date has been formed based on the principles of nature protection and conservation of areas having high ecological value. The functionally integrated network is aimed at maintaining high biological diversity (Verkhovna Rada Ukrainy, 2000). Future expansion of the Ukrainian ecological network implies taking account of innovative European concepts and approaches demonstrating importance not only for nature conservation, but also for socioeconomic aspects. Further development of a European MPA network and its Ukrainian component should therefore consider the specific natural features of marine ecosystems resulting from the interactions between coastal and offshore, pelagic and bottom ecosystems (which have a three-dimensional structure and function), together with physical, chemical, biological and ecological processes that underlie cells of ecosystem functioning (Boero, this volume). Socioeconomic issues are also important as a basis for regulatory mechanisms (Ojea *et al.*, this volume), and for forming and managing the objectives of ecological networks, which can have different purposes: connectivity, conservation, socioeconomic, geographic, collaborative, cultural and transnational (Beal *et al.*, this volume). Out of the various types of ecological networks, the Ukrainian MPA network could be classified chiefly as the conservation type. More attention should be paid in future to a more multi-faceted development of the national ecological network, strengthening socioeconomic, geographical, cultural and other aspects. Already, Ukrainian experts taking nature conservation principles as the basis are starting to pay more attention to the socioeconomic features of ecological networks, which can provide a strong foundation for 'blue' and 'green' economic growth in the country and regions (Harichkov and Nezdoyminov, 2013). Ukraine still has important marine and coastal areas not yet included in its ecological network, which together with their natural value, have high recreation and resource potential, including the possibility for setting up offshore wind-farms.

One of the legislative measures ensuring further development of the Ukrainian national MPA network is the listing of new areas and objects (Verkhovna Rada Ukrainy, 2004). The selection of promising new MPAs was based on criteria in the regulations (Verkhovna Rada Ukrainy, 1992) as well as approaches to ecological value assessment in line with the WFD and MSFD (EC, 2008) (Alexandrov *et al.*, 2010; Minicheva, 2013).

As the result of expert work, taking into account the new national and European principles of forming ecological networks, as well as new approaches in the determination of the ecological value of marine areas, it is presently proposed to include 12 new nationally protected areas in the existing Ukrainian MPA network. These sites have a combined area of 104 300 ha, which represents 17% of the current total area of Ukrainian MPAs (Figure 12.3).

Most of the MPAs in the north-west shelf are connected with the Crimean coast by the main cyclonic (counter-clockwise) Black Sea rim current (Öztürk *et al.*, this volume), which ensures population stability of many flora and fauna species by carrying their larvae and mature individuals downstream. To correct the misbalance of area coverage, 8 out of 12 of the new MPAs will be established on the Crimean Peninsula. These sites have high environmental status, socioeconomic potential and support reproduction of key plant and animal species both for Ukrainian MPAs and for the Black Sea in general. Implementation of plans for extending the Ukrainian MPA network and its integration within the European Coastal and Marine Ecological Network will unite the efforts of researchers and state officials responsible for ecological integration across Europe.

Conclusion

The total area of MPAs of international and national significance in the Ukrainian ecological network is over 6000 km^2. As a result, Ukraine ranks highest among the Black Sea countries for the overall extent of MPAs, and more than 82% of the area of all Black Sea MPAs are in Ukraine. The highest percentage cover of MPAs, 10.9%, occurs in the north-west shelf area of the Black Sea. However, in Ukraine the distribution of MPAs is very uneven: 99.8% of them are in the shelf area and only 0.2% around the Crimean Peninsula.

It is proposed to expand the Ukrainian MPA network to include 12 new sites covering more than $1040\,km^2$ (about 17% of the existing area). To help correct the misbalance of distribution between the MPAs in the shelf area and the Crimean Peninsula, eight of the proposed new MPAs lie in the coastal part of Crimea. The expansion of the Ukrainian MPA network takes account of such important natural characteristics as the main cyclonic Black Sea current and the influence of river discharges (as a main factor of eutrophication).

In order to integrate the Ukrainian MPA network into the European Coastal and Marine Ecological Network, a number of new methods of identifying MPAs were elaborated based on the requirements and standards of the WFD and MSFD. These methods and indicators should be incorporated into the Black Sea Integrated Monitoring and Assessment Programme (2015–2020) of Ukraine.

Acknowledgements

This chapter is based on research carried out with funding from the European Community's Seventh Framework Programme (FP7/2007–2013) under Grant Agreement No. 287844 for the project 'Towards coast to coast networks of marine protected areas (from the shore to the high and deep sea), coupled with sea-based wind energy potential' (CoCoNET), as well as the results of fundamental investigations on the themes of the National Academy of Sciences of Ukraine.

The authors would like to express their gratitude to the staff members of the Institute of Marine Biology, National Academy of Sciences of Ukraine: Marina Kosenko for her help with collecting information about the number and sizes of Black Sea MPAs, and Serhii Zaporozhets for drawing the maps.

References

Alexandrov, B. (2012) Black Sea marine protected areas and an approach to the creation of ecocorridors, in *Marine Nature Conservation and Management at the Borders of the European Union* (ed. D. Czybulka). Beitraege zum Landwirtschaftsrecht und zur Biodiversitaet 7. Nomos Verlag, Baden-Baden. pp. 121–135.

Alexandrov, B. and Zaitsev, Y. (1989) Man-made impact on the neuston of marine coastal waters and modes of its assessment. *Biologya Moria*, **2**, 56–60. (In Russian)

Alexandrov, B. and Zaitsev, Y. (1998) Black Sea biodiversity in eutrophication conditions, in *Conservation of the Biological Diversity as a Prerequisite for Sustainable Development in the Black Sea Region* (eds V. Kotlyakov, M. Uppenbrink and V. Metreveli). Kluwer, Dordrecht. pp. 221–234.

Alexandrov, B., Galperina, L., Groza, V. *et al.* (2010) *Strategy and Methodological Approaches to Building a Network of Marine Protected Areas in the Coastal Waters of the Ukrainian Part of the Black and Azov Seas.* Report on the project JWP-Ukr-2009–14 Living Black Sea: Strategy for expansion of marine protected areas in the Black Sea coastal waters of Ukraine funded by the Royal Embassy of the Netherlands. Odessa. (In Ukrainian)

Alexandrov, B., Gomoiu, M., Mikashavidze, E. *et al.* (2013) *Non-Native Species of the Black Sea.* Abstracts book of 4th Black Sea Scientific Conference 'Challenges towards good environmental status' (28–31 October 2013, Constanta). Editura Boldaş, Constanta. pp. 62–63.

Begun, T., Velikova, V., Muresan, M. *et al.* (2012) *Conservation and Protection of the Black Sea Biodiversity: Review of the existing and planned protected areas in the Black Sea (Bulgaria, Romania and Turkey) with a special focus on possible deficiencies regarding law enforcement and implementation of management plans.* Report for Marine Strategy Framework Directive (MSFD) Guiding Improvements in the Black Sea Integrated Monitoring System (MISIS). http://www.misisproject.eu/

Black Sea Environment Programme (2009) *Black Sea Red Data Book.* http://www.grid.unep.ch/bsein/redbook/index.htm

Commission on the Protection of the Black Sea Against Pollution (2007) *Black Sea Transboundary Diagnostic Analysis and Strategic Action Plan* (TDA). Istanbul. 269 pp. http://www.blacksea-commission.org/_tda2008.asp

Convention on Biological Diversity (2008) *COP Decision IX/20 Marine and Coastal Biodiversity.* https://www.cbd.int/decision/cop/default.shtml?id=11663

Harichkov, S. and Nezdoyminov, S. (2013) Ecological networks as a 'green' growth factor of the region's economy. *Economy: Realities Time,* **4** (9), 174–182. (In Ukrainian)

Lausche, B. (2011) *Guidelines for Protected Areas Legislation.* IUCN, Gland, Switzerland. xxvi + 370 pp.

Marushevsky, G. (ed.) (2003) *Directory of Azov–Black Sea Coastal Wetlands: Revised and Updated.* Wetlands International, Kyiv.

Minicheva, G. (1998) *The foundations of morphofunctional forming of the marine phytobenthos.* PhD thesis, Institute of Biology of the Southern Seas, NAS Ukraine, Sevastopol. (In Russian)

Minicheva, G. (2013) Use of the macrophytes morphofunctional parameters to assess ecological status class in accordance with the EU WFD. *Marine Ecological Journal,* **12** (3), 5–21.

Minicheva, G. (2014) Monitoring of wetlands ecological status on the basis of morphofunctional estimation of the bottom vegetation, in *Monitoring of Wetlands of*

International Significance (ed. O. Petrovich). DIA, Kyiv. pp. 19–23.

Minicheva, G., Zotov, A. and Kosenko, M. (2003) *Methodical recommendations for determining the complex of morpho-functional parameters of unicellular and multicellular forms of aquatic vegetation.* GEF Project for Renewal of the Black Sea Ecosystem. Odessa.

Minicheva, G., Maximova, O., Morushkova, N. *et al.* (2008) State of the environment of the Black Sea (2001–2006/7), in *State of the Environment of the Black Sea (2001–2006/7)* (ed. Temel Oguz). Commission on the Protection of the Black Sea Against Pollution, Istanbul, Turkey. pp. 247–272.

Minicheva, G., Afanasyev, D. and Kurakin, A. (2014) *Black Sea Monitoring Guidelines: Macrophytobenthos.* http://emblasproject.org/wp-content/uploads/2013/12/Manual_macrophytes_EMBLAS_ann.pdf

Ohio Environmental Protection Agency (1987) *Biological criteria for the protection of aquatic life: User manual for biological field assessment of Ohio surface water.* Ohio Environmental Protection Agency 2. Columbus, Ohio.

Orfanidis, S., Panayotidis, P. and Ugland, K. (2011) Evaluation index continuous formula (EEI-c) application: a step forward for functional groups, the formable and reference condition values. *Mediterranean Marine Science,* **12**, 199–231.

Panchenko, T. (2009) *Guidelines on Territorial Planning in Coastal Zone. Version 2.* Environmental Collaboration for the Black Sea Project (ECBSea). EuropeAid/120117/C/SV/Multi.

Toropova, C., Meliane, I., Laffoley, D. *et al.* (eds) (2010) *Global Ocean Protection: Present Status and Future Possibilities.* Agence des aires marines protégées, Brest, France; IUCN WCPA, Cambridge, UK; TNC, Arlington, USA; UNU, Tokyo, Japan.

Tsyban, A. (1971) Marine bacterioneuston. *Journal of the Oceanographical Society of Japan*, **27** (2), 56–66.

Verkhovna Rada Ukrainy (1992) *On the nature reserve fund of Ukraine.* Law 2456-XII, 16 June 1992. (In Ukrainian)

Verkhovna Rada Ukrainy (2000) *On national program of formation of the national ecological network of Ukraine for 2000–2015.* Law 1989-III, 21 September 2000. (In Ukrainian)

Verkhovna Rada Ukrainy (2004) *On the ecological network of Ukraine.* Law 1864-IV, 24 June 2004. (In Ukrainian)

Vernadsky, V. (1968) *The Biosphere.* Mysl, Moscow. (In Russian)

Yena, V., Yena, A. and Yena, A. (2004) *Tavridas' Protected Landscapes.* Business-Inform, Simferopol. (In Russian)

Zaitsev, Y. (1986) Contourobionts in ocean monitoring. *Environmental Monitoring and Assessment*, **7**, 31–38.

Zaitsev, Y. (1992) Ecological state of the Black Sea shelf zone near the Ukrainian coast (a review). *Hydrobiology Journal*, **28** (4), 3–18. (In Russian)

Zaitsev, Y. (2006) Littoral concentration of life in the Black Sea area. *Journal of the Black Sea/Mediterranean Environment*, **12**, 113–128.

Zaitsev, Y. (2015) On contour structure of the hydrosphere. *Hydrobiology Journal*, **51** (1), 3–27. (In Russian)

Zakutsky, V. and Vinogradov, K. (1967) *Macrozoobenthos.* Naukova dumka, Kiev. (In Russian)

13

Prospects for Marine Protected Areas in the Turkish Black Sea

Bayram Öztürk[1], Bettina A. Fach[2], Çetin Keskin[1], Sinan Arkin[2], Bülent Topaloğlu[1] and Ayaka Amaha Öztürk[1]

[1] *Faculty of Fisheries, Istanbul University, Beyazıt, Istanbul, Turkey*
[2] *Institute of Marine Science, Middle East Technical University, Erdemli, Mersin, Turkey*

Introduction

The Black Sea is one of the world's seas most isolated from major oceans, and the largest anoxic body of water on the planet (87% of its volume is anoxic). Even though the Black Sea is small in terms of volume compared to the Mediterranean Sea, it is very peculiar and has unique characteristics (Öztürk and Öztürk, 2005). It is connected to the Sea of Azov via the Kerch Strait in the north and to the Marmara Sea (which is connected to the Aegean Sea through the Canakkale (Dardanelles) Strait) via the Istanbul Strait (Bosporus) in the south-east. Moreover, the Black Sea is surrounded by six riparian countries whose socio-economic and political conditions differ greatly.

The unique basin-wide cyclonic boundary current (known as the rim current, Figure 13.1) is driven by prevailing winds and the large freshwater discharge from rivers, and is steered by the steep bottom topography around its periphery that consists of narrow shelves and a maximum depth of around 2200 m (Oguz et al., 2005). The cyclonic rim current encloses two cyclonic cells within the interior basin and separates the cyclonically dominated inner basin from the anticyclonically dominated coastal zone (Oguz et al., 1992). The anticyclonic eddies near the Istanbul Strait, Sakarya, Sinop, Kızılırmak and Batumi have been shown to be important for accumulation and transport of biota and fish larvae between the coastal zone and the open ocean (Oguz et al., 2002; Fach, 2014).

Since the late 1960s, a wide spectrum of anthropogenic influences on the Black Sea ecosystem has been apparent (Oguz and Velikova, 2010). Eutrophication has become the main issue, especially in the coastal sectors (Sapozhnikov, 1991; Mee, 1992; Zaitsev, 1993), due to large amounts of sediments, organic matter and pollutants discharged via large rivers especially the Danube, Dnieper and Dniester flowing into the north-western shelf of the Black Sea (Alexandrov et al., this volume). As Sur et al. (1996) state, eutrophication has increased significantly, influencing Secchi disc readings in the central Black Sea: from 20 m in the 1920s they had decreased to about 15 m by the mid-1980s and to 5–6 m in the early 1990s (Eremeev et al., 1992). In addition, a decline in the total stocks and species of fish has occurred, many organisms have disappeared from the region, and the Black

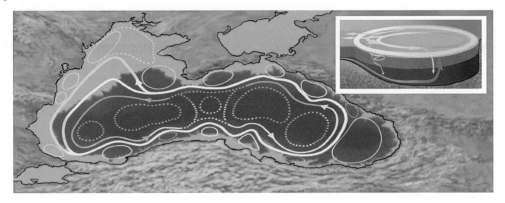

Figure 13.1 Circulation patterns in the Black Sea (see text for details). Artwork: Alberto Gennari.

Sea has been invaded by non-native opportunistic species (Zaitsev, 1991; Shiganova, 1998; Shiganova *et al.*, 2001; Kideys, 2002).

The coastal and marine biodiversity of the Turkish Black Sea is constantly under serious threat due to human pressures. Major threats are posed by the destruction of marine habitats and ecosystems, overexploitation of marine resources and the loss of coastal habitats through extensive urbanization. In addition, illegal, unreported and unregulated (IUU) fishing poses a serious threat for Black Sea marine biodiversity (Öztürk, 2013). Pollution by ships (e.g. oil spills and discharging bilge water), the intentional and/or accidental introduction of alien species (Zaitsev and Öztürk, 2001; Galil, this volume), marine litter (Topçu and Öztürk, 2012), and climate change are other threats of concern. Some commercial fish species such as *Thunnus thynnus*, *Acipenser sturio* and *Scomber colias* have been under pressure from overfishing during the last few decades and some species such as *T. thynnus*, *S. colias* and *S. scombrus* have even disappeared from the basin completely.

As stated in the Black Sea Transboundary Diagnostic Analysis 2007 (BSC, 2007) and confirmed in the Black Sea Strategic Action Plan 2008, Marine Protected Areas (MPAs) form a key element of the ecosystem-based approach to managing and safeguarding the Black Sea marine environment, including improving the sustainability of fisheries. The aim of this management regime is to manage the use and values of ecosystems with all stakeholders in order to maintain ecological integrity together with consideration for the uncertainty and ever-changing nature of ecosystems. This approach also contains precautionary safeguards to account for common problems such as lack of scientific data, the uncertainty of natural processes and lack of fisheries management. In the case of the Black Sea, establishing MPAs is an important way to exercise these precautionary principles, as well as protecting ecosystems where the single-species management for threatened species such as *A. sturio*, *Scophthalmus maximus*, monk seal *Monachus monachus* and cetaceans has failed.

Overview of the Regional Situation

According to Alexandrov *et al.* (this volume), 37 protected areas have been designated around the Black Sea which include marine waters, totalling 755 840 ha. However, more than half of this area is represented by Zernov's Phyllophora Field Botanical Reserve (in Ukraine), declared in November 2008, which covers 402 500 ha. Another major part is located in the Danube Delta Biosphere Reserve in Romania. There is at present not one MPA including offshore waters in the Turkish part of the Black Sea.

Turkish Perspectives and Rationale for Establishing MPAs

The length of the entire Turkish coastline is 8592 km (excluding coasts of islands), of which 1132 km are under protected designations such as National Parks, Ramsar Sites and Nature Parks. In addition, Special Protected Areas comprise 6.6% of all coasts. They have been designated to protect certain species such as the monk seal or for biodiversity objectives in the Aegean and Mediterranean Sea. The Black Sea coastline of Turkey is 1700 km long (Demirkesen *et al.*, 2008); there are many protected areas, but no specific MPA has been designated, and it has the least coverage of coastal protected areas, compared with other Black Sea countries (Alexandrov *et al.*, this volume). Several sites on the Turkish Black Sea coast are already recognized for their high ecological value, such as two internationally important wetlands: Kızılırmak Delta (designated in 1998 as a Ramsar Site) and the Yeşilırmak Delta, both of which are located in Samsun Province.

Recently, Öztürk *et al.* (2013) proposed five ecologically important sites for designation as MPAs along the Turkish coast of the Black Sea (Figure 13.2). These proposed sites comprise only 2% of the Turkish territorial water in the Black Sea. The largest site proposed covers the coastal waters from Şile to Kefken, and the smallest is the Mezgit Reef. The two deltas in Samsun Province mentioned above are also included as one

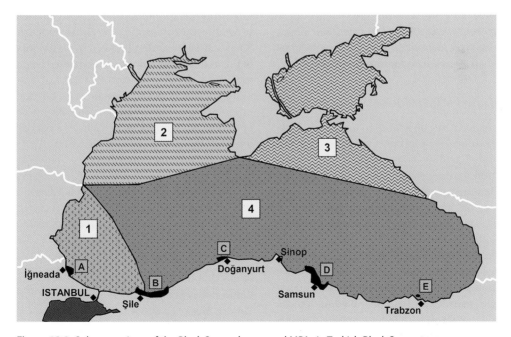

Figure 13.2 Sub-ecoregions of the Black Sea and proposed MPAs in Turkish Black Sea waters. 1, Pre-Bosphoric Region; 2, North-western Shelf; 3, Kerch Strait; 4, Southern Part. A, İğneada; B, Şile-Kefken; C, Doğanyurt; D, Samsun deltas; E, Mezgit Reef.

Table 13.1 Main threats identified for the proposed MPAs on the Turkish coast of the Black Sea.

Proposed area for MPA	Type of threat						
	Land-based and ship-originated pollution	Overfishing	Illegal sand extraction	Reed burning	Coastal erosion	Agriculture	Forestry
İğneada	Y	Y	Y	N	Y	Y	Y
Şile	Y	Y	N	N	Y	Y	Y
Doğanyurt	Y	Y	Y	N	N	Y	Y
Samsun deltas	Y	Y	Y	Y	Y	Y	Y
Mezgit Reef	Y	Y	N	N	N	N	N

Y, yes; N, no.

MPA. These areas were proposed by taking into account those criteria specified by the Convention on Biological Diversity (CBD, 2008), such as uniqueness; life history stages of species; importance for threatened, endangered species or habitats; vulnerability; fragility; sensitivity or slow recovery potential; and biological productivity.

Table 13.1 summarizes the threats these proposed areas currently face. Six major threats were identified, and apart from Mezgit Reef, all areas are under multiple threats.

Ecoregions of the Black Sea

Ecoregions are considered to be the smallest-scale units in Marine Ecoregions of the World (Spalding *et al.*, 2007). They show natural similarities and should be considered for nature planning and conservation. An ecoregion has a strong connection within itself and represents uniqueness, peculiar conditions and species diversity at a regional scale. Ecoregions are also connected to each other within wider geographical ranges. Although the Black Sea itself constitutes a single marine ecoregion (Spalding *et al.*, 2007), we suggest that four sub-ecoregions can be recognized within the Black Sea based on biodiversity characteristics as follows: 1. Pre-Bosphoric Region; 2. North-western

Shelf; 3. Kerch Strait; and 4. Southern Part, which contains the Turkish and Georgian waters (Figure 13.2).

Among these sub-ecoregions, the Pre-Bosphoric Region is under the influence of the Mediterranean–Black Sea interaction due to the presence of the Istanbul Strait, and thus contains a critical biotope for migratory fish, mammals, birds and species of Mediterranean origin. The North-western Shelf is shallow and influenced by sediments deposited by the Danube and other rivers, making it the richest area in terms of primary production. The Kerch Strait has the unique peculiarity of freezing during most of the coldest winters, which causes a barrier between the Black Sea and the Sea of Azov, especially for migratory species. The last ecoregion is Southern Part which contains Ponto-Caspian species such as relict Gobiid fish species. The five MPAs proposed by Öztürk *et al.* (2013) are located inside sub-ecoregions 1 or 4, which lie along the Turkish coast.

Connectivity Between the Proposed Turkish Black Sea MPAs

It has been found that larval dispersal by ocean currents and connectivity between different oceanic regions are crucial factors when designing MPAs (Cowen *et al.*, 2006; Lester *et al.*, 2009; Moffitt *et al.*, 2009). Connectivity also plays a major role in assuring population persistence in an MPA network (Moffitt *et al.*, 2011). Hence, a modelling study was carried out (not previously published) to assess the degree of connectivity between the five MPAs proposed above. The aim of this study was to identify basin-scale pelagic larval connectivity using an ecosystem-based approach (e.g. Coll *et al.*, 2012; Guidetti *et al.*, 2013) as opposed to focusing on one target species, such as the commercially important anchovy (Fach, 2014). The common trait of many of the pelagic fish species is that they have pelagic larval stages that stay in the water for different lengths of time, also referred to as pelagic larval duration (PLD).

Virtual pelagic larvae were released in the Black Sea surface current velocity fields for the years 2001–2003, obtained from the sbPOM model run for years 2000–2010, set up and validated for the Black Sea in the framework of the European FP7 OPEC

project (http://marine-opec.eu; Allen *et al.*, 2013). It was assumed that larval dispersal is dependent on the duration of larvae in the surface water (PLD), the timing of spawning and the circulation pattern. Particles were released in 10 different coastal areas using winter, spring and summer spawning times (1 January, 1 April and 1 July) as well as three different PLD times (30, 45 and 60 days) for the years 2001–2003 which are ecologically meaningful for a number of Black Sea organisms. In total, more than 3300 drifters were released every 2 km along the coast, up to 6 km offshore (Figure 13.3).

The particle drift study with a PLD of 45 days showed that Region (R) 1 where İğneada and Şile are located had a high level of connectivity (Figure 13.4) in all three years and at all spawning times. The area retained about 50% of the pelagic larvae starting there, while the other 50% were consistently transported downstream eastwards along the coast throughout all spawning times and years, mainly to Regions 2 (*c.*15%) and 3 (15%) as well as to R10 and open sea regions.

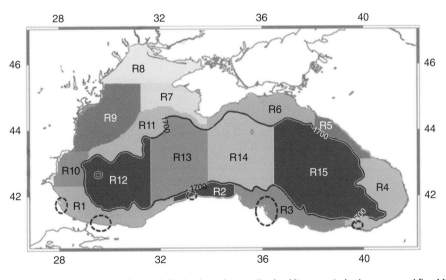

Figure 13.3 Sink regions for modelled pelagic larvae. Dashed lines encircle the proposed five MPAs for the southern Black Sea coast (see text). The thick black line marks the 1700 m isobath separating coastal regions from the open sea. Virtual larvae were released within the 6 km band surrounding the entire Black Sea coast.

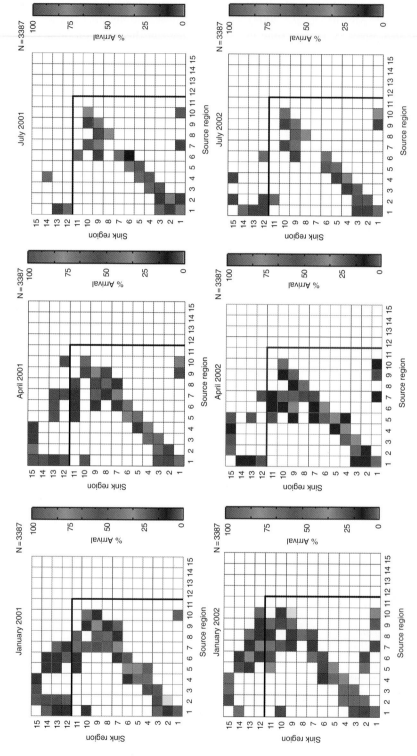

Figure 13.4 Connectivity matrices for modelled pelagic larvae released in coastal regions only (see Figure 13.2) on 1 January, 1 April and 1 July (first to third column) in each of the years 2001, 2002 and 2003 (first to third row) with a PLD of 45 days. Matrices indicate the probability (%) for larvae originating from a source region (x-axis) to be transported to a sink region (y-axis) estimated from individual 30-day trajectories. The thick black line separates shelf regions from open sea regions >1700 m deep (R12–15).

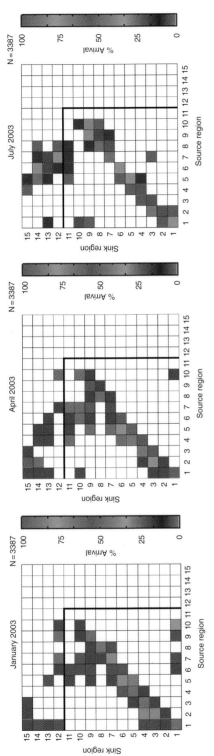

Figure 13.4 (Continued)

Region 1 also receives an inflow of pelagic larvae from R10 and the North-western Shelf (R7 and R9). Doğanyurt (R2) is connected downstream with R3 and the open sea R15 during spawning times in January and April of all years; in January 2002 and 2003 some drifters reach R4. Kızılırmak and Yeşilırmak (R3) have high retention rates: >80% during summer spawning times, 60–70% during spring spawning and approximately 50% in winter. The pelagic larvae that do leave are transported only as far as R4 and the adjacent open sea R15. Rather high retention rates also occur in R4, though not as much as in R3: the simulations show 70–80% retention in summer, 50–65% in spring and 20–40% in winter.

These transport patterns were broadly the same for smaller and larger PLD times, with generally higher retention during small PLD (30 days) and lower retention during longer PLD (60 days) as would be expected.

To illuminate the exact drift of the virtual pelagic larvae released within the five proposed MPAs along the Turkish coast, it is necessary to examine where they end up after the respective PLDs. Thus, when examining in detail the results of the model for July 2002 with a PLD of 45 days (Figure 13.5) it becomes clear that larvae originating in the İğneada region are not transported far at all but are retained or are merely transported a few kilometres downstream (Figure 13.5a). However, the larvae originating in the Şile region show much less retention and end up as far as R2 as well as far offshore in R12 (Figure 13.5a). Larvae originating in Doğanyurt are also transported long distances: there is no retention at all and larvae end up as far as Trabzon, close to Mezgit Reef (Figure 13.5b). This pattern is not surprising as this area is where the rim current flows close to shore and currents are fast and highly dynamic (Oguz *et al.*, 1992, 1993; Oguz and Besiktepe, 1999). On the other hand, pelagic larvae released in Kızılırmak and Yeşilırmak are very much retained in the area and though some larvae leave the immediate area, they cannot even reach Trabzon (Figure 13.5c). This is expected because the region comprises a big river delta where water is retained, known for serving as a nursery area for many species. Pelagic larvae released at Mezgit Reef are transported downstream up to the Rioni River delta (Figure 13.5d).

From the above, it was found that out of the five MPAs proposed for the Turkish coast, Şile is particularly well connected to upstream regions, at long PLD even all the way to the Kerch Strait. Similarly, the proximity of Doğanyurt to the strong rim current flow enables pelagic larvae originating there to travel downstream to distant regions. Hence the Şile and Doğanyurt sites are good locations for establishing MPAs, and because they are well connected, can also play an important role in maintaining a Black Sea MPA network. The other three proposed sites (İğneada, Kızılırmak and Mezgit Reef) exhibit more or less high retention rates of pelagic larvae and therefore need protection because of their localized biodiversity characteristics.

A Case Study of Şile Proposed MPA

Among the five proposed MPAs, Şile is of special interest due to its closeness to the Istanbul Strait which has crucial importance for migration of marine species between the Black and Mediterranean Seas (unfortunately including alien marine organisms, of which 19 species have been reported from Şile). In addition, Şile can be a success story because as well as its nature value and growing environmental concern, it has historical sites that attract tourists who can provide a source of revenue.

Şile is one of the smaller districts of Istanbul (with about 137 000 inhabitants),

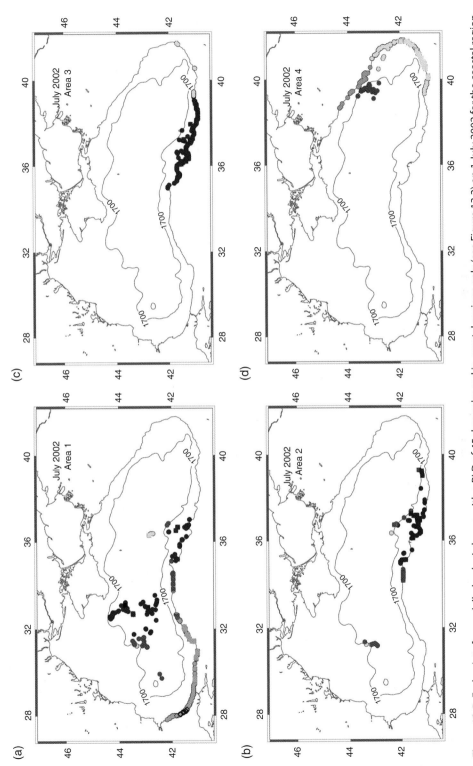

Figure 13.5 End points for modelled pelagic larvae with a PLD of 45 days released in coastal regions only (see Figure 13.2) on 1 July 2002 for the coastal region (a) 1, (b) 2, (c) 3 and (d) 4. Grey shades denote the different regions these larvae ended up in (see Figure 13.3). Black squares in (a), (b), (c) and (d) show the end points of pelagic larvae released in Şile, Doğanyurt, Kızılırmak and Mezgit Reef, respectively. Open black diamonds in (a) show the end points of those released in İğneada.

and only 70 km north of the city itself (Figure 13.2). It is one of the famous resort areas of the Black Sea, popular for its long sandy beach. In recent years, several hotels and many summer houses have been constructed for accommodating tourists.

Fishing

The Black Sea entrance of the Istanbul Strait and Şile are important areas for feeding, and for sheltering the larvae and eggs, of commercial fish such as *Engraulis encrasicholus*, *Sardina pilchardus*, *Sprattus sprattus*, *Scomber scombrus*, *S. colias*, *Merlangius merlangus* and *Trachurus trachurus* (Mater and Cihangir, 1990). Consequently, fishing is one of the major livelihoods of people in Şile (Table 13.2). In addition to 89 local fishermen, around 30 external fishermen arrive when the main fish migration period starts. They generally use artisanal methods such as set nets, gill nets and hand nets. The target species change seasonally, depending on the presence of migratory species such as anchovy, bluefish and horse mackerel. Demersal species (red mullet, turbot and whiting) are all fully or partially overfished. Most local fishermen complain about

Table 13.2 Fishing methods and number of local and external fishermen.

Fishing method	Local fishermen	External fishermen
Trawling	4	10
Purse seining	3	12
Rapana diving	2	8
Set nets	30	—
Hand nets	30	—
Gill nets	20	—
Total	**89**	**30**

Source: Unpublished data acquired from the Şile Fisheries Cooperative.

overfishing, pollution and disappearance of some of the commercially valuable species, such as *Scophthalmus rhombus*, *Xiphias gladius*, *S. scombrus*, *Pomatomus saltator* and *T. thynnus* in the Şile area. They also complain about fishermen coming from outside Şile. The total fish catch is estimated at 1000 tons and that of the Asian rapa whelk *Rapana venosa* as 750 tons. The latter is an alien species brought from the Sea of Japan and later commercially harvested in the Black Sea by diving, mostly in summer.

As small-scale fisheries are important around Şile, it can be expected that the local fishermen would benefit from the designation of an MPA in most of the area. The sandy shallow waters along the Şile coast are important nursery areas, especially for species like sand sole *Pegusa lascaris*, common sole *Solea solea* and turbot *Scophthalmus maximus*. In addition, this area is important for some fish species which are included in the IUCN Red List, such as common thresher shark *Alopias vulpinus*, spiny dogfish *Squalus acanthias*, thornback skate *Raja clavata*, long-snouted seahorse *Hippocampus guttulatus* and European sturgeon *Huso huso* (Anonymous, 2010). Turkey has been making efforts to protect several marine species in the Black Sea which are reflected in Fisheries Law 1380, which includes some restrictions on harvesting species found in the Şile area, such as seagrass *Zostera* spp., the mollusc *Cerithium vulgatum*, sturgeons, and seahorse *Hippocampus hippocampus*.

The Working Group on the Black Sea of the General Fisheries Commission for the Mediterranean (WGBS-GFCM, 2015) reported the status of the Black Sea turbot population as both 'overexploited' and 'in overexploitation'. Similarly, the Black Sea anchovy population was found to be 'in overexploitation'. The Black Sea horse mackerel stock was reported as 'overexploited', while the spiny dogfish population was considered to be depleted at the Black

Sea scale. The implementation of a recovery plan for both turbot and spiny dogfish as well as the reduction of fishing of both anchovy and horse mackerel was recommended.

The islands off Şile and Kefken include diverse habitats such as seagrass meadows, muddy bottoms, rocky bottoms, caves, reefs and biogenic formations. These habitats signify a rich fauna of fish and invertebrates in the proposed MPA which deserves more stringent measures to be introduced for its conservation. In particular, the MPA would allow ecosystem-based fisheries management to be introduced. Furthermore, Akbulut *et al.* (2011) reported that sturgeons need *in-situ* protection, but without holistic and ecosystem-based management, success will be limited.

Marine Mammals

There are three cetacean species found in the Black Sea: harbour porpoise *Phocoena phocoena relicta*, bottlenose dolphin *Tursiops truncatus ponticus*, and short-beaked common dolphin *Delphinus delphis ponticus*. While the harbour porpoise and bottlenose dolphin are listed as Endangered (Birkun and Frantzis, 2008; Birkun, 2012), the common dolphin is considered Vulnerable (Birkun, 2008) in the IUCN Red List of Threatened Species. Dolphins were once harvested throughout the Black Sea until Turkey finally banned it in 1983. Their populations in the Black Sea have started to recover, but due to their slow breeding rate, as well as the existence of many threats such as lack of prey fish, bycatch, pollution and epidemics, their recovery cannot be realized without protection measures. Bycatch is the most serious problem: Tonay and Öztürk (2003) reported that during the turbot season at least 3000 individuals of harbour porpoises were stranded due to entanglement in turbot set nets. In addition, cetaceans are transboundary species and concerted actions are needed for effective protection.

Around Şile, the bottlenose dolphin and harbour porpoise are the most commonly seen cetaceans. The coastal waters off Şile, Agva, Kerpe and Kefken are feeding and calving grounds for them, and calves of bottlenose dolphins have been observed there during the summer and autumn seasons (BÖ, unpublished data). Furthermore, some bycatches have been reported due to turbot set nets in spring. A proposed MPA can provide better protection of these cetaceans in terms of reducing bycatch, recruiting more prey fish, and securing feeding and calving grounds.

There is also one pinniped species, the monk seal, which is one of the most critically endangered species in the world. It was last seen in the Turkish Black Sea coast in the 1980s between Şile and Zonguldak. While it is highly likely that monk seals are completely extinct in the Black Sea, they still occur in the Sea of Marmara, so it is important to designate some areas with caves and beaches for potential monk seal re-colonization of the Black Sea coast.

Legal Framework Concerning MPAs in the Turkish Part of the Black Sea

There are several laws on the protection of coastal areas, environment, natural resources, national parks, and natural and cultural values. However, there is no appropriate legal mechanism for establishing MPAs and this constitutes an obstacle for their designation in Turkey. Moreover, even existing measures for protection of the marine environment or biodiversity are very weak and poorly enforced in terms of imposing fines or penalties. This is also another impediment for the conservation of the marine environment and nature protection as a whole. The legal instruments most relevant for MPAs are summarized below.

The purpose of the Coastal Law (number: 3621/3830, date: 1990/1992) is stated in Article 1 as 'to set out the principles for protection of the sea, natural and artificial lakes and river coasts and the shore buffer zones, which are extensions of these places and are under their influence, by paying attention to their natural and cultural characteristics and for their utilization towards the public interest and access for the benefit of society'. The Law defines the 'coastline' as 'the line along which water touches the land at the coasts of seas, natural or artificial lakes and rivers, excluding the inundation periods'.

The Environmental Law (of 9 August 1983, amended on 4 June 1986 and 3 March 1988) is administered by the Ministry of Environment and Urbanization. It covers environmental issues in general.

The Fisheries Law (22 March 1971, amended 15 May 1986) regulates the protection, exploitation, production and control of living resources. The responsible authority is the Ministry of Food, Agriculture and Husbandry (Nurlu and Erdem, 2002). It prohibits fishing certain species in certain areas, but does not designate particular protected areas. There are also the National Parks Law (9 August 1983); Law on the Protection of Cultural and Natural Wealth (21 July 1983); Council of Ministers' Decree (19 October 1989) for the establishment of an Agency for Specially Protected Areas (which is the legal base for special protected areas but not MPAs); the Coast Guard Security Force Law (9 July 1982); the Forestry Law (31 August 1956, amended 23 September 1983); the Law for the Protection of Cultural and Natural Values (Code No: 2863 of 1983); the Environmental Law (Code No: 2872 of 1983); and the National Parks Law (Code No: 2873 of 1993).

However, due to the lack of an appropriate law related to MPAs, we propose here to establish a specific law for the establishment and management of MPAs, independent from other laws.

Socio-economic Benefits of MPAs in the Turkish Part of the Black Sea

The Turkish part of the Black Sea coast is an area where a large number of human activities take place and several conflicts of interests exist between local people, fishermen, tourism operators, farmers and forestry. For local people, the coast is the area where they come into contact with the sea. One type of economic use of the coastal zone quite often denies opportunities for other activities: the construction of coastal highways limits the development of coastal tourism and wildlife reserves. The construction of hotels on the beach and in the immediate vicinity of the shore for tourism puts a burden of waste from human activities on the environment, and the quality of the beach and the coastal waters deteriorate even though wastewater discharge is forbidden by the Coastal Law.

The benefits of MPAs are generally accepted as natural capital for all stakeholders, but in particular for fishermen and the tourism industry. Tourism development is especially important for the Black Sea region where the most popular tourist destinations are the coastal areas, protected areas and historical settlements. If tourism is not sustainable, socio-economic and environmental problems will develop and pose extra stress for both coastal and marine environments. Accordingly, it is necessary to determine the carrying capacity and limiting factors for sustainable tourism. In recent years, coastal areas such as Şile have been subject to mass tourism, large-scale construction and infrastructure expansion, intensive land development and extensive urbanization, which have caused episodes of intense land-based and marine pollution during a very short period in summer.

However, tourism in protected areas is associated with appreciating and observing nature, scientific endeavour and education. This type of tourism is called ecotourism and associated with minimal development of infrastructure and small-scale interventions. Therefore, this kind of tourism is promising for the Black Sea, which is already facing many anthropogenic threats. Sand dunes, long coasts, reefs and caves can be attractive for ecotourism, as also are bird and dolphin watching. Local fishermen can also benefit from this development in tourism as the demand for fish increases when more tourists visit the area. Moreover, fishermen can rent out their boats for extra income when tourists wish to swim or snorkel and visit MPAs close by. The designation of an MPA in the Şile area would help in creating a plan for the sustainable use of natural resources, provide more income for fishermen, attract ecotourism investors and help raise the environmental consciousness of local people. Furthermore, within MPAs, control and surveillance measures for illegal fishing practices are generally more strict and this is an advantage for local fishermen, although their fishing grounds may be limited spatially. Nevertheless, in the long term, the benefits of MPAs for nature and all stakeholders are obvious.

Conclusion

The Turkish government should act to designate MPAs in the Black Sea before it is too late: most of the fish resources have already diminished since the mid-1970s (Kutaygil and Bilecek, 1976). Designation of transnational marine and coastal protected areas around the borders of Turkey with Bulgaria and Georgia would help to develop integrated protection measures in the entire southern portion of the Black Sea (indeed, Bulgaria already has an MPA at Strandja, close to the Turkish border). Designation of MPAs would also contribute to securing the biological corridor of the Istanbul Strait between the Sea of Marmara and Black Sea. The modelling study described above showed clearly that of all the proposed MPAs, the sites at Şile and Doğanyurt are the two areas that are most beneficial for establishing MPAs, because they have a high inflow of pelagic larvae from upstream areas and themselves ensure a high transport downstream to other areas. It may even be beneficial to establish another MPA along the western part of the south coast between Kızılırmak and Mezgit Reef to achieve a well-connected network of MPAs, assuring the exchange of pelagic larvae necessary for population persistence in the MPA network as detailed in Moffitt *et al.* (2011). Marine Protected Areas can also help to establish sustainable fisheries, rather than simply reducing the damage of the local fisheries or traditional fishing practices in the Black Sea, as they protect nursery grounds of many commercial fish species. Besides, poor fishing practices are more strictly controlled in MPAs, so that the fishermen who conduct 'legal' fishing activities in the region will be better protected (Öztürk, 2013).

According to the International Maritime Organization's MARPOL Convention, the Black Sea is designated as a special area because of its oceanographical and ecological conditions, and its level of sea traffic. Special areas require the adoption of mandatory methods by the relevant authorities for the prevention of marine pollution. In this regard, Uysal *et al.* (2002) reported that the Şile area has shown some signs of pollution and its benthic community is characterized by notable species enrichment. The Şile area is the only transition zone in the Black Sea under the influence of Mediterranean water due to its close geographical connection with the Sea of Marmara and Istanbul Strait, hence it needs special attention in terms of protection of marine biodiversity. The Sea of Marmara has a connection with the Black Sea in terms of maintaining some populations for breeding, over-wintering and/or migration, but it is not considered within the geographical scope of this chapter. For the designation of MPAs in the Black Sea, however, the Sea of Marmara should also be taken into account.

The European Union's goal of achieving Good Environmental Status (GES) in its seas by 2020 in accordance with the Marine Strategy Framework Directive (MSFD, 2008/56/EC) should be considered in parallel with Turkish initiatives for protecting the marine environment in the Black Sea, especially for the five proposed MPAs.

Turkey is a party of the CBD and one of the recent strategic goals is to improve the status of biodiversity by safeguarding ecosystems, species and genetic diversity. The CBD has set Aichi Targets in which by 2020 at least 17% of terrestrial and inland waters, and 10% of coastal and marine areas – especially those of particular importance for biodiversity and ecosystem services – should be conserved through effectively and equitably managed, ecologically representative and well-connected systems of protected areas and other effective area-based conservation measures, and integrated into the wider landscapes and seascapes. To reach this 2020 target, Turkey needs more MPAs, covering all Turkish waters, particularly in the Black Sea.

Finally, a robust, ecologically coherent network of MPAs in the Turkish part of the Black Sea as a whole will both contribute to, and depend on, the achievement of other conservation objectives concerning pollution reduction, sustainable fisheries management, improvement of legislation and enforcement, and capacity building as set out in the updated Black Sea Transboundary Diagnostic Analysis and Strategic Action Plan (BSC, 2007, 2009).

References

Akbulut, B., Zengin, M., Çiftçi, Y. *et al.* (2011) Stimulating sturgeon conservation and rehabilitation measures in Turkey: an overview on major projects (2006–2009). *Journal of Applied Ichthyology*, **27**, 415–419.

Allen, I., Arkin, S., Cossarini, G. *et al.* (2013) *Report on Reanalysis Hindcast Skill. Project Report for OPerational ECology: Ecosystem forecast products to enhance marine GMES applications.* EU Seventh Framework Programme. DG SPACE. 66 pp.

Anonymous (2010) http://www.iucnredlist.org/apps/redlist. Downloaded 23 August 2010.

Birkun, A. (2012) *Tursiops truncatus* ssp. *ponticus. The IUCN Red List of Threatened Species 2012.* e.T133714A17771698. http://dx.doi.org/ 10.2305/IUCN.UK.2012.RLTS.T133714A17771698.en. Downloaded 27 September 2015.

Birkun Jr., A.A. (2008) *Delphinus delphis ssp. ponticus. The IUCN Red List of Threatened Species 2008.* e.T133729A3875256. http://dx.doi.org/ 10.2305/IUCN.UK.2008.RLTS.T133729A3875256.en. Downloaded 27 September 2015.

Birkun Jr., A.A. and Frantzis, A. (2008) *Phocoena phocoena* ssp. *relicta. The IUCN Red List of Threatened Species 2008.* e.T17030A6737111. http://dx.doi.org/ 10.2305/IUCN.UK.2008.RLTS.T17030A6737111.en. Downloaded 27 September 2015.

Coll, M., Piroddi, C., Albouy, C. *et al.* (2012) The Mediterranean Sea under siege: spatial overlap between marine biodiversity, cumulative threats and marine reserves. *Global Ecology and Biogeography*, **21**, 465–480.

Commission on the Protection of the Black Sea Against Pollution (BSC) (2007) *Black Sea Transboundary Diagnostic Analysis and Strategic Action Plan* (TDA). Istanbul. 269 pp. http://www.blacksea-commission.org/_tda2008.asp

Commission on the Protection of the Black Sea Against Pollution (BSC) (2009) *Implementation of the Strategic Action Plan* for the Rehabilitation and Protection of the Black Sea (2002–2007)*. Istanbul. 252 pp.

Convention on Biological Diversity (CBD) (2008) *COP Decision IX/20 Marine and Coastal Biodiversity*. https://www.cbd.int/decision/cop/default.shtml?id=11663

Cowen, R.K., Paris, C.B. and Srinivasan, A. (2006) Scaling of connectivity in marine populations. *Science*, **311**, 522–527.

Demirkesen, A.C., Evrendilek, F. and Berberoğlu, S. (2008) Quantifying coastal inundation vulnerability of Turkey to sea-level rise. *Environmental Monitoring and Assessment*, **138** (1–3), 101–106.

Eremeev, V.N., Vladomiriov, V.L. and Krasheninnikov, B.N. (1992) Long-term variability of the Black Sea water transparency, in *Hydrophysical and Hydrochemical Studies of the Black Sea*. Marine Hydrophysical Institute the National Academy of Science of Ukraine, Sevastopol. pp. 28–30.

Fach, B.A. (2014) Modeling the influence of hydrodynamic processes on anchovy distribution and connectivity in the Black Sea. *Turkish Journal of Fisheries and Aquatic Sciences*, **14**, 353–365. doi:10.4194/1303-2712-v14_2_06

General Fisheries Commission for the Mediterranean Working Group on the Black Sea (WGBS-GFCM) (2015) *Report of the Fourth Meeting of the Ad hoc Working Group on the Black Sea, Tbilisi, Georgia, 9–11 March 2015*. http://www.gfcmonline.org/reports/technical

Guidetti, P., Notarbartolo-di-Sciara, G. and Agardy, T. (2013) Integrating pelagic and coastal MPAs into large-scale ecosystem-wide management. *Aquatic Conservation in Marine and Freshwater Ecosystems*, **23**, 179–182.

Kideys, A.E. (2002) Fall and rise of the Black Sea ecosystem. *Science*, **297**, 1482–1484.

Kutaygil, N. and Bilecik, N. (1976) Observations sur les principaux produits demersaux qui sont peches sur les cotes Turques de la mer Noire. *Rapport Commission Internationale pour l'Exploration Scientifique de la Méditerranée*, **23**, 75–77.

Lester, S.E., Halpern, B.S., Grorud-Colvert, K. et al. (2009) Biological effects within no-take marine reserves: a global synthesis. *Marine Ecology Progress Series*, **384** (2), 33–46.

Mater, S. and Cihangir, B. (1990) An investigation on the fish egg–larva distribution in the Black Sea entrance of the Bosphorus, in *Xth Ulusal Bioloji Kongresi 18–20 July 1990, Erzurum*. pp. 209–215. (In Turkish)

Mee, L.D. (1992) The Black Sea in crisis: a need for concerted international action. *Ambio*, **21** (4), 278–286.

Moffitt, E.A., Botsford, L.W., Kaplan, D.M. and O'Farrell, M.R. (2009) Marine reserve networks for species that move within a home range. *Ecological Applications*, **19**, 1835–1847.

Moffitt, E.A., White, J.W. and Botsford, L.W. (2011) The utility and limitations of size and spacing guidelines for designing Marine Protected Area networks. *Biological Conservation*, **144**, 306–318.

Nurlu, E. and Erdem, U. (2002) An overview of the Turkish coastal zone management within the framework of the European Union, in *Coastal Zone Management in the Mediterranean Region* (eds D. Camarda and L. Grassini). CIHEAM, Bari. pp. 107–111.

Oguz, T. and Besiktepe, S. (1999) Observations on the Rim Current structure, CIW formation and transport in the western Black Sea. *Deep-Sea Research I*, **46**, 1733–1753.

Oguz, T. and Velikova, V. (2010) Abrupt transition of the northwestern Black Sea shelf ecosystem from a eutrophic to an alternative pristine state. *Marine Ecology Progress Series*, **405**, 231–242.

Oguz, T., La Violette, P. and Unluata, U. (1992) Upper layer circulation of the southern Black Sea: its variability as inferred from hydrographic and satellite observations. *Journal of Geophysical Research*, **97** (C8), 12569–12584.

Oguz, T., Latun, V.S., Latif, M.A. *et al.* (1993) Circulation in the surface and intermediate layers of the Black Sea. *Deep-Sea Research I*, **40**, 1597–1612.

Oguz, T., Deshpande, A.G. and Malanotte-Rizzoli, P. (2002) The role of mesoscale processes controlling biological variability in the Black Sea coastal waters: inferences from SeaWIFS-derived surface chlorophyll field. *Continental Shelf Research*, **22**, 1477–1492.

Oguz, T., Tugrul, S., Kideys, A.E. *et al.* (2005) Physical and biogeochemical characteristics of the Black Sea. *The Sea*, **14** (33), 1331–1369.

Öztürk, B. (2013) Some remarks of illegal, unreported and unregulated (IUU) fishing in Turkish part of the Black Sea. *Journal of Black Sea/Mediterranean Environment*, **19** (2), 256–267.

Öztürk, B. and Öztürk, A.A. (2005) Biodiversity in the Black Sea: threats and the future, in *Mankind and Oceans* (eds N. Miyazaki, Z. Adel and K. Ohwada). UN University Press, Tokyo. pp. 155–172.

Öztürk, B., Topaloğlu, B., Kideys, A. *et al.* (2013) A proposal for new marine protected areas along the Turkish Black Sea coast. *Journal of Black Sea/Mediterranean Environment*, **19** (3), 365–379.

Sapozhnikov, V.V. (1991) Biohydrochemical barrier along the border of shelf waters of the Black Sea. *Oceanology*, **31** (4), 417–423.

Shiganova, T.A. (1998) Invasion of the Black Sea by the ctenophore *Mnemiopsis leidyi* and recent changes in pelagic community structure. *Fisheries Oceanography*, **7**, 305–310.

Shiganova, T.A., Mirzoyan, X.A., Studenikina, E.A. *et al.* (2001) Population development of the invader ctenophore *Mnemiopsis leidyi* in the Black Sea and in other seas of the Mediterranean basin. *Journal of Marine Biology*, **139**, 431–445.

Spalding, M.D., Fox, H.E., Allen, G.R. *et al.* (2007) Marine Ecoregions of the World: a bioregionalization of coastal and shelf areas. *BioScience*, **57** (7), 573–583.

Sur, H.I., Ozsoy, E., Ilyin, Y.P. and Unlata, U. (1996) Coastal/deep ocean interactions in the Black Sea and their ecological/environmental impacts. *Journal of Marine Systems*, **7**, 293–320.

Tonay, A.M. and Öztürk, B. (2003) Cetacean bycatches in turbot fishery on the western coast of the Turkish Black Sea, in *International Symposium of Fisheries and Zoology* (eds I.K. Oray, M.S. Çelikkale and G. Özdemir). Istanbul. pp. 131–138.

Topçu, E.N. and Öztürk, B. (2012) Abundance and composition of solid waste materials on the western part of the Turkish Black Sea seabed. *Aquatic Ecosystem Health and Management*, **13** (3), 301–306.

Uysal, A., Yüksek, A., Okuş, E. and Yılmaz, N. (2002) Benthic community structure of the Bosphorus and surrounding area. *Water Science and Technology*, **46** (8), 37–44.

Zaitsev, Y. (1991) Cultural eutrophication of the Black Sea and other South European Seas. *La Mer*, **29**, 1–7.

Zaitsev, Y. (1993) *Impact of Eutrophication on the Black Sea Fauna. Fisheries and Environment Studies in the Black Sea System. Part 2. Studies and Reviews.* General Fisheries Council for the Mediterranean No. 64. FAO, Rome. pp. 63–86.

Zaitsev, Y. and Öztürk, B. (2001) *Exotic Species in the Aegean, Marmara, Black, Azov and Caspian Seas.* Turkish Marine Research Foundation Publication No. 8. Istanbul, Turkey. 263 pp.

14

Marine Protected Areas and Offshore Wind Farms

Natalie Sanders, Thomas Haynes and Paul D. Goriup

NatureBureau, Newbury, UK

Introduction

European Union policy makes it clear that we need two things: one, a move towards a more sustainable source of energy from green developments, such as offshore wind farms (OWFs), in line with the EU Energy Directive (2009/28/EC); and two, an increase in Marine Protected Areas (MPAs) (EEA, 2015). The first is to safeguard our future energy demands and to mitigate the impacts of climate change. The second is to safeguard our marine biodiversity to ensure we have sustainable seas that are both diverse and support human needs. However, with a total extent of $338\,000\,km^2$, less than 6% of EU water was designated as an MPA by the end of 2012 (EEA, 2015). The Greater North Sea has the greatest proportion of the MPA network (17.9%) and it is also leading the way in OWF developments with large OWFs such as the Dogger Bank, Thanet and London Array.

In February 2016, the world's largest OWF was consented, Hornsey Project One (DONG Energy), which will start generating energy in 2019 and will consist of 172 turbines, producing 1.2 gigawatts of energy (DONG Energy, 2016). With more than 80 countries developing wind energy, there is enough installed wind power capacity worldwide to meet the residential needs of 380 million people at the European level of consumption (Wilkes *et al.*, 2014). The vast majority of wind turbines operating today are on land, but offshore wind developments are quickly increasing in number and size. Since 2006, offshore wind generating capacity has grown at a dramatic rate and the European Wind Energy Association expects this trend will continue, with Europe's offshore generating capacity reaching 150 GW by 2030 (Wilkes *et al.*, 2014). This amounts to 14% of the projected EU electricity demand and will avoid emitting 315 million tonnes of CO_2 per year. The EU already has an ambitious climate and energy target known as the '20-20-20 target', whereby it aims to reduce carbon emissions by 20%, produce 20% of energy from renewables and improve energy efficiency by 20%, all by the year 2020. But this is to be increased to have 30% of total energy generated to come from renewable sources by 2030 (EC, 2014). However, the most recent Intergovernmental Panel on Climate Change report (IPCC, 2014) states that climate change is a very real threat, and if we are to avoid the worst effects of it, we need to further reduce our dependency on coal and oil and at least

Management of Marine Protected Areas: A Network Perspective, First Edition. Edited by Paul D. Goriup.
© 2017 John Wiley & Sons Ltd. Published 2017 by John Wiley & Sons Ltd.

treble our production of energy from clean, renewable sources. Therefore, the need to construct offshore wind farms has never been so urgent.

The benefits of OWFs are numerous. With stronger and more frequent winds offshore than onshore, they generate more power per turbine than their onshore counterparts. The visual impacts are reduced the further out to sea the turbines are located, and there is less congestion on land in the construction phase with most transport being carried out at sea. Therefore, despite their high cost to construct, EU countries are keen to invest in OWFs. At present, the North Sea, Baltic Sea and to a lesser extent the Atlantic Ocean are the only European seas where OWFs are operating or under construction. Some offshore wind farm projects are planned in the Mediterranean, but with only a small proportion of them having reached consent stage. Given that OWFs often result in the reduction or exclusion of fishing due to risk of fishing gear entanglement (Inger *et al.*, 2009), it may seem intuitive that OWFs and MPAs go hand in hand. However, the situation actually turns out to be more complicated.

Marine Protected Areas to date have usually been assigned to areas that are known to be either important for charismatic biodiversity, or because a particularly important species or habitat is present. Although there is no legally binding definition of an MPA, the International Union for Conservation of Nature (IUCN, 1994) defines an MPA as

> a clearly defined geographical space, recognised, dedicated and managed, through legal or other effective means, to achieve the long-term conservation of nature with associated ecosystem services and cultural values.

According to this definition, the purpose of an MPA is to minimize any potential risks or threats to these key nature conservation features and prevent exploitation (including fisheries) and damaging activities. The construction of an OWF is in itself a damaging process that will impact the benthic environment and water column, and potentially affect a vast array of organisms. However, once operational, the impacts are reduced and in some cases OWFs have even been known to enhance biodiversity (Lindeboom *et al.*, 2011) since due to the absence of fishing, they can become quasi-MPAs.

There are cases where proposed OWFs are located very close to important existing MPAs. In the UK, the Wash and North Norfolk coastline has many important habitats and species, including mudflats, sandbanks and reefs (listed in Annex 1 of the Habitats Directive, 92/43/EEC), many important migrating bird species, the harbour porpoise *Phocoena phocoena* and the common seal *Phoca vitulina*, and has been designated as a Site of Community Importance (North Ridge SCI). This area is also the site of multiple OWF developments, including the Lincs, Lynn and Inner Dowsing, Racebank and soon-to-be Hornsea Project One. Further north in the UK, in the Firth of Forth in Scotland, there are also OWF proposals – the Neart na Gaoithe OWF and the 700-turbine Firth of Forth development – yet the area is important for many species and also has a Special Protection Area (Birds Directive, 2009/147/EC) for birds. With more and more OWFs being planned in areas important for biodiversity it is important to understand how the two interests interact.

In order to establish whether OWFs and MPAs can in fact ever co-exist, this chapter will discuss:

- The potential negative impacts that an OWF can have on marine biodiversity
- The ways an OWF can increase species abundance, richness and biodiversity
- The future prospects for OWF development that may reduce the initial impact, thereby enhancing their co-existence.

This chapter will also consider examples where OWFs and MPAs exist in the same place as well as examples of where the existence of an MPA has prohibited or stalled the construction of an OWF.

Offshore Wind Farms

European legislation necessitates a detailed understanding of the impacts of large developments on the marine and coastal environments. National planning projects, such as wind energy plans, are subject to an initial Strategic Environmental Assessment (SEA) under EU Directive 2001/42/EC. At this stage, governments are required to assess how a project may affect biodiversity on a wide scale. The cumulative impacts of projects and developments are assessed and inappropriate areas and plans are excluded. Large projects, such as OWFs, are subject to Environmental Impact Assessments (EIAs) under EU Directive 85/337/EEC. Such EIAs are applied to public and private projects and the individual and cumulative effects on the immediate and surrounding environment are assessed. This includes, amongst other things, assessments of the impacts on biodiversity, hydrology and the seabed. Under the Habitats Directive (92/43/EEC), projects that may compromise the integrity of Natura 2000 sites are required to undergo further appropriate assessments. If it cannot be concluded that the site(s) will remain uncompromised by the development, a licence will not be granted unless, in exceptional circumstances, there is deemed to be an overriding public interest. Therefore if an MPA is already established, there are additional steps and consideration that need to be taken in order to gain approval for the construction of an OWF. However, this does not mean that OWFs can never be constructed in or near to an existing MPA.

Offshore wind farms generally occur in water depths of 60 m or less (Simmonds and Dolman, 2008) and most have been located up to 5 km out to sea (Simmonds and Dolman, 2008). Further out to sea, the winds are stronger and so more energy can be generated. However, due to the requirement of more advanced technology, the costs are higher. Recent developments have seen wind farms placed in deeper waters and further out to sea, such as the Thanet wind farm which is 12 km off the coast of Kent in the UK (Vattenfall, 2013) and the London Array which is 20 km off the Kent and Essex coasts. As technology develops and wind farms move further offshore into deeper waters, there is the potential for different habitats and wildlife populations to be affected. The transmission of noise, one of the main impacts generated in construction and operation, is largely dependent on the foundations of the wind farms and also on the local seafloor bathymetry and oceanographic features. Therefore, each wind farm could transmit noise in a unique way, which is why it is essential to undertake an EIA for each proposed OWF. It is important to have adequate baseline studies to understand both the impacts and the mitigation strategies which can be put in place to prevent or limit environmental damage (Evans, 2008).

There are three main stages in the life of an OWF: construction, operation and decommissioning. All three are known to have some impacts on the marine environment although it is generally considered that the construction phase is the most damaging to marine life (Inger *et al.*, 2009). Offshore wind farms are a relatively new technology and therefore knowledge of their long-term effects on the marine environment is in its infancy (Simmonds and Dolman, 2008), but post-construction monitoring that has been done to date is providing important information on the ability of marine life to recover. The three main stages of an OWF and their associated risks to marine life will be briefly set out below (summarized in Table 14.1) and then discussed in greater detail for each biological group of concern.

Table 14.1 A summary of the main effects on marine life during the construction and operational stages of OWFs.[a]

Stage	Biological group	Effects	Response
Construction	Benthic organisms	Decreased food supply due to increased turbidity and reduced light penetration limiting vegetation growth	Negative
		Smothering of organisms due to dredging resulting in sediment re-suspension	Negative
		Impacts of re-suspended pollutants in sediment	Negative
		Displacement and initial loss of habitat due to cable laying and cable protection (mattressing, boulders)	Negative
		Loss of habitat in the immediate area where turbine bases are placed	Negative
	Fish communities	Smothering of eggs affecting recruitment due to re-suspended sediments and re-suspended pollutants	Negative
		Possible effects for magneto-sensitive fish and elasmobranchs due to EMF produced from cables (likely to be negligible)	Negative
		Avoidance and displacement from OWF area due to noise and vibrations from pile-driving. Potential for physical injury	Negative
	Marine mammals	Ship strikes and collisions (physical damage and mortality) due to increased construction vessels on water	Negative
		Masking effects (interruption of communication affecting group cohesion) due to noise from pile-driving and other construction activities	Negative
		Behaviour changes (e.g. orientation to sound, cessation of feeding or social interaction) due to increased noise from pile-driving and other construction activities	Negative
		Behaviour changes (attraction to OWF zone due to increased food supply)	Positive
		Behaviour changes (displacement and habitat abandonment due to intense pile-driving)	Negative
		Hearing loss and injury (TTS, PTS) due to intense pile-driving	Negative
	Birds	Displacement from construction zone due to the presence of construction vessels deterring birds from the area	Negative
Operational	Benthic	Increase in colonization rates due to new substrates with turbine bases and cable protection structures acting as artificial habitats	Positive
		Shifting sediments/modified habitat due to altered hydrological conditions around the bases	Negative
	Fish	Increase in abundance (and potentially diversity) due to reduced fishing effort in the OWF area	Positive
		Increased abundance or higher production of fish with increased structures acting as Fish Aggregating Devices (artificial reef effect)	Positive
	Mammals	Behaviour changes (decreased energy expenditure when hunting by using buried cables for direction between turbines and increased prey)	Positive
		Behaviour changes (displacement and habitat abandonment)	Negative
	Birds	Disturbance (time/energy spent avoiding structures)	Negative
		Disturbance (altered flight paths)	Negative
		Displacement from and avoidance of OWF zone due to visual impact, noise and vibrations from turbines and maintenance traffic	Negative
		Displacement due to habitat change	Negative
		Increase in availability of prey fish species	Positive
		Collisions with rotating blades	Negative

a) The decommissioning stage has been omitted due to lack of information.

Construction Stage

The noise generated during construction is periodic but can have relatively severe adverse effects (Rye *et al.*, 2008). Pile-driving is the process of inserting the supporting pile into the foundation structure, which can result in extremely high noise levels which may be damaging to marine life (Prior and McMath, 2008; Simmonds and Dolman, 2008). Each pile can take between two and three hours to drive (Evans, 2008) with several piles driven per day during construction. Due to limited weather windows, construction may only occur in certain months of the year and may be further limited to avoid key biological processes. In addition to the noise generated during pile-driving, there will be increased boat traffic and associated noise, due to the large number of cargo ships that are required to transport the turbines to the offshore location (Wilson *et al.*, 2007). Large areas of seabed are also disturbed in the construction phase in order to lay and bury all the cables that will transmit the electricity from the turbines to the landfall site and converter station. This will increase sediment loads in the water column which can impact pelagic species and also displace or smother benthic species. Therefore the construction phase is considered to be the most important aspect when considering the impacts on marine life.

Operational Stage

Once operational, the movement of the turbine blades produces a low frequency noise which can result in some marine organisms avoiding the area either temporarily before habituation to the noise or permanently (Prior and McMath, 2008; Simmonds and Dolman, 2008). As turbines get larger the noise generated increases, having the potential to create disturbance effects over several kilometres (Prior and McMath, 2008). During the operational phase, there will be a small amount of increased boat traffic due to routine inspections and maintenance trips, as well as on-going EIA inspections taking place on small research vessels. However, it is likely this will be offset by reduced numbers of fishing vessels in the area around the wind farm. Electromagnetic fields (EMF) produced by the buried cables that carry the generated electricity can interfere with some sensitive species such as elasmobranches (sharks and rays) and migratory species that use the Earth's magnetic field for orientation (Gill and Bartlett, 2010). The rotating blades can have a detrimental impact on some birds that can either collide with the blades resulting in mortality or are displaced from the area due to avoidance behaviour (Furness *et al.*, 2013).

Decommissioning Stage

At the end of their commercial life, the turbines need to be dismantled and returned to land for disposal. Removal of bases can involve activities which generate a lot of noise, such as cutting, drilling and in extreme cases use of explosives (Prior and McMath, 2008). Explosives have the potential to be very damaging for any cetaceans that might be in the local vicinity as well as to benthic species and epifauna that are growing on or around the turbine pillars. Similarly to the construction phase, there will be increased boat traffic to the area due to cargo ships returning the dismantled turbines to the shore. Only one OWF in the world, consisting of only five turbines, has so far been decommissioned and that is the Yttre Stengrund in Kalmar Sound, Sweden (Vattenfall, 2016). The OWF has had the turbine bases cut back to the level of the seabed but the actual bases and cables are still in place. These were due to be removed in the summer of 2016. Therefore the full impacts of decommissioning are not yet understood but studies from this site will be invaluable in understanding how long it takes for marine life to return to pre-OWF levels.

Potential Impacts of OWFs on Marine Life

Impacts on Benthos

It is inevitable that the construction of an OWF will have an impact on the benthic community due to artificial structures being placed on and in the seafloor. However, although the overall size of the wind farm might be large, the actual area of seafloor which is disturbed is relatively small. The actual size will depend on the type of turbine foundation used, with gravity bases affecting the largest area. The increased turbidity of the water during construction can also decrease the light penetration and therefore have a negative impact on vegetation. These negative impacts are thought to be slight and short-lived (Boesen and Kjaer, 2005). Benthic communities are complex, but can be considered within two broad groups. Infauna are the organisms such as worms and clams that live in the sediment, and epifauna are those, such as oysters and mussels, that live on the surface and attach to a hard substrate.

Construction can result in noise and vibrations from pile-driving that can disturb the seafloor and the associated organisms. During the construction phrase, dredging of sediments can re-suspend pollutants into the water column and displace the benthic communities that were living in the dredged area. There are large amounts of cables associated with OWFs in order to carry the electricity generated from the turbines to an onshore converter station. These cables need to be buried within the substrate to minimize the impacts of the EMF, reduce snagging (e.g. on fishing nets or anchors) and also to ensure the cables stay in place and are not dragged around by currents. Mattressing is the process of using concrete slabs along the cable route to prevent cables from being exposed. Other methods include the use of rock boulders being placed over the cables. In both cases, this means that a large amount of the area surrounding the OWF will be altered, from the turbine bases, between turbines and then back to the landfall site. Sometimes the cable route may pass through several different habitats and therefore cable protection can impact a large number of species. Consideration should be given to vulnerable species such as the Ross worm *Sabellaria spinulosa* not just in the OWF site but also along the cable route back to the landfall site. Conversely, the turbine bases and the boulders/mattresses used in cable protection can act as additional substrate adding habitat complexity in what could otherwise be a barren muddy seafloor. This therefore increases the heterogeneity of the environment in and around the OWF (Lindeboom *et al.*, 2011). Consequently, species diversity, abundance, community structure as well as functional properties such as nutrient cycling and bioturbation can be altered (Coates *et al.*, 2013). Whilst colonization of turbine bases can increase local biodiversity, this is often not considered beneficial by developers and so the turbine bases are frequently coated with anti-fouling paint to prevent such colonization. After construction, there may be altered hydrographic conditions in and around the pile bases which can also impact the benthic communities. Therefore, it is likely that there will be some degree of alteration and degradation of the benthic environment, especially during the construction period, and the actual footprint of the turbine bases will be permanently altered.

Surveys from pre- and post- construction in the UK have shown that overall benthic communities do recover from the construction of an OWF and there are minimal long-lasting impacts (Boesen and Kjaer, 2005; Lindeboom *et al.*, 2011). Moreover, changes in the benthic community may be different for each component, with a decrease in the infauna and an increase in the epifauna in some cases. This results in high food abundance for other species such as fish, which can lead to greater biodiversity (Köller *et al.*, 2006). However, the epibenthic communities that form can be different to the native biota and have different patterns of succession (Boesen and Kjaer, 2005). At the Egmond aan Zee OWF in the Netherlands, the overall biodiversity of benthic communities increased in the area after construction (Lindeboom *et al.*, 2011). In contrast, the benthic communities at the North Hoyle wind farm (off the coast of Wales, UK), which has also been designated as an MPA, had still not returned to pre-construction levels after 12 months, demonstrating that even at sites that are fully protected, time is required for recovery (May, 2005). However, whilst research to date suggests that there are minimal long-term impacts on benthic communities, the impacts will be different for each site and so monitoring must occur with each OWF development.

Impacts on Fish

As with the benthic environment, cable-laying and pile-driving can have negative impacts on the fish population inhabiting the water column in and around an OWF. Cable-laying will disturb and redistribute sediments and may re-suspend pollutants that were previously buried. Re-suspended sediments may smother fish eggs which can lead to reduced survival and recruitment, especially if construction occurs near fish spawning grounds (Gill, 2005). It is possible that the EMF generated by electric currents in the submarine cables could be detrimental to fish, in particular for species that are magneto-sensitive and use geomagnetic fields for information. There are two groups that might therefore be affected by the EMF: those that use geomagnetic information for locating food such as elasmobranchs (Collin and Whitehead, 2004) and fish species which migrate long distances, such as the European eel *Anguilla anguilla* and the

Atlantic salmon *Salmo salar* (Boehlert and Gill, 2010). Several field studies have observed responses of fish to cables (Marra, 1989; Gill and Taylor, 2001); however, none have yet shown any effect at the population level. The EMF can be reduced by burying the cable (Gill *et al.*, 2009) or by sheathing it (CMACS, 2003).

During the construction phase, the noise from pile-driving can be detected by fish from a considerable distance and can result in fish leaving or deter them from entering the zone of construction. However, this is a behavioural response and not due to physiological damage (Wahlberg and Westerberg, 2005). Avoidance of the zone can have a significant impact for sensitive species such as Atlantic herring *Clupea harengus* and Atlantic cod *Gadus morhua* which can detect pile-driving as far as 80 km away (Thomsen *et al.*, 2006). Along with behavioural effects such as avoiding the area during pile-driving, evidence suggests that some species may be susceptible to physical damage (internal and external injuries) and if in close proximity, pile-driving can even result in mortality (Thomsen *et al.*, 2006). Even after construction, the noise generated by the rotating blades can be detected by herring and cod up to 4 km away, and dab *Limanda limanda* and Atlantic salmon can detect it up to 1 km away (Thomsen *et al.*, 2006). Some studies do indicate that once operational, the noise from the rotating blades can be detected and fish may avoid the turbine bases themselves by up to 4 m, but it is unlikely that they suffer any destructive hearing effects during the operational stage of an OWF (Inger *et al.*, 2009).

Whilst noise during construction may result in fish leaving and avoiding the OWF area, studies show that the abundance of both pelagic and demersal fish actually increases during the operational phase (Lange *et al.*, 2010). However, no impacts on diversity were reported, suggesting that fish that already inhabit that area will increase in number rather than more species being attracted to it (Lange *et al.*, 2010). Long-term monitoring at Burbo Bank OWF (UK) reported that fluctuations in fish abundance were unrelated to the OWF (CEFAS, 2009). But there have been reports that OWFs result in an increase in abundance of some fish species such as Atlantic cod and flounder *Platichthys flesus* due to improved food conditions after construction (Pedersen and Leonhard, 2006). Some fisheries can also benefit as fishing is generally prohibited within a wind farm area, therefore making it a de facto MPA.

Man-made structures such as oil rigs, and now OWFs, may serve as artificial reefs and actually attract fish to the area, a phenomenon known as the Artificial Reef Effector or a type of Fish Aggregating Device (FAD) (Jensen *et al.*, 1994; Inger *et al.*, 2009). However, it is not always clear if it is a case of attraction of fish to the turbines (i.e. the same number of fish just in a higher concentration around the bases or within the OWF area) or actual increased production with greater recruitment. At the Lillgrund wind farm in Sweden, fish have been shown to aggregate around the turbine bases (Bergström *et al.*, 2013). Fish Aggregating Devices have been used by fishermen for centuries to increase their catch efficiency (Inger *et al.*, 2009) but their role for conservation purposes remains unclear. Early research suggests that OWFs may act as artificial reefs and provide a safe, fishing-free zone which will ensure higher recruitment and survival rates. This in turn can increase the abundance of fish outside the OWF zone through a spillover effect (Garcia-Rubies *et al.*, this volume). Offshore wind farms may also increase connectivity between areas as each turbine is close enough for many organisms to move from one to the other which can help increase production (Krone *et al.*, 2013). However, production increases can only occur if fishing is prohibited within the wind farm

area, which is not always the case. If fishing is not prohibited (or not sufficiently managed), the effects of the OWF as a FAD may in fact have a negative impact on biodiversity and fisheries as it attracts fish into a concentrated area, increasing the chance of overfishing and then stock depletion. But if fishing is prohibited, Inger *et al.* (2009) believe that it is possible that the overall effects of turbines can be positive. Interestingly, all OWFs in the Belgian part of the North Sea are closed to fisheries and the debate of 'attraction vs. production' has been the subject of studies. Results of these studies indicate that there is actual production and therefore fish are not just being attracted to and aggregating around the turbines, but increasing in abundance (Reubens and Degraer, 2014). Therefore, OWFs may be more than just simple FADs, but actually help improve the local abundance of fish.

Impacts on Marine Mammals

Marine mammals can be affected by the construction of an offshore wind farm, primarily through the sound generated by pile-driving. Toothed marine mammals, odontocetes, rely on sound for hearing and echolocation to communicate and locate their prey. Any detrimental impacts on their ability to echolocate can have serious implications for their ability to inhabit a given area. There is a direct negative link between acoustic emissions and harbour porpoise survival, as these animals have very acute hearing (Lucke, 2008). Both the frequency and level of noise generated during pile-driving need to be considered. The communication signals of odontocetes fall mainly in the medium to high frequencies (1–20 kHz), with echolocation occurring at high and very high frequencies (20–150 kHz). On the other hand, baleen whales or mysticetes are primarily sensitive to low and medium frequencies (12 Hz – 8 kHz) which can leave them more vulnerable to low frequency pile-driving noise. While it is generally agreed that a noise level of 180 dB can cause irreversible damage (Southall *et al.*, 2007), this level is often exceeded during the construction phase, with pile-driving reaching 250 dB. Long-term studies of the effects of noise pollution from OWFs on marine mammals have so far focused on harbour porpoises (Simmonds and Dolman, 2008). Harbour porpoises have a wide hearing range, and therefore can be affected by sounds ranging from low frequencies (250 Hz) to the ultrasonic range (160 kHz) (Diederichs *et al.*, 2008). However, there is evidence of variable susceptibility to noise not just between different species, but between individuals of the same species (Thomsen *et al.*, 2006). This means the full effects of underwater noise from OWFs are unclear and unpredictable. In particular, the chronic effect of long-term exposure (Inger *et al.*, 2009) needs further investigation. Some of the known effects of OWF noise on marine mammals include masking effects, behaviour changes, temporary hearing loss and permanent tissue damage: these are discussed further below.

Masking Effects

Noise pollution resulting from any stage in an OWF life cycle can create a masking effect preventing cetaceans from detecting biologically relevant sound signals (Southall *et al.*, 2007; Inger *et al.*, 2009). This can result in marine mammals not being able to detect the presence of a school of fish or a predator, or prevent communication between individuals. Given that cetaceans are highly communicative and often feed in groups, masking effects interfering with communication between group members can be highly detrimental to their social structure and group survival. The effects of masking on marine mammals may differ greatly between species and within species and even age of individuals (Erbe *et al.*, 2014). However, masking effects are greatest

during the construction phase and do not appear to cause long-lasting effects.

Behaviour Changes

There are a wide range of unusual behaviours that can be seen during periods of high noise levels, such as orientation or attraction to the noise source, increased alertness, cessation of feeding or social interaction, habitat abandonment, panic and, in severe cases, stranding (Southall *et al.*, 2007). Some behaviour changes may be small and subtle; however, wind farm construction can lead to avoidance reactions resulting in entire populations of marine mammals being displaced from the wind farm vicinity. Therefore, careful consideration needs to be given in terms of wind farm location so as not to displace a resident population of marine mammals. Consideration must also be given to the location of alternative areas which are suitable for cetacean feeding grounds in case they are displaced by the noise and need to hunt elsewhere. Alternative food sources need to be close by so as not to result in a large-scale displacement. The disturbance from noise can reduce time spent foraging, resting or socializing and can also result in an increase in time spent travelling between good foraging sites. This can have negative energetic repercussions, affecting the overall fitness of a population if the disturbance is sustained for an extended period of time. During the construction phase, it has been shown that porpoises and seals display avoidance behaviour and leave the area of the OWF. However, in most studies, the mammals returned two years after completion to near baseline levels (Prior and McMath, 2008; Teilmann *et al.*, 2008). Therefore the construction phase may have an effect on a large number of individuals but for a limited time period (Teilmann *et al.*, 2008). The dependence on the local area also appears to affect the rate of return of mammals. For example, at the Horns Rev OWF (in the Wadden Sea area of the eastern North Sea) the area is of great importance to the marine mammal population and so porpoises have been shown to tolerate a certain degree of disturbance and returned to the site after the construction phase. On the other hand, the Nysted OWF (south-west Baltic) is not particularly important for the local porpoise population and so they simply avoided the area during construction (Teilmann *et al.*, 2008). At the Egmond aan Zee OWF in the Netherlands, marine mammals were sighted more often post-construction than pre-construction (Lindeboom *et al.*, 2011), potentially due to increased prey abundance in the OWF area.

Buried cables have also been shown to affect the behaviour of marine mammals. Both the grey seal *Halichoerus grypus* and the common seal have displayed the ability to use the EMF emitted from the cables to guide them between turbines. Using a systematic grid-like approach, seals at the Sheringham Shoal OWF in Norfolk moved from turbine to turbine foraging along the cables and turbine bases (Russell *et al.*, 2014). It is thought that by following the cables, the seals were able to take advantage of the increase in prey species aggregating around the turbines to maximize their foraging and minimize their energy expenditure.

Hearing Loss and Injury

Noise pollution can result in damage to auditory tissues and lesions in parenchymatous organs (Piantadosi and Thalmann, 2004). Acoustic damage can also cause nitrogen bubbles to form in the blood (known as the 'bends'), affect the vestibulocochlear nerves of the cochlea (Degollada *et al.*, 2003) and cause lesions. Lesions are most often reported in the lungs and tissues associated with the transmission of sound such as the jaws and inner ear. Given that marine mammals have a sensitive and wide hearing range, hearing damage can occur at relatively low noise exposure so that the effects of even low noise pollution generated

from an OWF can have a wide zone of impact. Damage to tissues resulting from pile-driving can lead to temporary threshold shift (TTS) and permanent threshold shift (PTS). Permanent threshold shift is considered to be an auditory injury (Lucke, 2008) caused when the noise source reaches a level that permanent auditory impairment is inflicted. Moreover, noise pollution can also affect non-auditory tissues. With hearing loss, marine mammals may not be able to find food, reach breeding and feeding grounds, locate mates or locate other individuals from within their group or maintain their balance (André *et al.*, 2003). This can have long-term impacts on the viability of the population if they are not able to breed successfully.

It has been reported that the impacts of increased noise from OWF construction are less severe in areas which are already subject to high ambient noise such as a high degree of boat traffic (Bergström *et al.*, 2014). This implies that marine mammals that are already habituated to high noise levels are less likely to be impacted by the construction or presence of an OWF. It is therefore important to consider the ambient noise levels in the planning stages. If the OWF is to be constructed in an area with low ambient noise levels, which is likely in or near MPAs, the construction of an OWF may have disproportionately negative impacts on any marine mammals that rely on that area.

Mitigation Measures for Mammals

Standard mitigation measures for marine mammals include the use of a soft-start and ramp up of drilling activities. This is where pile-driving starts at a lower energy to allow mammals time to move out of the zone before higher energies are used which could cause damage. However, the effectiveness of this measure is unclear. Other mitigation measures, recommended by the Marine Mammal Observer Association, include the use of Marine Mammal Observer and Passive Acoustic Monitoring (PAM) to detect mammals in the vicinity of the construction zone. Acoustic Mitigation Devices (AMDs) emit a frightening or adverse sound to encourage marine mammals to move out of the construction zone (SMRU Ltd, 2007). New techniques such as Big Bubble Curtains (BBC) are being tested: the system creates a shield of bubbles around the pile-driving device to reduce the distance harmful noise can travel, which may prove to protect marine mammals more and reduce the area from which they are displaced (Reyff, 2009).

Impacts on Birds

Conducting an EIA for seabirds is problematic as very little is known about their behaviour and it is more difficult to conduct surveys for birds at sea than it is on land. Impacts on birds involve disturbance, displacement and collision, which can result from a single OWF (placed near a breeding colony or in a wintering area) or from the cumulative risk of encountering several wind farms over a wide geographic area as the birds forage or migrate along the coast.

Disturbance

Disturbance to birds caused by the presence of OWFs has been defined as birds expending extra time and/or energy to avoid structures (Furness *et al.*, 2013). Some species are known to deliberately fly around OWFs, rather than fly directly through them (Desholm and Kahlert, 2005) and depending on the location of the OWF, this has the potential to affect both migration pathways and local flight paths (Drewitt and Langston, 2006). The level of disturbance and associated energy costs depend on a variety of factors including species behavioural traits, size of the OWF and level of disturbance compared to 'normal' flight patterns (Fox *et al.*, 2006; Pereksta, 2013). Radar tracking of common eider *Somateria mollissima* at

Danish OWFs followed the migration routes of over 200 000 individuals (Masden *et al.*, 2009). It was found that, while the birds adjusted their flight path to avoid the OWF, the extra distance amounted to about 500 m, which is trivial compared to the total migration distance of 1400 km. Similarly, studies modelling the cumulative impacts of disturbance by OWFs suggest that disturbance caused by increased travel distance are minor (Topping and Petersen, 2011).

Displacement

Many species show spatial responses to new OWFs during both the construction and operational stages (Drewitt and Langston, 2006). The responses can be attributed to factors caused by the turbines themselves (visual impact, noise, vibration) as well as maintenance traffic (Drewitt and Langston, 2006). Studies carried out to assess spatial responses to OWFs have found some species to be more sensitive than others. Post-construction monitoring at Horns Rev OWF found that the northern gannet *Morus bassanus*, common scoter *Melanitta nigra*, common guillemot *Uria aalge* and razorbill *Alca torda* were all fewer in number than expected (Petersen *et al.*, 2004). However, post-construction monitoring at the same site showed increased numbers of gulls, terns and the great cormorant *Phalacrocorax carbo* (Kahlert *et al.*, 2004). Other studies have shown that red-throated divers *Gavia stellata* completely avoid OWFs post-development and long-tailed duck *Clangula hyemalis* numbers are of a lower density in OWFs than is expected (Pereksta, 2013). Displacement, where it occurs, is generally linked to habitat availability as the reduction of a species from an area suggests that a suitable habitat has either been lost or is no longer accessible, even though the direct loss of habitat due to the footprint of an OWF is typically only 2–5% (Fox *et al.*, 2006).

Collision

Collision mortality is believed to be the greatest risk posed to seabirds by OWFs. Mortality can arise from direct collision with a pylon, collision with stationary or rotating blades, and the impact of pressure vortices created by rotating blades (Fox *et al.*, 2006). In this regard, the height at which birds typically fly and their manoeuvrability are the most important factors when assessing seabird vulnerability to collision (Furness *et al.*, 2013). However, passerines migrating at night can also be affected (Kerlinger, 2001; Evans, 2008), especially when structures are illuminated (Manville, 2001) and during poor weather conditions (Richardson, 2000; Erickson *et al.*, 2001; Hueppop *et al.*, 2006). On the other hand, studies of migrating geese have shown that they tend to avoid OWFs (even at night) and less than 1% flew close enough to be at risk of collision (Desholm and Kahlert, 2005). Nevertheless, seabirds tend to have low productivity and slow maturation rates so that populations could be negatively impacted by even relatively low collision rates (Drewitt and Langston, 2006).

Generally, collision risk (let alone actual collisions) in seabirds is difficult to quantify and to monitor (Pereksta, 2013). While most seabirds fly below 20 m and therefore well below rotor height (Krijgsveld *et al.*, 2011), some species do fly higher and have a greater risk (Furness *et al.*, 2013). Models used to estimate collision risk are chiefly based on species abundance and flight height (Johnston *et al.*, 2014). The models tend to show a lower than previously assumed risk of collision when taking account of a heterogeneous, rather than homogeneous, distribution of flight heights (Johnston *et al.*, 2014), and it may be that reduced seabird abundance in the vicinity of an OWF is more due to avoidance behaviour rather than collisions, as birds display the ability to recognize turbines and learn to avoid OWFs (Inger *et al.*, 2009).

In the UK alone, there are 25 species of offshore seabirds, 13 of which receive protection through designated Natura 2000 sites. Some wind farm applications have been refused consent due to potential impacts on birds. For example, Docking Shoal, a 500 MW OWF in the outer Wash area of Lincolnshire and Norfolk, was proposed by Centrica in 2009. The proposed wind farm was predicted to result in the death of over 90 sandwich terns *Thalasseus sandvicensis* each year, so the application was refused in 2012, with a loss of several million pounds of preparation costs to Centrica.

Rejected OWF Proposals

The main reasons for rejection of OWF proposals to date are unacceptable risks to wildlife and objections from local people over visibility and tourism impacts. In addition to the Docking Shoal proposal mentioned above, a proposal by Eneco for two OWFs in the North Sea (one for 101 turbines off the coast of Callantsoog and another for 137 turbines near Zuid-Holland) was rejected by a court in the Netherlands over concerns for the safety of the great black-backed gull *Larus marinus*, a protected bird species (Windpower Intelligence, 2011). A proposed OWF in Navitus Bay in Dorset, UK was refused owing to its proximity to the Jurassic Coast World Heritage Site and two Areas of Outstanding Natural Beauty. The 970 MW OWF would have consisted of 194 turbines, and created as many as 1700 jobs in the local area. However, the potential visual impacts on the World Heritage Site and Areas of Outstanding Natural Beauty were considered too great, especially as tourism was an important economic activity for this region and its presence might deter visitors (Business Green, 2015). However, Rampion OWF (being constructed by E.ON) is located 13 km off the coast of Brighton, UK and when finished it will be visible from Brighton and Worthing. The construction also includes laying 14 km of cables through the South Downs National Park. In order to receive permission for installing the OWF, E.ON agreed to reduce the number of turbines from 195 to 116 to reduce the impacts of the development (E.ON, 2016).

Potential of Floating Platforms

The impacts of conventional OWFs on marine life described in this chapter can result in developers avoiding areas in or near existing MPAs due to the risks of consent being denied and money lost. If the same amount of energy can be generated but with less impact on the environment, consent is more likely to be granted, which would encourage more OWF development. Given that the main source of detrimental impact on the marine environment is pile-driving during the construction stage, removing the need for turbine piles would be a huge advantage. Floating wind turbines were first designed in 1972 by W.E. Heronemus (Wang *et al.*, 2010); however, it is only relatively recently that this concept has been given more attention. Traditional pile-driven turbine bases can normally be used in depths of up to 50 m, whereas floating turbines can be placed in water up to 700 m deep (Leung and Yang, 2012). Stronger winds occur further offshore in deeper waters, but traditional pile-driven bases cannot withstand them. Deploying floating wind turbines would therefore minimize the impact on marine life whilst exploiting the stronger winds.

Floating turbines simply consist of a wind turbine on top of a floating structure anchored to the seabed by tension cables which are designed to allow movement with the waves during extreme weather conditions. They have underground cables to transport the generated electricity to the

national grid in a similar manner as traditional OWFs. Therefore, whilst the impact will be reduced due to omission of the pile-driving stage, there will still be some environmental impact from cable-laying and protection (mattressing and sediment disturbance).

There are currently three designs for floating platforms, although some designs involve a mix of at least two, like the Dutch tri-floater (Leung and Yang, 2012). The aim of all three designs is to withstand all weather conditions, like the deep water oil rigs on which the concept and designs are based (Inger *et al.*, 2009); the design must resist the motion of the sea and minimize pitch, roll and yaw (the way the platform would rotate in various ways around its central axis) whilst still maintaining the weight of the turbine. Therefore they need to be tethered to the seafloor but in a way that allows stable movement.

- The *ballast stabilized platform* achieves stability by having ballast weights below a buoyancy tank which creates the righting moment and high inertial resistance to pitch and roll. There is enough draft to offset heave motion. The anchors are embedded into the seafloor.
- The *mooring line stabilized platform* relies on mooring line tension for righting stability. The tension legs have suction pile anchors.
- The *buoyancy stabilized platform* uses distributed buoyancy to achieve stability. It takes advantage of the weighted water plane area for righting moment. From a bird's eye view, it is triangular on the surface.

The Dutch tri-floater is a mixture of the mooring line and buoyancy platform. The distributed buoyancy tanks are attached to the central tower by truss arms and this achieves stability by a weighted water plane area but also the mass of the steel tanks and truss structure.

Whether this new design of offshore wind turbines can be installed in MPAs is important to consider. As discussed, OWF can be beneficial to marine organisms, particularly in the operational stage due to reduced fishing effort. With reduced noise disturbance as a result of the elimination of pile-driving, negative impacts may be mitigated allowing the benefits of OWF developments to outweigh the disadvantages (Koschinski and Lüdemann, 2013).

Floating platforms may pose a risk to pelagic fish in terms of collisions with mooring chains and cables (Wilson *et al.*, 2007). The anchoring as well as the mooring floats and weights would also be a possible risk. Moreover, despite the anchoring chains and cables being smaller and having a reduced effect on water flow, their movement may cause an 'acoustic strumming', or simple entanglement. Nevertheless, floating platforms are likely to also act as FADs and increase local abundance of fish populations in a similar manner to traditional turbines (Inger *et al.*, 2009). Mooring lines underneath the water may pose a threat to diving seabirds such as gannets. However, given the distances and depths offshore, the risks are likely to be minimal.

Conclusions

From the review of OWF effects on marine life presented here, the potential for an MPA and a conventional OWF to work in synergy is limited. The very nature of traditional pile-driven turbines means that there are likely to be negative impacts on benthic habitats and associated fauna, as well as marine mammals and birds, even if only temporary. Therefore, extreme caution would be needed before an OWF was placed inside or close to an MPA in order to prevent such damage. However, once an OWF is operational, there is great potential for the site to deliver benefits for restoring fish populations and increasing habitat heterogeneity

(Inger *et al.*, 2009). It should be noted that environmental impacts of an OWF are discrete in space and time, whereas the impacts of certain fishing techniques such as bottom trawling can be much longer lasting and devastating (Inger *et al.*, 2009).

Recent development of floating turbines may provide solutions to reduce the impact of OWFs on marine life. However, it would still be advisable to avoid areas already designated as MPAs, particularly if the area is of high importance for marine mammals, fish spawning grounds or bird migration routes. Monitoring and continued efforts to minimize the risks to marine life should be paramount in the planning of an OWF, but the prospects for collaboration between marine conservation organizations and energy developers can allow a move towards both clean renewable energy and protected marine life.

References

André, M., Supin, A., Delory, E. *et al.* (2003) Evidence of deafness in a striped dolphin, *Stenella coeruleoalba. Aquatic Mammals*, **29** (1), 3–8.

Bergström, L., Sundqvist, F. and Bergström, U. (2013) Effects of an offshore wind farm on temporal and spatial patterns in the demersal fish community. *Marine Ecology Progress Series*, **485**, 199–210.

Bergström, L., Kautsky, L., Malm, T. *et al.* (2014) Effects of offshore wind farms on marine wildlife: a generalised impact assessment. *Environmental Research Letters*, **9**, 12.

Boehlert, G.W. and Gill, A.B. (2010) Environmental and ecological effects of ocean renewable energy development: a current synthesis. *Oceanography*, **23**, 69–81.

Boesen, C. and Kjaer, J. (2005) *Review Report 2004. The Danish Offshore Wind Farm Demonstration Project: Horns Rev and Nysted Offshore wind farms, environmental impact assessment and monitoring.* Elsam Engineering and ENERGI E2. 135 pp.

Business Green (2015) *Ministers reject Navitus Bay offshore wind farm.* http://www.businessgreen.com/bg/news/2425618/amber-rudd-rejects-navitus-bay-offshore-wind-farm. Accessed 19 October 2016.

Centre for Environment, Fisheries and Aquaculture Science (CEFAS) (2009) *Strategic review of offshore wind farm monitoring data associated with FEPA licence conditions: Fish.* Contract ME117, 17 July 2009.

Centre for Marine and Coastal Studies (CMACS) (2003) *Cowrie Phase 1 Report: A baseline assessment of electromagnetic fields generated by offshore wind farm cables.* COWRIE Report EMF.

Coates, D.A., Deschutter, Y., Vincx, M. and Vanaverbeke, J. (2013) Enrichment shifts in macrobenthic assemblages in an offshore wind farm area in the Belgian part of the North Sea. *Marine Environmental Research*, **95**. doi:10.1016/j.marenvres.2013.12.008

Collin, S.P. and Whitehead, D. (2004) The functional roles of passive electroreception in non-electric fishes. *Animal Biology*, **54**, 1–25.

Degollada, E., André, M., Arbelo, M. and Fernández, A. (2003) *Preliminary findings on the inner ear structures of beaked whales after the 2002 mass stranding in the Canary Islands.* 17th Annual Conference ECS, Las Palmas de Gran Canaria, 8–12 March 2003.

Desholm, M. and Kahlert, J. (2005) Avian collision risk at an offshore wind farm. *Biology Letters*, **1**, 296–298.

Diederichs, A., Nehls, G., Dähne, M. *et al.* (2008) *Methodologies for measuring and assessing potential changes in marine mammal behaviour, abundance or distribution arising from the construction, operation and decommissioning of offshore wind farms.* Report to COWRIE Ltd, Crown Commissioners, UK.

DONG Energy (2016) http://www.hornseaprojectone.co.uk/en/news/articles/worlds-largest-ever-offshore-wind-farm-to-be-built-by-dong-energy. Accessed 16 March 2016.

Drewitt, A.L. and Langston, R.H.W. (2006) Assessing the impacts of wind farms on birds. *Ibis*, **148**, 29–42. doi.wiley.com/10.1111/j.1474-919X.2006.00516.x

E.ON (2016) https://www.eonenergy.com/About-eon/our-company/generation/planning-for-the-future/wind/offshore/rampion-offshore-wind-farm. Accessed 16 March 2016.

Erbe, C., Williams, R., Sandilands, D. and Ashe, E. (2014) Identifying modelled ship noise hotspots for marine mammals of Canada's pacific region. *PLoS ONE*, **9** (3), e89820. doi:10.1371/journal.pone.0089820

Erickson, W.P., Johnson, G.D., Strickland, M.D. and Young Jr., D.P. (2001) *Avian collisions with wind turbines: a summary of existing studies and comparisons to other sources of avian collision mortality in the United States.* National Wind Coordinating Committee, c/o RESOLVE, Inc., Washington, DC.

European Commission (EC) (2014) *A policy framework for climate and energy in the period from 2020 to 2030.* Communication from the Commission. COM/2014/015.

European Environment Agency (EEA) (2015) *Marine Protected Areas in Europe's seas: an overview and perspective for the future.* European Environment Agency Report 3/2015. doi:10.2800/99473. 40 pp.

Evans, P. (2008) Offshore wind farms and marine mammals: impacts and methodologies for assessing impacts. Proceedings of the ASCOBANS/ECS Workshop. *ECS Special Publication Series*, 49.

Fox, A.D., Desholm, M., Kahlert, J. *et al.* (2006) Information needs to support environmental impact assessments of the effects of European marine offshore wind farms on birds. *Ibis*, **148**, 129–144.

Furness, R.W., Wade, H.M. and Masden, E. (2013) Assessing vulnerability of marine bird populations to offshore wind farms. *Journal of Environmental Management*, **119**, 56–66.

Gill, A.B. (2005) Offshore renewable energy: ecological implications of generating electricity in the coastal zone. *Journal of Applied Ecology*, **42**, 605–615. doi:10.1111/j.1365-2664.2005.01060.x

Gill, A.B. and Bartlett, M. (2010) *Literature review on the potential effects of electromagnetic fields and subsea noise from marine renewable energy developments on Atlantic salmon, sea trout and European eel.* Scottish Natural Heritage Commissioned Report No. 401.

Gill, A.B. and Taylor, H. (2001) *The potential effects of electromagnetic fields generated by cabling between offshore wind turbines upon elasmobranch fishes.* Countryside Council for Wales Science Report No. 488.

Gill, A.B., Huang, Y., Gloyne-Phillips, I. *et al.* (2009) *COWRIE 2.0 Electromagnetic Fields (EMF) Phase 2: EMF-sensitive fish response to EM emissions from subsea electricity cables of the type used by the offshore renewable energy industry.* COWRIE Ltd, Crown Commissioners, UK.

Hueppop, O., Dierschke, J., Exo, K.M. *et al.* (2006) Bird migration studies and potential collision risk with offshore wind turbines. *Ibis*, **148**, 90–109.

Inger, R., Attrill, M., Bearhop, S. *et al.* (2009) Marine renewable energy: potential benefits to biodiversity? An urgent call for research. *Journal of Applied Ecology*, **46**, 1145–1153.

Intergovernmental Panel on Climate Change (IPCC) (2014) *Climate changes 2014. Impacts, adaptation and vulnerability. Part A: Global and sectoral aspects.* Working Group II contribution to the Fifth Assessment Report of the Intergovernmental Panel on Climate Change.

International Union for Conservation of Nature (IUCN) (1994) *Guidelines for Protected Area Management Categories.* IUCN, Cambridge, UK and Gland, Switzerland.

Jensen, A.C., Collins, K.J., Lockwood, A.P.M. *et al.* (1994) Colonisation and fishery potential of a coal-ash artificial reef, Poole Bay, United Kingdom. *Bulletin of Marine Science*, **55**, 1263–1276.

Johnston, A., Cook, A., Wright, L. *et al.* (2014) Modelling flight heights of marine birds to more accurately assess collision risk with offshore wind turbines. *Journal of Applied Ecology*, **51**, 31–41.

Kahlert, J., Petersen, I.K., Fox, A.D. *et al.* (2004) *Investigations of birds during construction and operation of Nysted Offshore Wind Farm at Rødsand.* Annual Status Report 2003. http://uk.nystedhavmoellepark.dk

Kerlinger, P. (2001) *Avian issues and potential impacts associated with wind power development of nearshore waters on Long Island, New York.* Unpublished report for B. Bailey, AWS Scientific. 20 pp. http://www.winergyllc.com/reports/report_16.pdf

Köller, J., Köppel, J. and Peters, W. (eds) (2006) *Offshore Wind Energy: Research on Environmental Impacts.* Springer, Berlin.

Koschinski, S. and Lüdemann, K. (2013) *Development of noise mitigation measures in offshore wind farm construction.* Federal Agency for Nature Conservation, Berlin. 102 pp.

Krijgsveld, K.L., Fijn, R.C., Japink, M. and Dirksen, S. (2011) *Effect studies Offshore Wind Farm Egmond aan Zee: Final report on fluxes, flight altitudes and behaviour of flying birds.* Bureau Waardenburg Report, 10–219. Culemborg, Netherlands.

Krone, R., Gutow, L., Brey, T. *et al.* (2013) Mobile demersal megafauna at artificial structures in the German Bight: likely effects of offshore wind farm developments. *Estuarine, Coastal and Shelf Science*, **125**, 1–9.

Lange, M., Burkhard, B., Garthe, S. *et al.* (2010) *Analyzing coastal and marine changes: offshore wind farming as a case study.* Zukunft Küste – Coastal Futures Synthesis Report. LOICZ Research and Studies No. 36. GKSS Research Center, Geesthacht. 212 pp.

Leung, D., and Yang, T. (2012) Wind energy development and its environmental impact: a review. *Renewable and Sustainable Energy Review*, **16**, 1031–1039.

Lindeboom, H.J., Kouwenhoven, H.J., Bergman, M.J.N. *et al.* (2011) Short-term ecological effects of an offshore wind farm in the Dutch coastal zone; a compilation. *Environmental Research Letters*, **6** (3). doi: 10.1088/1748-9326/6/3/035101

Lucke, K. (2008) Auditory studies on harbour porpoises in relation to offshore wind turbines. Proceedings of the Workshop on Offshore Wind Farms and Marine Mammals: Impacts and Methodologies for Assessing Impacts. *ECS Special Publication Series*, **49**, 17–20.

Manville, A.M., II. (2001) The ABCs of avoiding bird collisions at communication towers: next steps, in *Avian Interactions with Utility and Communication Structures* (ed. R.L. Carlton). Proceedings of a Workshop held in Charleston, South Carolina, 2–3 December 1999. EPRI Technical Report, Concord, CA. pp. 85–103.

Marra, L.J. (1989) Sharkbite on the SL submarine lightwave cable system: history, causes, and resolution. *IEEE Journal of Ocean Engineering*, **14**, 230–237.

Masden, E.A., Hayden, D.T., Fox, A.D. *et al.* (2009) Barriers to movement: impacts of wind farms on migrating birds. *ICES Journal of Marine Science*, **66**, 746–753.

May, J. (2005) *Post-construction results from the North Hoyle offshore wind farm*. Paper for the Copenhagen Offshore Wind International Conference, Project Management Support Services Ltd. 10 pp.

Pedersen, J. and Leonhard, S.B. (2006) *The Danish Monitoring Programme: Final results –Electromagnetic fields*. Conference material, Wind Farms and the Environment 2006.

Pereksta, D.M. (2013) *Birds and offshore wind: studying and assessing effects*. Bureau of Ocean Energy Management, Department of the Interior. http://tinyurl.com/hou8tkb

Petersen, I.K., Clausager, I. and Christensen, T.K. (2004) *Bird numbers and distribution in the Horns Rev offshore wind farm area*. http://www.hornsrev.dk/Miljoeforhold

Piantadosi, C.A. and Thalmann, E.D. (2004) Comment on: Pathology: whales, sonar and decompression sickness. *Nature*, **428**, 1 p. following 716; discussion 2 pp. following 716.

Prior, A. and McMath, M. (2008) Marine mammals and noise from offshore renewable energy projects: K developments. Proceedings of the Workshop on Offshore Wind Farms and Marine Mammals: Impacts and Methodologies for Assessing Impacts. *ECS Special Publication Series*, **49**, 12–16.

Reubens, J. and Degraer, S. (2014) The ecology of benthopelagic fishes at offshore wind farms: a synthesis of 4 years of research. *Hydrobiologia*, **727**, 121–136.

Reyff, J.A. (2009) Reducing underwater sounds with air bubble curtains. *TR News*, **262**, 31–33.

Richardson, W.J. (2000) Bird migration and wind turbines: migration timing, flight behaviour, and collision risk, in *Proceedings of National Avian – Wind Power Planning Meeting III, San Diego, California, May 1998*. Prepared for the Avian Subcommittee of the National Wind Coordinating Committee by LGL Ltd, King City, Ontario. pp. 132–140.

Russell, D., Brasseur, S., Thompson, D. *et al.* (2014) Marine mammals trace anthropogenic structures at sea. *Current Biology*, **24** (14), 638–639.

Rye, J., Gilles, A. and Verfuβ, U.K. (2008) Linking wind farms and harbour porpoises: a review of methods. Proceedings of the ASCOBANS/ECS Workshop. *ECS Special Publication Series*, **49**, 21–26.

Simmonds, M. and Dolman, S. (2008) All at sea: renewable energy production in the context of marine nature conservation. Proceedings of the Workshop on Offshore Wind Farms and Marine Mammals: Impacts and Methodologies for Assessing Impacts. *ECS Special Publication Series*, **49**, 6–11.

SMRU Ltd (2007) *Assessment of the potential for acoustic deterrents to mitigate the impact on marine mammals of underwater noise arising from the construction of offshore windfarms*. COWRIE Ltd, Crown Commissioners, UK.

Southall, B., Bowles, A., Ellison, W. *et al.* (2007) Marine mammal exposure criteria: initial scientific recommendations. *Aquatic Mammals*, **33** (4), 411–509.

Teilmann, J., Tougaard, J. and Carstensen, J. (2008) Effects from offshore wind farms on harbour porpoises in Denmark. Proceedings of the ASCOBANS/ECS Workshop. *ECS Special Publication Series*, **49**, 50–59.

Thomsen, F., Lüdemann, K., Kafemann, R. and Piper, W. (2006) *Effects of offshore wind farm noise on marine mammals and fish*. COWRIE Ltd, Crown Commissioners, UK.

Topping, C. and Petersen, I.K. (2011) *Report on a red-throated diver agent-based model to assess the cumulative impact from offshore wind farms*. Report commissioned by the Environmental Group. Aarhus University, Danish Centre for Environment and Energy (DCE). 44 pp.

Vattenfall (2013) *Summary of environmental statement, Thanet Offshore Windfarm*. http://corporate.vattenfall.co.uk/projects/operational-wind-farms/thanet/. Accessed 10 February 2016.

Vattenfall (2016) *The first decommission in the world of an offshore wind farm is now complete*. http://corporate.vattenfall.com/press-and-media/press-releases/2016/the-first-decommission-in-the-world-of-an-offshore-wind-farm-is-now-complete. Accessed 15 March 2016.

Wahlberg, M. and Westerberg, H. (2005) Hearing in fish and their reactions to sounds from offshore wind farms. *Marine Ecology Progress Series*, **288**, 295–309.

Wang, C.M., Utsunomiya, T., Wee, S.C. and Choo, Y.S. (2010) Research on floating wind turbines: a literature survey. *The IES Journal Part A: Civil and Structure Engineering*, **3** (4), 267–277. doi:10.1080/19373260.2010.517395

Wilkes, J., Pineda, I. and Corbetta, G. (2014) *Wind energy scenarios for 2020*. A report for the European Wind Energy Association, July 2014.

Wilson, B., Batty, R., Daunt, F. and Carter, C. (2007) *Collision risks between marine renewable energy devices and mammals, fish and diving birds*. Report to the Scottish Executive. Scottish Association for Marine Science, Oban, Scotland.

Windpower Intelligence (2011) *The Netherlands: Court supports rejection of Eneco's offshore wind farm proposal in North Sea*. http://www.windpowerintelligence.com/article/0UjfVeh3yHE/2011/01/05/the_netherlands_court_supports_rejection_of_enecos_offshore_/. Accessed 19 October 2016.

Index

Page numbers in *italics* have a figure and those in **bold** have a table.

Management of Marine Protected Areas: A Network Perspective, First Edition. Edited by Paul D. Goriup.
© 2017 John Wiley & Sons Ltd. Published 2017 by John Wiley & Sons Ltd.